STUDIES IN MULTIDISCIPLINARITY   VOLUME 4

Memory Evolutive Systems

Hierarchy, Emergence, Cognition

STUDIES IN MULTIDISCIPLINARITY

SERIES EDITORS

Laura A. McNamara *Sandia National Laboratories, Albuquerque, New Mexico, USA*

Mary A. Meyer *Los Alamos National Laboratory, Los Alamos, New Mexico, USA*

Ray Paton[†] *The University of Liverpool, Liverpool, UK*

On the cover:
The artwork is by Jean-Paul Vanbremeersch and represents his view of the Archetypal Core which is at the root of the self in the Memory Evolutive Neural System modeling the mind. It filters the incoming information and dynamically interprets it by integration into reverberating circuits, to trigger the emergence of an adapted response

STUDIES IN MULTIDISCIPLINARITY   VOLUME 4

Memory Evolutive Systems
Hierarchy, Emergence, Cognition

Andrée C. Ehresmann
Université de Picardie Jules Verne
Amiens, France

and

Jean-Paul Vanbremeersch
Centre Médical St. Roch – Tivoli
Amiens, France

ELSEVIER

Amsterdam – Boston – Heidelberg – London – New York – Oxford
Paris – San Diego – San Francisco – Singapore – Sydney – Tokyo

Elsevier
Radarweg 29, PO Box 211, 1000 AE Amsterdam, The Netherlands
Linacre House, Jordan Hill, Oxford OX2 8DP, UK

First edition 2007

Notice
No responsibility is assumed by the publisher for any injury and/or damage to persons
or property as a matter of products liability, negligence or otherwise, or from any use
or operation of any methods, products, instructions or ideas contained in the material
herein. Because of rapid advances in the medical sciences, in particular, independent
verification of diagnoses and drug dosages should be made

Library of Congress Cataloging-in-Publication Data
A catalog record for this book is available from the Library of Congress

British Library Cataloguing in Publication Data
A catalogue record for this book is available from the British Library

ISBN: 978-0-444-52244-3
ISSN: 1571-0831

For information on all Elsevier publications
visit our website at books.elsevier.com

Printed and bound in The Netherlands

07 08 09 10 11   10 9 8 7 6 5 4 3 2 1

Series Dedication

Studies in Multidisciplinarity is dedicated to the memory of Ray Paton.

> Sure, he that made us with such large discourse,
> Looking before and after, gave us not
> That capability and god-like reason
> To fust in us unused.
>
> – William Shakespeare, Hamlet

Table of Contents

Introduction   1

PART A.   HIERARCHY AND EMERGENCE

CHAPTER 1: NETS OF INTERACTIONS AND CATEGORIES   21

1. Systems Theory and Graphs   22
    1.1. Objects and Relations   22
    1.2. Definition of a Graph   23
    1.3. Supplementary Properties.   24

2. Categories and Functors   25
    2.1. Some History   26
    2.2. Definition of a Category   27
    2.3. Some More Definitions   29
    2.4. Functors.   31
    2.5. Universal Problems   32

3. Categories in Systems Theory   32
    3.1. Configuration Categories of a System   33
    3.2. Fields of an Object   35
    3.3. Some Examples.   37

4. Construction of a Category by Generators and Relations.   37
    4.1. The Category of Paths of a Graph   37
    4.2. Comparison between Graphs and Categories.   38
    4.3. Labelled Categories.   39
    4.4. A Concrete Example   40

5. Mathematical Examples of Categories   42
    5.1. Categories with at Most One Arrow between Two Objects   43
    5.2. Categories with a Unique Object   44
    5.3. Categories of Mathematical Structures   45

CHAPTER 2: THE BINDING PROBLEM   49

1. Patterns and Their Collective Links   49
    1.1. Pattern in a Category   50

1.2. Collective Links   53
1.3. Operating Field of a Pattern   55

2. Colimit of a Pattern   56
2.1. The Mathematical Concept of a Colimit   57
2.2. The Binding Problem.   59
2.3. Decompositions of an Object   61

3. Integration vs. Juxtaposition   62
3.1. Comparison of the Sum to the Colimit   62
3.2. Some Concrete Examples.   64
3.3. Some Mathematical Examples   66

4. Interlude: A Transport Network   68
4.1. Characteristics of the Network   68
4.2. Local Area Nets, Selected Nets.   68
4.3. Central Node   69
4.4. Creation of a Central Node   70
4.5. Inter-Area Nets.   71

CHAPTER 3: HIERARCHY AND REDUCTIONISM   73

1. P-Factors of a Link Towards a Complex Object   73
1.1. Links Mediated by a Pattern   74
1.2. Link Binding a Perspective.   76
1.3. Examples of Perspectives   78
1.4. Field of a Pattern   79

2. Interactions between Patterns: Simple Links   80
2.1. Clusters   80
2.2. Composition of Clusters   82
2.3. Simple Links.   83

3. Representative Sub-Patterns   85
3.1. Comparison Link   86
3.2. Representative Sub-Pattern   87

4. Multiplicity Principle   89
4.1. Patterns with the Same Colimit   89
4.2. Connected Patterns   90
4.3. Multiplicity Principle and Complex Links   92

5. Hierarchies.   94
5.1. Hierarchical Categories   95

5.2. Examples 97
5.3. Ramifications of an Object and Iterated Colimits. 99

6. Complexity Order of an Object: Reductionism 102
   6.1. Obstacles to a Pure Reductionism. 102
   6.2. Reduction of an Iterated Colimit 104
   6.3. Complexity Order of an Object 105
   6.4. Examples 106
   6.5. n-Multifold Objects 107

CHAPTER 4: COMPLEXIFICATION AND EMERGENCE. 109

1. Transformation and Preservation of Colimits 110
   1.1. Image of a Pattern by a Functor 110
   1.2. Replacement of a Category by a Larger One 111
   1.3. Preservation and Deformation of Colimits. 113
   1.4. Limits 114

2. Different Types of Complexifications 117
   2.1. Options on a Category 117
   2.2. Complexification 118
   2.3. Examples 120

3. First Steps of the Complexification. 121
   3.1. Absorption and Elimination of Elements. 121
   3.2. Simple Adjoining of Colimits 122
   3.3. 'Forcing' of Colimits 123

4. Construction of the Complexification 125
   4.1. Objects of the Complexification 126
   4.2. Construction of the Links 126
   4.3. End of the Construction 128

5. Properties of the Complexification 128
   5.1. Complexification Theorem 128
   5.2. Examples 129
   5.3. Mixed Complexification. 130

6. Successive Complexifications: Based Hierarchies 131
   6.1. Sequence of Complexifications 132
   6.2. Based Hierarchies 134
   6.3. Emergent Properties in a Based Hierarchy. 136

7. Discussion of the Emergence Problem 138
   7.1. Emergentist Reductionism 138
   7.2. The Emergence Problem 139
   7.3. Causality Attributions 140

PART B.   MEMORY EVOLUTIVE SYSTEMS

CHAPTER 5: EVOLUTIVE SYSTEMS 147

1. Mechanical Systems vs. Living Systems 147
   1.1. Mechanical Systems. 147
   1.2. Biological and Social Systems. 148
   1.3. Characteristics of the Proposed Model 149

2. Characteristics of an Evolutive System 150
   2.1. Time Scale 150
   2.2. Configuration Category at Time t. 151
   2.3. Transition from t to t' 152

3. Evolutive Systems. 153
   3.1. Definition of an Evolutive System 153
   3.2. Components of an Evolutive System. 155
   3.3. Boundary Problems. 158

4. Hierarchical Evolutive Systems and Some Examples 161
   4.1. Hierarchical Evolutive Systems. 161
   4.2. The Quantum Evolutive System and the Cosmic Evolutive
       System 162
   4.3. Hierarchical Evolutive Systems Modelling Natural
       Systems 162

5. Stability Span and Temporal Indices 163
   5.1. Stability Span 164
   5.2. Complex Identity 165
   5.3. Other Spans 167
   5.4. Propagation Delays. 170

6. Complement: Fibration Associated to an Evolutive System 171
   6.1. Fibration Associated to an Evolutive System. 171
   6.2. Particular Cases 173
   6.3. The Large Category of Evolutive Systems 173

CHAPTER 6: INTERNAL REGULATION AND MEMORY
EVOLUTIVE SYSTEMS 175

1. Regulatory Organs in Autonomous Systems 176
   1.1. General Behaviour 176
   1.2. The Co-Regulators 177
   1.3. Meaning of Information 178

2. Memory and Learning 180
   2.1. Several Types of Memory 180
   2.2. Different Properties of Memory 181
   2.3. Formation and Development of the Memory. 182

3. Structure of Memory Evolutive Systems 183
   3.1. Definition of a Memory Evolutive System. 183
   3.2. Function of a Co-Regulator. 186
   3.3. Propagation Delays and Time Lags 187

4. Local Dynamics of a Memory Evolutive System 188
   4.1. Phase 1: Construction of the Landscape (Decoding). 189
   4.2. Phase 2: Selection of Objectives 192
   4.3. Phase 3: Commands (Encoding) of the Procedure, and Evaluation . . . 194
   4.4. Structural and Temporal Constraints of a Co-Regulator. 198

5. Global Dynamics of a Memory Evolutive System 200
   5.1. Conflict between Procedures. 200
   5.2. Interplay Among the Procedures 201

6. Some Biological Examples. 203
   6.1. Regulation of a Cell 204
   6.2. Gene Transcription in Prokaryotes 204
   6.3. Innate Immune System 205
   6.4. Behaviour of a Tissue 206

7. Examples at the Level of Societies and Ecosystems 207
   7.1. Changes in an Ecosystem. 207
   7.2. Organization of a Business. 208
   7.3. Publication of a Journal. 210

CHAPTER 7: ROBUSTNESS, PLASTICITY AND AGING 213

1. Fractures and Dyschrony 213
   1.1. Different Causes of Dysfunction. 213
   1.2. Temporal Causes of Fractures 214
   1.3. Dyschrony 216

2. Dialectics between Heterogeneous Co-Regulators 217
   2.1. Heterogeneous Co-Regulators 217
   2.2. Several Cases 221
   2.3. Dialectics between Co-Regulators. 222

3. Comparison with Simple Systems. 224
   3.1. Classical Analytic Models 224
   3.2. Comparison of Time Scales 225
   3.3. Mechanisms vs. Organisms. 226

4. Some Philosophical Remarks. 228
   4.1. On the Problem of Final Cause 228
   4.2. More on Causality in Memory Evolutive Systems 229
   4.3. Role of Time 230

5. Replication with Repair of DNA 231
   5.1. Biological Background. 231
   5.2. The Memory Evolutive System Model 232

6. A Theory of Aging. 234
   6.1. Characteristics of Aging. 234
   6.2. Theories of Aging at the Macromolecular Level. 237
   6.3. Level of Infra-Cellular Structures 239
   6.4. Cellular Level 240
   6.5. Higher Levels (Tissues, Organs, Large Systems). 241

CHAPTER 8: MEMORY AND LEARNING 245

1. Formation of Records. 245
   1.1. Storage and Recall 245
   1.2. Formation of a Partial Record 247
   1.3. Formation and Recall of a Record 249
   1.4. Flexibility of Records 250

2. Development of the Memory. 250
   2.1. Interactions between Records. 251
   2.2. Complex Records 251
   2.3. Examples 252

3. Procedural Memory 253
   3.1. Effectors Associated to a Procedure 254
   3.2. Basic Procedures. 256
   3.3. Construction of the Procedural Memory 257

4. Functioning of the Procedural Memory 259
   4.1. Activator Links. 259

4.2. Recall of a Procedure 260
4.3. Generalization of a Procedure 261

5. Selection of Admissible Procedures. 262
   5.1. Admissible Procedures. 262
   5.2. Procedure Associated to an Option. 263
   5.3. Selection of a Procedure by a Co-Regulator 264

6. Operative Procedure and Evaluation 265
   6.1. Interplay among the Procedures. 266
   6.2. Formation of New Procedures 266
   6.3. Evaluation Process and Storage in Memory. 267

7. Semantic Memory. 269
   7.1. How Are Records Classified? 270
   7.2. Pragmatic Classification with Respect to a Particular Attribute 270
   7.3. Formation of an E-Concept. 272
   7.4. Links between E-Concepts. 275
   7.5. Semantic Memory. 277

8. Some Epistemological Remarks. 280
   8.1. The Knowledge of the System 280
   8.2. Acquisition of Knowledge by a Society. 281
   8.3. The Role of the Interplay among Procedures. 282
   8.4. Hidden Reality. 283

PART C.   APPLICATION TO COGNITION AND CONSCIOUSNESS

CHAPTER 9: COGNITION AND MEMORY EVOLUTIVE NEURAL
SYSTEMS. 287

1. A Brief Overview of Neurobiology. 287
   1.1. Neurons and Synapses. 287
   1.2. Coordination Neurons and Assemblies of Neurons 288
   1.3. Synchronous Assemblies of Neurons. 289
   1.4. Binding of a Synchronous Assembly. 290

2. Categories of Cat-Neurons 292
   2.1. The Evolutive System of Neurons. 292
   2.2. Cat-Neurons as Colimits of Synchronous Assemblies 293
   2.3. Interactions between Cat-Neurons 295

3. The Hierarchical Evolutive System of Cat-Neurons. 297
   3.1. Higher Level Cat-Neurons. 297
   3.2. Extended Hebb Rule. 299
   3.3. The Evolutive System of Cat-Neurons. 300

4. The Memory Evolutive Neural System 302
    4.1. The Memory of an Animal 303
    4.2. The Memory as a Hierarchical Evolutive System 304
    4.3. A Modular Organization: The Net of Co-Regulators 305

5. Development of the Memory via the Co-Regulators 307
    5.1. Storage and Retrieval by a Co-Regulator 307
    5.2. Formation of Records 309
    5.3. Procedures and their Evaluation 312

6. Applications 315
    6.1. Physiological Drives and Reflexes 315
    6.2. Conditioning 316
    6.3. Evaluating Co-Regulators and Value-Dominated Memory 318

CHAPTER 10: SEMANTICS, ARCHETYPAL CORE AND
CONSCIOUSNESS 321

1. Semantic Memory 321
    1.1. Perceptual Categorization 322
    1.2. Concept with Respect to an Attribute 323
    1.3. The Semantic Memory 326
    1.4. Recall of a Concept 330

2. Archetypal Core 332
    2.1. The Archetypal Core and Its Fans 332
    2.2. Extension of the Archetypal Core: The Experiential Memory 334

3. Conscious Processes 337
    3.1. Intentional Co-Regulators 338
    3.2. Global Landscape 339
    3.3. Properties of the Global Landscape 341
    3.4. The Retrospection Process 342
    3.5. Prospection and Long-Term Planning 343

4. Some Remarks on Consciousness 344
    4.1. Evolutionary, Causal and Temporal Aspects of Consciousness 344
    4.2. Qualia 345
    4.3. The Role of Quantum Processes 346
    4.4. Interpretation of Various Problems 347
    4.5. Self-Consciousness and Language 347

5. A Brief Summary 348
    5.1. Basic Properties of Neural Systems 348
    5.2. Interpretation in Our Model 350

Appendix. 353

Bibliography 361

List of Figures 379

Index 383

Introduction

This is a presentation of what we call *memory evolutive systems*. We offer these as mathematical models for autonomous evolutionary systems, such as biological or social systems, and in particular, nervous systems of higher animals. Our work is rooted in category theory, which is a particular domain of mathematics. We have spent some 20 years developing the concepts involved in memory evolutive systems, and over that time have presented them in a series of articles and several conferences. This book is a synthesis of these two decades of research.

1. Motivations

One of us, Jean-Paul Vanbremeersch, is a physician who specializes in gerontology. He has long been interested in explaining the complex responses of organisms to illness or senescence. The second of us, Andrée C. Ehresmann, is a mathematician, whose research areas have included analysis, optimization theory and category theory, in collaboration with her well-known mathematician husband, late Charles Ehresmann. In 1980, she organized an international conference on category theory in Amiens, France, in memory of her husband, who died in 1979. In doing so, she asked Vanbremeersch for assistance in writing an explanation of category theory for non-mathematicians. It was during these initial interactions that he first suggested that categories might have applications for problems related to complexity.

   This is how our study of memory evolutive systems began. Our subsequent examination of the literature revealed that there had not yet been any real work done on this subject. Although Rosen (1958a) had promoted the use of category theory in biology, he considered only its basic notions and not its more powerful constructions. Hence, we decided to combine our interests and pursue research in this direction.

1.1. How Can Complexity Be Characterized?

During the late 1970s and early 1980s, there was a great deal of excitement around the question of 'complexity', with researchers discussing non-linear systems, chaos theory, fractal objects and other complex analytical constructs. We quickly realized that category theory could provide tools to study concepts germane to complexity, such as the following.

(i) *The binding problem*: how do simple objects bind together to form a 'whole that is greater than the sum of its parts'?
(ii) *The emergence problem*: how do the properties of a complex object relate to the properties of the more elementary objects that it binds?
(iii) *The hierarchy problem*: how may we explain the formation of increasingly complex objects, beginning with elementary particles that form atoms, which in turn form molecules, up through increasingly complicated systems such as cells, animals and societies?

We considered these three problems in our first joint paper (Ehresmann and Vanbremeersch, 1987), in which we defined a model called *hierarchical evolutive systems*, based on the categorical concept of colimit, and the process of complexification.

1.2.   Self-Regulation

In our 1987 paper, however, we did not introduce those characteristics of living systems that allow for autonomy through self-regulation; namely, some type of internal regulation systems, as well as a capacity to recognize, innately or through learning, those environmental characteristics that require the system to develop adequate and appropriate responses.

Our work in hierarchical evolutive systems had to be enriched to take these characteristics into account, and we did so in subsequent papers. Initially we introduced the concept of a single regulatory organ (Ehresmann and Vanbremeersch, 1989). However, soon we realized that it was not possible to have only a single regulatory organ, because of differences in laws and time scales across various levels of the hierarchy. Hence, we introduced (in Ehresmann and Vanbremeersch, 1990, 1991, 1992a) the concept of a net of such regulatory organs, individually called *co-regulators* (CR). To function, these co-regulators must rely on a central internal archive, a kind of 'memory'. Such a memory would not be rigid, like that of a computer, but would instead be flexible enough to allow for successful adaptation to change over time, and the formation, possibly, of increasingly better adapted behaviours. From this work, we developed the model which we call a memory evolutive system.

1.3.   Cognitive Systems

In 1989, we sketched some applications of memory evolutive systems to the nervous system and to cognition. In that same year, Gerald Edelman published *The Remembered Present: A Biological Theory of Consciousness*

(1989). We were amazed to see that Edelman's ideas corroborated many of the concepts we had arrived at through applying the methods of category theory to problems of cognition and consciousness. In particular, Edelman insists on a notion of degeneracy, which is readily modelled by what we now call the *multiplicity principle*, and which we place at the basis of emergent properties. Edelman's book also encouraged us to develop our study of semantics and higher cognitive processes within the framework of memory evolutive systems, and in particular to attempt to model consciousness. The issue of consciousness has been central in some of our recent articles (Ehresmann and Vanbremeersch, 1999, 2002, 2003), in which we have singled out some of its characteristics, and shown how they rely on the development of a personal memory, called the *archetypal core*, which forms the basis of the self.

2. Why Resort to a Model?

From whence came the interest in designing such models in the first place? Through memory evolutive systems, we propose a mathematical model that provides a framework to study and possibly simulate natural complex systems. Indeed, since their beginnings, the dream of philosophy and of science has been to give an explanatory account of the universe. In seeking deeper explanations for life and consciousness, scholars in many fields have become increasingly aware of the problem of complexity in biological systems. Computational science has played a very important role in pushing these understandings forward, but the pursuit requires increasingly elaborate mathematical tools. Our hope is that an adequate mathematical model will shed some light on the characteristics of complex evolutionary systems, on what distinguishes them from simple mechanisms or straightforward physical systems, and on the development of complex systems over time, from birth to death.

Moreover, the behaviour of such a system depends heavily on its experiences. In a memory evolutive system, we posit that the system may remember these experiences for later use. A model that represents a system over a certain time period, one that accounts for the system's responses to various situations that it encounters, might be able to anticipate the system's later behaviour and perhaps even predict some developmental alternatives for the system. This dream of developing a computational forecasting ability, which is rather like seeking a modern Pythia, has been considerably stimulated by the increasing power of computers, which makes it possible to deal with very large numerical and non-numerical data sets. However, computation also has its limits. Thus the role of a mathematical model is twofold: theoretical, for comprehending the fundamental nature of complex systems; and practical, for applications in biology, medicine, sociology, ecology, economics, meteorology and other fields that trade in complexity.

2.1. Different Types of Models

There are many ways of designing models. For example, the traditional models in physics (*e.g.* those inspired by the Newtonian paradigm, or that are well known in thermodynamics, electromagnetism and quantum mechanics) generally use a representation based on 'observables' that satisfy systems of differential equations, which translate the laws of physics into a quantitative language.

Some of these traditional models include chaotic behaviour and have been imported into such fields as biology and ecology. The values (real numbers or vectors) of the observables are obtained empirically. Over the past five decades, such analytical models have assumed an increasingly important role in many scientific fields, as advances in computational science led to the development of powerful data processing systems, capable of handling large systems of equations with many parameters.

Another kind of model is the black box model, which does not try to reproduce the internal behaviour of a system. Rather, this kind of model takes into account only the inputs, the outputs and the change-of-state rules. These rules are formal, as in a Turing machine; or as in cellular automata, introduced by von Neuman (1966), one of the main architects of the modern digital computer. Black box models can be used to help develop decision trees that operate on variables issued from databases, according to usual Boolean logic: and/or; if … then; not and so on. Such trees are useful in expert systems; for example, in those used in the diagnosis and treatment of diseases.

Cybernetics is a field comprising another class of mathematical models. The term was defined by Wiener (1948) to mean 'the entire field of control and communication theory, whether in the machine or in the animal' (p. 19). Its models use in an essential way the concept of feedback, and at times Shannon's information theory (Shannon and Weaver, 1949). Cybernetics advanced throughout the 1940s, 1950s and 1960s, thanks to the collaboration of specialists in biology, neurobiology and economics, who compared their individual approaches, and found great similarity in the structure and the evolutionary modes of the systems they studied.

It is also in this multi-disciplinary environment that systems theory developed. Although it is related to cybernetics, systems theory focuses more on modelling the relations among the components of a system. As defined by von Bertalanffy (1926), a system is a set of interacting elements organized to achieve a particular goal. Today, in engineering or science, a system is generally defined as any item consisting of two or more components linked together. The idea is to model the dynamics of the system, *i.e.* its actions and interactions over time, as represented by the evolution of its state function.

The science of complexity integrates methods of cybernetics and systems theory to study complex systems or systems with complex dynamics. It includes a wide range of approaches to studying and modelling complexity. Examples include artificial intelligence, neural systems, catastrophe theory (Thom, 1974), chaos and dissipative structures (Prigogine and Stengers, 1982), fractals (Mandelbrot, 1975), autopoietic systems (Maturana and Varela, 1973) and anticipatory systems (Rosen, 1985b).

2.2. *The Limitations of Classical Models*

The models we have described above have proved useful in a variety of applications, ranging from pattern recognition, to weather prediction, to financial analysis and more. The model results are valid locally, as when a weather model gives reliable forecasts over the short term. However, the simulations are limited to a specific range of complexities and energies: for one thing, locally valid laws govern each level of behaviour in the system being modelled. For example, in a cellular system, the molecules that make up the cells follow laws different from those at the level of cells, even though both molecules and cells are part of the same whole. Moreover, each part operates on its own time scale, and these temporal variations play an essential part in the evolution of a system.

As a general rule, it is impossible to extrapolate the properties of the whole from local features, *i.e.* to understand the entire system just by analysing its parts one by one. The interactions of specific objects in a system might be well understood. However, the interactions of a very great number of these units create new forms of spatial, energetic and temporal organization, the existence of which can only be considered at a higher level of analysis, if we are to ever understand their functioning. It is as difficult to unify quantum physics and macro-physics, as it is to understand the functioning of the mind from the finest analyses of individual neurons. Likewise, we are challenged to understand how evolution could begin from simple macromolecules mingled in a primitive soup, that become simple cellular organisms, which develop gradually into autonomous systems, capable of adapting to changing environmental conditions while successfully reproducing themselves.

The prevalence in nature of complex systems is the reason why many authors have explored the concept of hierarchical systems in a very wide range of fields; for example, in physics (Reeves, 1988; Ullmo, 1993), biology (von Bertalanffy, 1973; Jacob, 1970; Monod, 1970), neurology (Changeux,

1983; Jeannerod, 1983; Laborit, 1983), evolutionary theory (Dobzhansky, 1970; Mayr, 1976; Teilhard de Chardin, 1955), ethology (Lorenz, 1973; Tinbergen, 1951) and the social sciences (Koestler, 1965; Le Moigne, 1990; Morin, 1977; Piaget, 1967). However, it is the passage from a lower level to a higher level separated from it by what Farre (in his preface to Schempp, 1998) calls an 'energy gap' that remains obscure.

The passage from level to level in a complex system leads to the consideration of what researchers describe as the *emergent properties* of a system. As an example, a brick heap is nothing more than the sum of all its bricks, with some random spaces between the individual bricks. The heap is characterized by its weight, which is the sum of the weights of the bricks, and its volume, which is the sum of the volumes of the bricks and of the spaces between them. The form of a brick pile is more or less arbitrary. In contrast, a dam made of bricks is, strictly speaking, also a pile of bricks, but now this pile of bricks is purposeful, insofar as there is a specific relationship between the pile and a river. In a dam, each brick participates, in collaboration with the other bricks and with the mortar that joins them, in a concerted way to block the flow of water. What emerges from the pile of bricks, the mortar, and the river itself is the unified operation of the dam that holds back the river.

3.   Two Representative Examples

3.1.   *Higher Vertebrates Nervous System*

In the case of the nervous system, the standard models—in particular the connectionist neural networks (following Hopfield, 1982)—do give rather reliable results for local analyses: starting with a small number of neurons and a set of basic rules, it is possible to simulate the neurons' collective behaviour over a short period of time. However, the problem becomes intractable when we extend the model to simulate the operation of a larger collection of neurons in an organ as complex as the brain. Even creating an adequate simulation of a single function, like visual recognition, requires much more detail that even the most complete set of local analyses can provide. Attempting to describe visual recognition through modelling the behaviour of individual neurons is comparable to attempting to describe the workings of a multinational corporation by modelling the activities of all its employees.

The organization of a system like that for visual recognition is complex, with a structure that can be described as hierarchical. At the most basic level, multiple neurons in the optic nerve sense an image against the retina of the eye. This event stimulates activity in groups of neurons in the brain, for example, the neuronal groups that make up the visual areas of the brain conduct more or less parallel analyses of the various features of an observed object, while the neurons in the brain's memory centres are activated by these visual stimuli, and support recognition of the image of the object as a particular thing. Visual recognition also requires placing the object in context: simply classifying an object as a bottle tells one little about what kind of bottle it is or what it contains. To this end, the neuronal groups of the associative areas integrate the image with the meaning of the perceived image stored in the semantic memory. One may then recognize the bottle as a wine bottle, and the liquid it contains as wine of a particular colour, origin and vintage.

In short, visual recognition is representative of a class of processes that become increasingly complex as more and more neurons are activated in a synchronous way. The processes follow the laws of cellular neurophysiology (involving observables such as the firing (spiking) frequencies of neurons, the transmission delays and the synaptic strengths), as well as the functional relationships among the specific neuronal groups involved in vision, and between these groups and other areas of the brain. Indeed, a local assembly of various neuronal groups can spontaneously organize to behave as a completely new entity, when provided with the appropriate stimuli. It is clear that the properties of complex systems such as the brain are different at each hierarchical level, each one functioning with a different time scale. For example, the firing frequency of a retinal neuron (*i.e.* the period encompassing the reception of the signal, its transmission along the axon and through the synapse, and the subsequent refractory period) is not of the same order of magnitude as the oscillation frequency of the larger-scale neuronal systems involved in facial recognition.

Let us give a clinical example of what is probably a kind of epileptic process, which illustrates this temporal variability. The process is characterized by a feeling of acceleration, with each gesture, each thought, each action seeming to be conveyed at a record speed, like in a speeded-up film. However, in spite of this disturbing sensation, the subject can attempt to analyse the real rhythm of things, as if he were looking from the outside at his 'accelerated self'. It is then as if the subject had two levels of consciousness: a first level extremely and abnormally fast, and a second, normal one. We could analyse this phenomenon as an unusual eruption at the conscious level of a multitude of fragments of actions or thoughts. This fragmentation is normally unconscious, and the fragments reflected at the conscious level only after they have been united into a continuous process; but in the pathological case, they become conscious as such, each fragment being at the origin of a discontinuity, of a rupture. Instead of a continuous flow, at the normal pace of thoughts and general activities, this phenomenon introduces a split flow, with decomposition of each complex movement into its successive components, of each thought into bits of sentences, or even into words, put end to end.

3.2.   A Hierarchical Business Enterprise

Another example is a hierarchically organized manufacturing company, one that includes several groups of workers, each of which has a well defined function within the manufacturing process, that function involving specific tools and raw materials. One group of workers produces parts, which are passed on to another set of workers trained to assemble them, along with parts made by other groups, into a particular product, such as a car. In the same factory, we can imagine that still other groups of workers might be in charge of controls, packaging, forwarding and other activities that occur in a manufacturing environment. All these activities take place under the supervision of foremen, who are in charge of making sure that operations on the manufacturing floor run smoothly. These foremen, in turn, work under the supervision of a group of middle managers, responsible for coordinating the activities of a complex, which is a highly differentiated hierarchy of specialized personnel. This level of management, in turn, operates under the supervision of directors of various departments, who themselves report to the president of the company and, ultimately, to a board of directors (*cf.* Fig. 6.8).

   The workers, the foremen, the managers, the president and the board of directors are all part of the same system. All are necessary to keep the company active, but each operates within a specific level of organization within this system. A workman assembling parts on the company's manufacturing floor does not deal with the same objects, workspaces, tools, people or responsibilities, that a member of the board of directors does. Executives have to deal with a particular set of decision contexts and temporal horizons, a set that is quite different from the one that shapes the actions of the shop floor worker. The actions the executives take in their daily work are correspondingly different from those of a foreman or a manager. The selection of strategies and procedures is different at the level of the workman, who is dealing with events occurring on a much shorter time scale than those faced by a company executive. The workman's procedures may be less complex, while the parameters of the workman's job may be far more circumscribed than those of the executive board. On the other hand, strategies that the executive develops—for example, conceptualizing a new mode of production or deciding to pursue a new market, developing new distribution systems and arranging for new capital financing— represent a complex set of responsibilities with long-term repercussions for the entire company. The impacts of the actions of the decision maker may be felt at the shop floor level only much later, as the repercussions gradually spread from the uppermost echelons of the company to its workforce.

So, what happens when the system encounters a challenge in its environment—for example, a delay in the acquisition of raw materials? The shortage is sensed differently at each level of the company. A worker on the shop floor might perceive the problem more immediately than an upper level executive. There may be a delay in activities on the shop floor as the executives attempt to source a replacement material, perhaps while engaging in an analysis of the causes of the shortage and the possible consequences, and deciding upon an optimum strategy for the foreseeable future. In this case also, the duration of the steps is very variable, as is their quality—in other words, their nature with respect to the number and the complexity of the parameters to be analysed, and the number and complexity of the choices of procedures to be applied.

4. How Does the Model Function?

The preceding examples are representative of the kinds of systems we are interested in studying: autonomous, evolutionary systems that manage their interactions with their environment, and which remember their experiences to anticipate future challenges, develop appropriate responses, and so adapt more effectively to changes in the environment. Following Rosen (1985a), we may refer to these systems as 'organisms', a term that comprises not only biological entities, from cells to complex living creatures, but also ecological, social, cultural and economic systems. What main concepts are necessary if we are to model such systems as memory evolutive systems?

4.1. *Evolutive Systems*

The configuration of an evolutionary system at a given time t is a snapshot of the system, showing its components and the interactions among them around that particular moment. These components and their interactions may be geometric, *i.e.* determining the shape of the system; or functional, *i.e.* transfers of information among components, or constraints. There may be a high degree of variability in the levels of complexity of these components, in their horizontal and vertical interconnections, their possible relations with other objects external to the system and in their specific types of function within the system. There is also a quantitative variability in the observables that measure the strengths and durations of the interactions. The structure and the organization of the system, like its components, are not invariant, but change in the course of time, as a result of exchanges between components and with the environment. We model the different configurations of a system, and the changes between them, by what we call an evolutive system.

The changes can maintain homeostasis: the total structure remains stable, although the components that comprise the system are renewed and repairs are carried out. For example, the position of accountant may be enduring within a given company, even though different individuals may carry out the activities of accounting. However, changes often lead to loss or acquisition of information, energy and matter, dissociation of certain components, the formation of more complex components, as well as links that will acquire new, emergent properties. This will be modelled by the process of *vertical complexification*.

The brain provides examples of these two situations. The regular renewal of the macromolecules of a neuron in the human brain is one example of the maintenance of stability while carrying out repairs. In contrast, an encounter with an unknown object may start an activity of learning that forms entirely new synaptic links in the brain. Similar examples may be found in society as well: we have the regular election of representatives, but this can lead to political modifications of a more fundamental and ideological nature, if a change of majority occurs. Vertical complexification may involve an enrichment of the hierarchy of the system, in particular with the formation of higher levels of organization that enable the system to capture and remember more complex experiences, their successive modifications and their immediate or foreseeable consequences.

4.2.   Net of Internal Regulators and Memory

Natural evolution is internally regulated. Because the components and groups of components that make up the different levels of a complex system are highly variable, a central regulating mechanism cannot exist, unless we posit a divine process present everywhere, at all moments and all levels, that oversees and calibrates every rhythm and activity. Instead, for defining memory evolutive systems we posit a whole parallel distributed network of internal regulatory organs, called *co-regulators*, which observe, analyse, evaluate and possibly make decisions. These co-regulators act in parallel, and integrate all levels of the system throughout the hierarchy, both horizontally and vertically. Each co-regulator, in turn, comprises a small group of components of a certain level of complexity, operating together in a stepwise mode at a specific time scale, using loops, feedback or feed-forward processes. Its specificity of function imparts a specificity to the procedures (or strategies), which it can select at a given moment.

A co-regulator selects a procedure by taking account of the partial information received from other parts of the system or the environment (which all together form what we call its *landscape*), of the structural and temporal constraints which it must respect, and of the results of its former experiences, which have been remembered, and so allow for better adaptation. The analysis of the situation will be very different depending on the co-regulators. In particular, their time scales are by nature heterogeneous: if we think of a biological system from its atoms, or even its sub-atomic particles, to the molecules, macromolecules, organelles, cells, organs and finally the entire organism, each one of these hierarchical levels functions at its own pace. It is thus natural that each co-regulator has its own procedures; the behaviour of an insect does not take account of the long-term evolution of the ecosystem.

However, the co-regulators are not independent and must all function coherently: we know the paradox of the wings of a butterfly in New Guinea, which in a certain way may contribute to the formation of a hurricane on the coasts of Florida. To explain such global effects, it should be understood that the procedures chosen by the various co-regulators must be realized not on their respective landscapes, but on the system itself, where they compete for common resources. Thus, an equilibration process, which we call the *interplay among the procedures*, must be carried out; this process is not directed centrally, but depends on the respective strengths and time scales of the different co-regulators. If all the chosen procedures are compatible, they are all put into effect, so that the global system maintains at least a certain homeostasis, and possibly develops new capacities. If not, their interplay will eliminate some of them, causing what we call a *fracture* for the corresponding co-regulators. This fracture will have to be repaired later on; if it cannot be overcome, it will result in a synchronization error, and thus a loss of homeostasis. In particular, what we call a 'dialectics' is generated between co-regulators of very different complexity levels and time scales. There is then the risk of a cascade of what we term *re-synchronizations* at increasingly higher levels, which we have proposed as characteristic of the aging of an organism (Chapter 7).

Moreover, a memory evolutive system is able to learn and adapt, thanks to the development of an internal long-term *memory*. This memory records the successive experiences of the system, as well as the choices of procedures by the co-regulators and their results. Each co-regulator takes part in the development of the memory, and resorts to it when making decisions. Moreover, in the case of more complex systems (*e.g.* the nervous system of a higher animal), there will be a classification of the records retained in the memory, leading to the formation of a *semantic memory*. This, in turn, allows the development of a personal memory, which we call the archetypal core (Chapter 10), encompassing and intertwining the different kinds of experiences of the animal, quickly activated, and maintaining self-activation over a long time. This archetypal core is the basis of the development of at least a primary consciousness. One characteristic of such a consciousness is the formation, when an arousing event occurs, of a global landscape, on which a process of retrospection allows a search for the nature of the present event, and on which a prospection process allows to select and program long-term well-adapted responses.

4.3.   *With What Tools?*

To study these problems, we make recourse to category theory, a sub-domain of mathematics introduced by Eilenberg and Mac Lane (1945). This theory has a unique status, at the border between mathematics, logic and meta-mathematics. It was introduced to relate algebraic and topological constructs, and later its foundational role in mathematics and logic was emphasized by several authors: for example, in the theory of *topos* developed by Lawvere and Tierney (Lawvere, 1972) and in the *sketch theory* developed by Ehresmann (1968), which provides a single setting for the main operations performed by the 'working mathematician' (a phrase taken from the title of Mac Lane's (1971) well-known book on the subject). As a language unifying many domains of mathematics, category theory makes a general concept of structure possible, and indeed it has been described as mathematical structuralism.

In fact, category theory, seen as an analysis of the manner in which the mathematician thinks, reflects some of the main capacities of the brain. Our basic idea is that the evolution of living systems, and in particular of the human brain, rests on a small number of prototypical operations, which are exactly those that category theory can model: the formation, dissolution, comparison and combination of relations between objects. Together, these allow for the transfer and analysis of information, and the synthesis of complex objects from more elementary ones (formation of colimits, Chapter 2). Equally, they allow for: analysis, through the decompositions of complex objects (Chapter 3); optimization processes (universal constructions, Chapter 1); formation of hierarchies of objects (complexification, Chapter 4); detection of the continuous identity of a composite in spite of suppression or addition of some elements (progressive transformation of a colimit, Chapter 5); recording of new objects and their later recognition (formation of, and comparison with, a colimit, Chapter 8); classification of objects into invariance classes, leading to the definition of concepts (Chapter 8).

These are the operations we use to model organisms, via memory evolutive systems. Thanks to their primarily relational and qualitative nature, they offer a method which covers at the same time the local, general, evolutionary and temporal aspects of complex autonomous systems.

5. Plan of the Book

This book is divided into three parts: Hierarchy and Emergence; Memory Evolutive Systems; Application to Cognition and Consciousness.

5.1. *Part A: Hierarchy and Emergence*

Here we introduce the main mathematical tools that we use to describe a hierarchical system, and we explain how such a system evolves via the emergence of increasingly complex components which bind patterns of more elementary components, acting coherently via some set of distinguished links.

More specifically, in Chapter 1, we explore the notion of a category, and use it to model a system. Then, in Chapter 2, we focus on an important concept, that of a colimit in a category. We explain how this concept describes the binding process in a system. In Chapter 3, we focus on the problem of mathematically modelling a hierarchical system. The key idea is to define a hierarchical category, as a category in which objects are partitioned into levels, with an object of one level being the colimit of a pattern of linked objects of strictly lower levels. Of particular importance is the *reduction theorem*, which sets the conditions for an object of one particular level to be reducible to a lower level of complexity. As a corollary, we show that a necessary and sufficient condition for the existence of objects of strictly increasing levels is what we call the multiplicity principle. This formalizes the degeneracy property singled out by Edelman (1989; see also Edelman and Gally, 2001; Tononi *et al.*, 1999) for neural system. We use the reduction theorem and the multiplicity principle to discuss the *reductionism problem* (Chapter 3).

Chapter 4 describes the important process at the root of the emergence of higher order objects, namely, the *complexification* of a category with respect to an *option*. In this formulation, an option has certain objectives, among them to bind patterns of linked objects, in order to integrate them into a higher order object with new, emergent properties. The main result is the explicit construction of the adequate relations between the emerging objects. With this consequence: a sequence of complexifications of a category satisfying the multiplicity principle cannot be reduced to a unique complexification—thus confirming the reality of the emergence of objects of increasing orders of complexity. We analyse the philosophical consequences for the emergence problem, and propose an alternative formal basis for the concept of *emergentist reductionism* advanced by Bunge (1979).

5.2.   Part B: Memory Evolutive Systems

Chapter 5 introduces evolutive systems, which model the evolution of a natural system. The idea is to represent the successive configurations of a system by categories, and changes of configuration by partial functors between them. The problem is then to identify the components of the system through their successive configurations, and to explain how complex components can maintain their identity in spite of the renewal of their own lower order components. The notion of a memory evolutive system is introduced in Chapter 6, with its net of local internal regulatory organs, the co-regulators and a memory which helps them to perform their specific function. We first study both the local dynamics (one step of a co-regulator), and then the global dynamics, directed by the interplay among the procedures of the various co-regulators, with a risk of a fracture to some of them. Numerous examples are given, both in biology and in sociology.

The competition between the co-regulators is studied more deeply in Chapter 7. In particular, we prove how what we call a 'dialectics' emerges between heterogeneous co-regulators, which may cause cascades of fractures and re-synchronizations at higher and higher levels. This process is exemplified in the replication, with DNA repair, of a bacterium. An important application is given to a *theory of aging*, and we show how numerous physiological aging theories can be incorporated into this scheme. Chapter 8 describes the development of a flexible memory, in which the records can be recalled under several forms, allowing for generalization. Two important sub-systems of the memory are analysed. The first is the procedural memory, consisting of procedures modelling behaviours, in particular those which determine the function of a co-regulator. The second is the semantic memory, in which records are classified into invariants.

5.3.   Part C: Application to Cognition and Consciousness

Chapter 9 gives an application of memory evolutive systems to cognitive systems. The nervous system of an animal is modelled by the evolutive system of neurons. Its successive complexifications give rise to a hierarchy of components, called *category-neurons* (or, more briefly, cat-neurons), which model more and more complex mental objects, in a robust though flexible way. They form a memory evolutive system, called the *memory evolutive neural system* (MENS). It is shown how its properties agree with experimental data, for instance on the formation of memories and the learning process. Higher order cognitive processes are studied in Chapter 10. Higher animals are able to develop a semantic memory, which allows the formation of the archetypal core, a personal memory of the animal which is at the root of the notion of self, and of the formation of conscious processes.

5.4. *Appendix*

A brief appendix describes two generalizations of the concept of a colimit. The multi-colimits (Diers, 1971) and their generalization, the local colimits (Ehresmann, 2002), may be used to model some particular processes, such as ambiguous images. The hyperstructures of Baas (1992) could provide a more general framework within which many constructions done in memory evolutive systems could be adapted (*cf.* Baas *et al.*, 2004).

Lastly, there are a few notes about mathematical notations. To make the text understandable by non-mathematicians, in each chapter we briefly recall all the mathematical notions used in the chapter, and provide a number of examples in biology and sociology to illustrate them. Some more mathematical parts (in particular proofs of the theorems) are included to provide a rigorous treatment, but are not necessary to understand the main ideas, and so can be omitted by the non-mathematician.

Acknowledgements

We are most grateful to Ray Paton, who passed away while this volume was being completed; he constantly encouraged us and, without his insistence, we doubt we would have had the courage to write it.

We thank Michael Kary who has very attentively read our text, corrected its English, and provided a large number of stimulating comments which have helped us to improve many points; naturally he is not responsible for the errors which might remain. We also thank Laura McNamara for her careful editing of our manuscript.

Among all those who have stimulated and encouraged us along the years, we more particularly thank Jerry Chandler, George Farre and Brian Josephson for many lively exchanges; and Sylvie, as well as Jean-Paul's children, for their patience.

PART A. HIERARCHY AND EMERGENCE

Chapter 1   Nets of Interactions and Categories

1   Systems Theory and Graphs
   1.1   Objects and Relations
   1.2   Definition of a Graph
   1.3   Supplementary Properties
2   Categories and Functors
   2.1   Some History
   2.2   Definition of a Category
   2.3   Some More Definitions
   2.4   Functors
   2.5   Universal Problems
3   Categories in Systems Theory
   3.1   Configuration Categories of a System
   3.2   Fields of an Object
   3.3   Some Examples
4   Construction of a Category by Generators and Relations
   4.1   The Category of Paths of a Graph
   4.2   Comparison between Graphs and Categories
   4.3   Labelled Categories
   4.4   A Concrete Example
5   Mathematical Examples of Categories
   5.1   Categories with at Most One Arrow between Two Objects
   5.2   Categories with a Unique Object
   5.3   Categories of Mathematical Structures

Chapter 2   The Binding Problem

1   Patterns and Their Collective Links
   1.1   Pattern in a Category

    1.2   Collective Links
    1.3   Operating Field of a Pattern
2   Colimit of a Pattern
    2.1   The Mathematical Concept of a Colimit
    2.2   The Binding Problem
    2.3   Decompositions of an Object
3   Integration vs. Juxtaposition
    3.1   Comparison of the Sum to the Colimit
    3.2   Some Concrete Examples
    3.3   Some Mathematical Examples
4   Interlude: A Transport Network
    4.1   Characteristics of the Network
    4.2   Local Area Nets, Selected Nets
    4.3   Central Node
    4.4   Creation of a Central Node
    4.5   Inter-Area Nets

Chapter 3   Hierarchy and Reductionism

1   P-Factors of a Link Towards a Complex Object
    1.1   Links Mediated by a Pattern
    1.2   Link Binding a Perspective
    1.3   Examples of Perspectives
    1.4   Field of a Pattern
2   Interactions between Patterns: Simple Links
    2.1   Clusters
    2.2   Composition of Clusters
    2.3   Simple Links
3   Representative Sub-Patterns
    3.1   Comparison Link
    3.2   Representative Sub-Pattern
4   Multiplicity Principle
    4.1   Patterns with the Same Colimit
    4.2   Connected Patterns
    4.3   Multiplicity Principle and Complex Links
5   Hierarchies
    5.1   Hierarchical Categories
    5.2   Examples
    5.3   Ramifications of an Object and Iterated Colimits

6   Complexity Order of an Object: Reductionism
    6.1   *Obstacles to a Pure Reductionism*
    6.2   *Reduction of an Iterated Colimit*
    6.3   *Complexity Order of an Object*
    6.4   *Examples*
    6.5   *n-Multifold Objects*

Chapter 4   Complexification and Emergence

1   Transformation and Preservation of Colimits
    1.1   *Image of a Pattern by a Functor*
    1.2   *Replacement of a Category by a Larger One*
    1.3   *Preservation and Deformation of Colimits*
    1.4   *Limits*
2   Different Types of Complexifications
    2.1   *Options on a Category*
    2.2   *Complexification*
    2.3   *Examples*
3   First Steps of the Complexification
    3.1   *Absorption and Elimination of Elements*
    3.2   *Simple Adjoining of Colimits*
    3.3   *'Forcing' of Colimits*
4   Construction of the Complexification
    4.1   *Objects of the Complexification*
    4.2   *Construction of the Links*
    4.3   *End of the Construction*
5   Properties of the Complexification
    5.1   *Complexification Theorem*
    5.2   *Examples*
    5.3   *Mixed Complexification*
6   Successive Complexifications: Based Hierarchies
    6.1   *Sequence of Complexifications*
    6.2   *Based Hierarchies*
    6.3   *Emergent Properties in a Based Hierarchy*
7   Discussion of the Emergence Problem
    7.1   *Emergentist Reductionism*
    7.2   *The Emergence Problem*
    7.3   *Causality Attributions*

Chapter 1

Nets of Interactions and Categories

'Man cannot live in chaos' writes Eliade (1965, p. 32). To put order into our perceptions of the world which surrounds us, we humans must distinguish the objects in our surroundings and the relations between them. These words 'object' and 'relation' are taken here in a very general and vague sense. A first idea would be to think of an object as a physical body located in space; this is why some philosophers consider space to be the primitive datum (that will not be our point of view). However, the objects we will consider can also be more tenuous, like a musical tone, an odour or an internal feeling. The word *phenomenon* (used by Kant, 1790) or *event* (in the terminology of Whitehead, 1925) would perhaps be more appropriate, since we would like to use the term object in its most general sense, to refer to anything which can be apprehended, whether body, property, event, process, conception, perception or sensation, even if it be only temporary, such as the shape of a passing cloud.

Here we run into a dilemma: do we want to describe everything that exists at every given moment, *i.e.* the world in its constant flux; or, instead, to extract some more permanent aspects from the world? It is obvious that it will be necessary to do the latter if we want to recognize our environment and the dangers that it conceals, to avoid them, and to develop actions allowing for a better adaptation. In other words, it will be necessary to organize the world that we perceive, without presuming in which sense such an organization is real. What counts is to describe an organization that helps to plan successful action, and which, for humans, is translatable into language. These conditions already force us to differentiate more or less recognizable objects, so that we need recall and name only a reasonable number of them.

Moreover, the world cannot be understood as consisting solely of objects. The objects are not isolated but interact in various ways, and we need to take into account the more or less temporary relations among them. Long ago, the Taoists imagined the universe as a dynamic web of relations, whose events constitute the nodes; each action of a living creature modifies its relations with its environment, and the consequences gradually propagate to the whole of the universe. The Taoist philosophy of inaction comes from this. It does not imply pure passivity, but rather a search for more harmonious existence with nature.

Modern science is becoming more conscious of the need to consider interactions among objects. For example, in ecology, it is now better understood that local alterations have repercussions which may far exceed the sought after effects: the application of an insecticide, intended to destroy the insects of a field, may kill the birds which eat the insects; and perhaps even more dramatically the equilibrium of the entire area might be compromised.

In the model we propose for natural systems, the successive configurations of a system, as defined by its components and the relations among them around a given time, will be represented by *categories*; the changes among configurations by *functors*. The evolution of the system will mostly depend on the interactions between agents of various levels of complexity, acting with different time scales.

This chapter recalls the notion of a category, which refines the notion of a graph, and which is an essential tool in the subsequent chapters. Examples prove the ubiquity of categories: they range from groups to ordered sets and from 'large' categories of mathematical structures to the 'small' categories that we will use to model natural systems.

1. Systems Theory and Graphs

Systems theory was initiated by von Bertalanffy in 1926. It proposes to describe a system as: '... a set of unities with relationships among them' (von Bertalanffy, 1956, p. 1). It allows study of the most varied structures, as several authors have emphasized (*e.g.* Piaget, 1967; Klir, 1969, 1985; Jacob, 1970; Morin, 1977; Atlan, 1979; Laborit, 1983; Hofstadter, 1985). This theory is not a unified corpus, but rather a compendium of various models which have been applied to the study of natural and artificial systems of any kind, including physical systems, biological organisms, social groups and organizations, cognitive systems and works of engineering, from bridges to robots.

1.1. *Objects and Relations*

The first difficulty we encounter when we want to study a system (even a simple one) is determining what are the objects and which relations to consider between them. This all the more since what can be considered the objects, and what the relations, can sometimes overlap.

In the examples we will consider, the objects correspond to the various components of the system at a given time, of whatever complexity level; for instance in an organism its atoms, molecules, cells, tissues and organs. The relations between them can be of different types:

- spatial, indicating for example if the objects are close or distant, adjacent or not;
- causal, meaning that one event is a factor in the occurrence of another;
- informational, corresponding to information transmitted from one object to another;
- energetic, supporting a flow of energy between two objects; and
- constraints of any kind.

Many other types could be distinguished, but the important fact is that the objects interact, this interaction being completed more or less instantaneously, or else after a delay (as is typically the case in cause–effect relations).

To model a system as described above by objects and relations, it seems natural to use the notion of a graph: the objects become the nodes of the graph, and the relations the edges between them. Graphs have been used this way in the most varied settings to give a visual representation by which the structure of the system is better apprehended. For instance, the roads of a county can be represented by a graph in which the objects model the towns, and the relations model the various routes between them. Let us define the terms more precisely.

1.2.   Definition of a Graph

In mathematics, the study of relations is rather recent. It is the introduction of set theory, developed by Cantor at the end of the 19th century (see the correspondence of Cantor and Dedekind, 1937), which opened the way for the modern theory of structures and relations, as developed in the mid 20th century, and for the development of an axiomatic theory of graphs. However, the word graph has been used with several meanings, and we must specify which one we have in mind.

Here a *graph* will always be a composite item, consisting of a class of objects (its vertices), and a class of arrows between them; therefore, it is a 'directed' graph. Moreover, there can be several arrows from an object A to an object B, as well as 'closed' arrows from A to A (many authors speak of a multi-graph, reserving the word graph if there is at most one arrow between two vertices).

Definition. A *graph* G (also called a directed graph, or a *diagram scheme*) consists of a set of objects, called its *vertices* (or nodes), which we denote by|G|, and a set of (directed) *edges* (or *arrows*) from a vertex A to a vertex B, denoted by f: A $\rightarrow$ B. We call A the *source* of the arrow, and B its *target*.

There may exist several arrows with the same source and the same target (these are said to be parallel) and closed arrows are also accepted (Fig. 1.1).

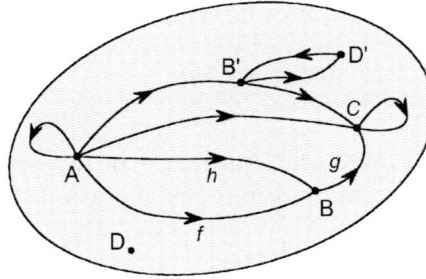

Fig. 1.1   A graph.
The vertices of the graph are represented by points, in this case A, B, B', C, D and D'.
The edges are represented by arrows such as *f*. Not all the arrows in the figure are
labelled. *f* is an arrow with source A and target B, as is *h*; these arrows are therefore
considered parallel. *g* is an arrow from B to C. The graph contains D as an isolated
vertex. There is a closed arrow (or loop) from A to A and one from C to C. Between B'
and D' there are two arrows in opposite directions.

Two arrows *f* and *g* are *successive* if the target of the first one is also the
source of the second one, say: *f*: A→B and *g*: B→C; they form a *path* (*f*, *g*)
of length 2 from A to B. More generally, a *path of length n* from A to B is
a sequence $(f_1, f_2, \ldots, f_n)$ of *n* successive arrows (Fig. 1.2), say:

$$f_1 : A \rightarrow A_1, \quad f_2 : A_1 \rightarrow A_2, \ldots, \quad f_n : A_{n-1} \rightarrow B$$

A *sub-graph* G' of a graph G is a graph whose set of vertices is contained
in |G| and such that each arrow *f* from A to B in G' is also an arrow from
A to B in G (but the converse is not necessarily true). For instance a path
of G (with its edges and its vertices) defines a particular sub-graph of G.

Since we are interested in the relations between objects, and a graph in
itself can be taken as an object, the question arises: what are the key
relations between two graphs? They are the ones which relate graphs having
the same basic structure, and are defined as follows:

Definition. If G and K are two graphs, a *homomorphism* from G to K is a
rule *p* which associates to every vertex A of G a vertex *p*A of K, and to every
arrow *f* from A to B in G an arrow *p*(*f*) from *p*A to *p*B in K.

1.3.   *Supplementary Properties*

Though a graph seems a good representation of a system—taking into
account both its components and the web of relations between them—it is
not sufficient for what we want to do, and it needs to be strengthened by
imposing some constraints on this web.

Any modelling rests on a certain approach towards the object to be
modelled: what features should be retained, and how those features should

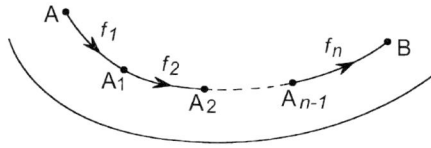

Fig. 1.2 A path.
A path of a graph G from vertex A to vertex B is a finite sequence of successive arrows of the graph, the first one beginning at A and the last ending at B. Here we have f_1 from A to A_1, f_2 from the target A_1 of f_1 to A_2, ..., f_n from A_{n-1} to B.

be represented. Our approach will be illustrated by the example of a cell. The objects that we consider model the components of the cell at a given time t: its atoms, molecules and macromolecules, but also larger differentiated sub-cellular parts, such as its organelles. These objects are interdependent, with various relations between them: contiguity, attachment of a molecule to an organelle, chemical interaction around t and so on. We consider that these objects and relations together constitute the internal organization, or con-figuration, of the cell at the given time t. These various objects are not all of the same level of complexity: from above, a molecule is itself a composite object, admitting a lower level internal organization formed by its system of atoms and their relations (*e.g.* bonding, spatial configuration); while from below, it is a 'simple' component of an organelle.

How then may we describe such a hierarchical structure with different complexity levels internally to a graph, so that an object of a certain level has an internal organization under the form of a lower level sub-graph of the graph (or, more generally, of a pattern in the graph, *cf.* Chapter 2)? In particular, what special properties must a sub-graph (or a pattern) P have in order to correctly represent the internal organization of a higher object C, which means that the relations of C to any other object A are correctly determined by the collective relations of P to A? As we shall see in Chapter 2, to answer this question, we have to recognize when two paths of a graph are functionally equivalent. For this, the graph must be endowed with a richer structure, that of a category, which ensures a kind of transitivity of relations by ensuring that each path of the graph be internally composed into a unique arrow. Then, two paths will be functionally equivalent if they have the same composite. Let us give a more precise definition.

2. Categories and Functors

A *category* can be defined as a composite item consisting of a graph and an internal law which associates an arrow of the graph to each path of the graph, called its composite, and which satisfies some axioms given further on. Categories form the basis of the model for natural systems which we develop.

2.1. Some History

The theory of categories has been much developed, both for itself and its applications in the most varied fields of mathematics. Even if parts of its autonomous development have been sometimes criticized as 'abstract non-sense' (as explained by Mac Lane, 1997, p. 5983), category theory is now recognized as a powerful language to develop a universal semantics of mathematical structures.

Eilenberg and Mac Lane introduced categories in the early 1940s (Eilenberg and Mac Lane, 1945) as a tool for studying difficult problems connecting topology and algebra, and to make calculation possible in topology, in particular to compute the homology and cohomology of a topological space. In fact, a particular case of categories, the groupoids, had been defined much earlier by Brandt (1926) as a generalization of groups, and was independently used in the 1940s by Charles Ehresmann in his important work on fibred bundles and the foundations of differential geometry (reprinted in Ehresmann, 1980–1982, part I).

There has been considerable debate regarding the nature and status of category theory and its role in mathematics (*e.g.* Bell, 1981; Landry, 1998). For Mac Lane (1986), category theory has a major organizing role within mathematics because, as he explicitly argues (Mac Lane, 1992), mathematics is not so much about things (objects) as about form (patterns or structures). Similarly, Charles Ehresmann, who made many essential contributions to category theory, stressed its unifying role. In a paper entitled 'Trends Toward Unity in Mathematics', he explains that, after the development of a large variety of new mathematical structures in the 19th century and the beginning of the 20th century,

> ... the necessity of unification was deeply felt: without a unifying theory following a period of rapid expansion, the mathematicians would fatally tend to use divergent, incompatible languages, like the builders of the tower of Babel. ... this theory of categories seems to be the most characteristic unifying trend in present-day Mathematics. (Ehresmann, 1967, p. 762–763)

Categories have also been extensively used in logic via the theory of topos. Grothendieck introduced the topos of presheaves for problems in algebraic geometry (Grothendieck and Verdier, 1963–1964), and Lawvere and Tierney (Lawvere, 1972) abstracted its properties in the general concept of an *elementary topos*, which can be thought of as a generalization of set theory allowing for an intuitive logic, into which a translation of most mathematical concepts is possible (*e.g.* natural number object, synthetic differential geometry, and so on.). More generally, many authors, following Lawvere (1966), think that category theory provides an alternative foundation to set theory. Others have argued that it is just a complement to set theory (*e.g.* Marquis, 1995), possibly imposing some supplementary conditions on it (Muller, 2001). We will not enter into this debate for we will essentially consider 'small' categories, *i.e.* those which are well defined in usual set theory.

Applications of categories have been developed in other domains, in particular in automata theory (Arbib and Manes, 1974; Eilenberg, 1974) and computer science. Since the 1980s, following the lead of Barr and Wells (1984, 1999), categorical techniques are used in the study of data types and of semantics of programming languages. Lawvere defended categories for physics as soon as 1980 (Lawvere, 1980; Lawvere and Schanuel, 1980), and recently, category theory has begun to be applied in string theory and in quantum physics (Aspinwall and Lawrence, 2001; Aspinwall and Karp, 2003; Abramsky and Coecke, 2004; Baez, 2004). Rosen introduced categories to biology as early as 1958, in his attempt to develop a relational biology (*cf.* Section 3 below).

2.2. Definition of a Category

In the following section, we will present only the notions of category theory which will be used in this book. For a complete theory, we refer to the well-known book of Mac Lane (1971). We define a category as a graph on which there has been given an internal law to compose successive arrows, satisfying some axioms. This presentation of categories follows that given in Ehresmann (1965). It is not the most usual one, but it has the advantage of emphasizing the diagrammatic and geometric nature of the theory, without suppressing its logical setting.

Definition. A *category* (Fig. 1.3) is the pair of a graph (called its *underlying graph*) and an internal composition law on this graph. The composition associates with each path (f, g) of length 2 from A to C an arrow of the graph from A to C, called the *composite* of the path and denoted by fg, so that the following axioms be satisfied:

(i) *Associativity*. If (f, g, h) is a path of length 3, the two composites $f(gh)$ and $(fg)h$ are equal (they can therefore be denoted unambiguously as fgh). It follows that a unique composite is also associated to any path (invariance of the route).

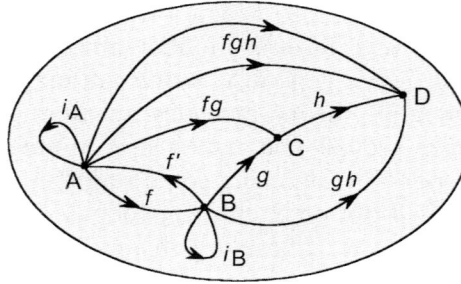

Fig. 1.3   A category.
A category is a graph in which any two successive arrows have a composite; for example, the arrows *f* from A to B and *g* from B to C have a composite arrow *fg* from A to C. Moreover, each vertex has a closed identity arrow, such as i_A for vertex A (the identities of C and D are not drawn). The composition is associative, whence giving a unique composite *fgh* for a path (*f, g, h*). The composite of an identity with another arrow is always that arrow. For the category as drawn, *i.e.* if no arrows have been omitted from the depiction, then *f'* must be the inverse of *f*, because i_A would have to act as the composite of *f* and *f'*.

(ii) *Identities*: To each vertex A there is a closed arrow i_A from A to A, called the *identity of* A, whose composite on the right or on the left with an arrow is equal to this other arrow.

Throughout the text, we denote the composite of (*f, g*) by *fg* (in the same order); most often it is denoted by *gf* (in the opposite order, by analogy with the notation for the composite of two maps). The vertices of the graph are called the *objects* of the category and its arrows the *morphisms* (we will often simply call them its *links*). An arrow *f* is an *isomorphism* if there exists an arrow, called its *inverse* and denoted by f^{-1}, such that the composites ff^{-1} and $f^{-1}f$ are defined and reduce to identities (this inverse is then unique). A category in which all the arrows are isomorphisms is called a *groupoid*.

Thus a category is formed by objects and arrows linking them (as in a graph), but in a category we also have an internal rule to compose successive arrows. The arrows and their composition play an essential role, for they determine the behaviour of an object A with respect to the other objects, as it is characterized by the class of arrows which arrive at, or go away from, A. What is taken into account is not the nature of an object, nor its construction, nor its internal structure independent of the context, but the manner in which it interacts with the other objects of the category through the arrows. Contrary to the more classical mathematical theories where the emphasis is put on the objects, here the morphisms are privileged over the objects (*cf.* Section 5.3 for an example).

To better understand how categories differ from and enrich graphs, let us give a simple example.

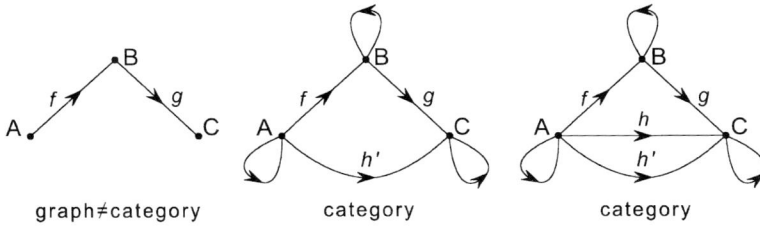

Fig. 1.4 Difference between a graph and a category.
The first figure represents a graph which does not underlie a category, since the path *(f, g)* has no composite and there are no identities. This graph is included in a smallest category (represented in the central figure) obtained by adding a link *h'* from A to C which becomes the composite *fg* of *f* and *g*, and identities for A, B and C. The third graph may underlie two different categories, depending on which arrow, *h* or *h'*, is taken as the composite of *f* and *g*.

We consider the three following graphs (Fig. 1.4): The first, which has only two successive arrows *f* and *g*, cannot underlie a category since the path *(f, g)* they form cannot have a composite in the graph (there is no arrow from A to C), and since there are no identity arrows. The second, obtained by adding one arrow from A to C and closed arrows, underlies a category in one and only one way, by taking the unique arrow *h'* from A to C to be the composite of *f* and *g*, and the closed arrows to be identities. The third graph, in which there are 2 arrows *h* and *h'* from A to C, underlies one category if we choose *h* to be the composite of *f* and *g*; but it underlies a different category if we choose *h'* to be the composite of *f* and *g*.

This shows that the structure of a category imposes more constraints than that of a graph, since it consists of a graph, and in addition, a composition law, specifying a rule to internally compose the paths of the graph. And the same graph can underlie several categories, or none.

2.3. *Some More Definitions*

Let us recall the following definitions:

- A category is *finite* if it has only a finite number of objects and arrows; this will be the case for the majority of the categories which we will use.
- Given a category, one obtains a category known as the *opposite* (or dual) *category* by keeping the same objects but by reversing all the arrows.
- If K is a category, a *sub-category* of K is a category H whose underlying graph is a sub-graph of K and whose composition law is a restriction of that of K; thus the identities and the composites in H are the same as in K. In other terms a sub-graph H defines a sub-category of K if it contains: with a vertex in H, its identity in K; and with a path of H, its composite in K.

In particular the set of isomorphisms of K defines a sub-category which is a groupoid.
• A sub-category H is said to be *full* if, for every two objects in H, it contains all the arrows that are between them in K.
• A sub-graph G of a category K is also called a *diagram in K*; it is said to be *commutative* if two paths of G with the same extremities have the same composite in the category. For instance: a commutative triangle consists of three arrows *f, g, h* such that *fg* = *h*; and a commutative square consists of 4 arrows *f, g, f',g'* such that *fg* = *f'g'* (Fig. 1.5).

Though category theory can be considered as a domain of pure algebra, its diagrammatic nature also lends itself to consideration as a geometrical domain. Often the proof of a categorical theorem amounts to displaying an appropriate sequence of diagrams and exploiting their commutativity; for instance, by computing the diagonal of a square in the two different possible ways. It is usually called 'chasing along the diagrams' (Mac Lane, 1971, p. 200). Naturally it is possible to do this by writing the corresponding formulas, but it is more intuitive simply to look at the figure. Moreover, a diagrammatic representation also allows one to imagine motion along it. For instance, paths in a graph are a way of jumping from one object to another. The composition rule of a category becomes a way of determining when two paths represent the same global motion. As Guitart writes:

> A diagram is a net of oriented segments between points (taken as abstract positions), the truth of which is found by moving along it, by selecting a good route, and by inserting this net in other nets. To do this, diagrams can be written inside the diagram, or the diagram itself can be taken as an abstract point in another diagram. (translated from Guitart, 2000, p. 93)

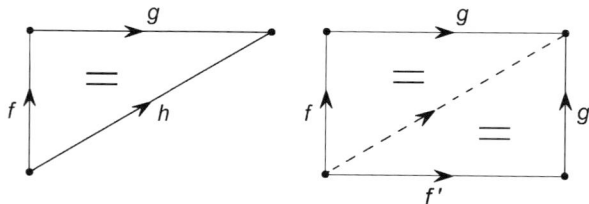

Fig. 1.5   Commutative diagrams.
Example of two simple commutative diagrams in a category. A commutative triangle has two successive arrows *f* and *g* and their composite *h*. A commutative square has four arrows forming two paths (*f, g*) and (*f', g'*) with the same extremities and the same composite, so that *fg* = *f'g'*.

2.4. Functors

We have said above that category theory privileges relations over the objects. At a higher level, we can consider a category itself as an object (of a particular type); then (as for graphs above) it is natural to ask, How do we appropriately define relations among categories? These relations are the *functors*, homomorphisms for the graph structure which preserve the composition law and the identities. More explicitly:

Definition. Let K and K′ be two categories. A *functor* from K to K′ is a graph homomorphism p from K (considered as a graph) to K′ which satisfies the following conditions:

(i) It maps a composite fg of two arrows f and g onto the composite $p(f)p(g)$ of their images by p:

$$p(fg) = p(f)p(g)$$

(ii) For each object A of K, it maps the identity of A onto the identity of pA.

Geometrically, the first condition means that a functor transforms a commutative triangle into a commutative triangle (Fig. 1.6). More generally, a functor transforms a commutative diagram into a commutative diagram.

If H is a sub-category of K, the *insertion* functor from H to K maps an object or arrow of H to the same element of K. In particular, if H = K, we have the *identity functor* from K to K. Functors can be composed. The *composite* of the functor p from K to K′ with a functor $p′$ is defined only

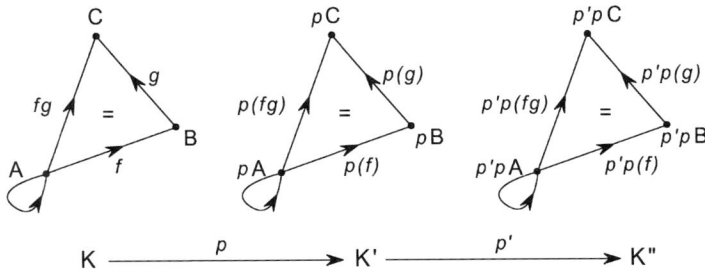

Fig. 1.6 Functors.
A functor p from K to K′ maps an object A of K to an object pA of K′, an arrow f from A to B to an arrow $p(f)$ from pA to pB, a commutative triangle in K to a commutative triangle in K′, and an identity to an identity. The composite of p from K to K′ with the functor $p′$ from K′ to K″ is a functor $p′p$ from K to K″ which maps A to $p′pA$ and f to $p′(p(f))$.

if p' is a functor from K' to some category K'', and then it is the functor from K to K'' which maps an object A of K to $p'pA$, and an arrow f from A to B in K to the arrow $p'(p(f))$ from $p'pA$ to $p'pB$ (Fig. 1.6).

A functor may map several objects of K to the same object of K' and several arrows to the same arrow; that is, it is not limited to one-to-one relationships. Moreover, it may not be 'onto' (the whole of) K', which means that there may exist objects or arrows of K' which are not the image of objects or arrows of K. A functor p from K to K' which comprises a one-to-one relationship and is onto K' is called an *isomorphism*. In this case, there exists a functor p^{-1} from K' to K, called the *inverse of p*, such that the composite of p^{-1} with p on the right is the identity of K and on the left the identity of K'.

2.5.   *Universal Problems*

A most significant aspect of category theory is its unifying way of handling *universal problems*, those which require an optimal solution for a given problem (in a sense to be made explicit in each case). An abstract formulation of universal problems has been given by Samuel (1948). Translated into the categorical setting it has led to the important notion of adjoint functors (Kan, 1958). In this book, we consider several universal problems, in particular the construction of colimits, the complexification process, and the construction of concepts. Though it will not always be done explicitly, they can be brought back to the following important situation.

Definition. Let p be a functor from K to K' and A' an object of K'. We say that A' generates a *free object* A *with respect to p* if A is an object of K and if there exists an arrow g' from A' to pA in K' satisfying the 'universal' condition: (UC) If B is an object of K and f' an arrow from A' to pB, then there exists a unique arrow f from A to B in K such that $f' = g'\ p(f)$ (Fig. 1.7).

If each object A' of K' generates such a free object, there is a functor from K' to K which maps A' to its free object; this functor is called an *adjoint of p*.

As we are essentially interested in applications of categories to model natural systems, in the next section we explain how they arise in this context, and more mathematical examples are given only later on.

3.   **Categories in Systems Theory**

Several authors have proposed using categories in systems theory and in biology. Rosen was probably the first to develop an abstract setting for a relational biology, in the terminology of Rashevsky, based on category

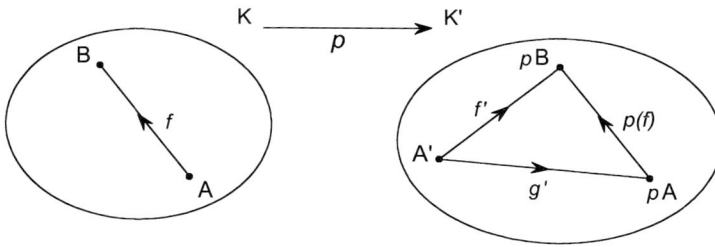

Fig. 1.7 A free object.

We have a functor p from K to K' and an object A' of K'. The object A' generates a free object, denoted by A, with respect to p if there is an arrow g' from A' to the image pA which satisfies the 'universal property': for each pair of an object B of K and of a link f' from A' to the image pB of B, there exists a unique arrow f from A to B such that f' is the composite $g'p(f)$ of g' with the image $p(f)$ of f.

theory. In his early papers (Rosen, 1958a,b, 1959, 1969), he described the category of (M, R)-systems, which are abstract models of the metabolism and repair activities of a cell (Arbib, 1966; Warner, 1982). He also gave a categorical version of the organismic systems of Rashevsky (1967, 1969, 1972; Baianu, 1970, 1971; Kainen, 1990). Later he introduced the category of morphogenetic networks (Rosen, 1981), and the category of natural systems (Rosen, 1978; Louie, 1983). However, in these works, the role of categories is most often purely descriptive, and deep results of category theory are not exploited. On the contrary, in our model where we use categories to represent the successive configurations of a system, we make use of fundamental constructions, to give an internal analysis of the structure and the dynamics of the system.

3.1.   Configuration Categories of a System

The internal organization, or *configuration*, of a natural system at a given time t, will be modelled by a category K, called the *configuration category at t*: the objects of K represent the components of the system which exist at t, and the arrows of K (which we simply call *links*) represent the interactions between them around t that define the present organization of the system. These links can be of different natures: more or less invariant structural links, such as spatial relations (*e.g.* as formed by desmosomes between contiguous cells); causal relations; energetic or informational relations; or those imposing constraints of any kind. The links can also represent labile connections, corresponding to a temporary interaction (chemical reaction). In any case, the links correspond to interactions between objects, which can be more or less extended in time.

Here we note an *important convention*: throughout this book, we employ a 'dynamic' vocabulary with respect to objects and links (and, in following chapters, sub-systems) of a category modelling a natural system: *e.g.* an object acts or observes, a link transmits information or constraints and so on. However, we do not intend to convey intentionality on the part of the objects or the links.

Applying the preceding convention within a configuration category, we say that an object A modelling a component of the system plays a double role:

(i) It acts as a causative agent or as an emitter, by way of its links toward other objects, which represent either its actions upon them, information it sends them, or constraints it applies to them.

(ii) It becomes a receptor, or an observer through the links which arrive at it, which correspond to aspects it observes, messages it receives, or constraints imposed upon it.

The identity of A models its 'self', in particular in a biological system. The composition law of the category models the transitivity of the interactions. For instance, if the object A transmits information to, or imposes a constraint on, B via f, and B relays that to C via g, then the composite fg of the two links represents the combined process by which the information or constraint gets from A to C. This does not prevent A from also having a direct channel to C, for example, to transmit information without passing through B, so as to accelerate the transmission and not to distort it, or to transmit confidential data. The composition being associative, more general paths of links are obtained, with several relays, and two such paths represent *functionally equivalent* interactions (of whatever type) if and only if they have the same composite in the category. Thus composition amounts to characterize classes of functionally equivalent paths of interactions between objects.

As the components of the system and their interactions may change over time, the category models only the configuration of the system at a given time t. A change of configuration, say from t to t', will be modelled by a functor which keeps track of the transformations of the various objects and interactions. However, if some objects no longer exist at t', the functor is no longer defined on them, and becomes instead a partial functor:

Definition. A *partial functor from* K *to* K′ is a functor from a sub-category of K to K′.

The evolutive systems, defined in Chapter 5, will model the entire system by its successive configuration categories, with partial functors between them modelling their changes of configurations.

3.2. Fields of an Object

Categories are useful for representing knowledge in a wide variety of fields because of the wealth of information that links convey about the objects they connect, whatever the nature and the anatomy of these objects. In the preceding discussion, we have explained how to interpret the arrows to and from an object; it suggests associating to each object A of a category (*e.g.* a configuration category of a system), another category which sums up all the information that A can internally collect, through the links it receives from other objects. This category is called the *field* (or, more precisely, the *perception field*) of A. Roughly, the objects of this field are the links to A, and its arrows are the commutative triangles with vertex A.

Definition. The (*perception*) *field* of an object A in a category K is the category defined as follows (Fig. 1.8):

(i) Its objects are all the links h with target A in K;
(ii) An arrow from h: C→A to h': C′→A is defined by a link k: C→C′ correlating h and h' in the sense that $h = kh'$ (so that h, h', k forms a commutative triangle with vertex A); it is denoted by k: $h→h'$;
(iii) The composite of k: $h→h'$ with k': $h'→h''$ is kk': $h→h''$.

There is a '*base*' functor from the field of A to K which associates C to the link h: C→A and k: C→C′ in K to the link k: $h→h'$. This functor measures the difference between the internal model that A can form of the system (namely, its field) and the system itself. It is an isomorphism (so that A has a

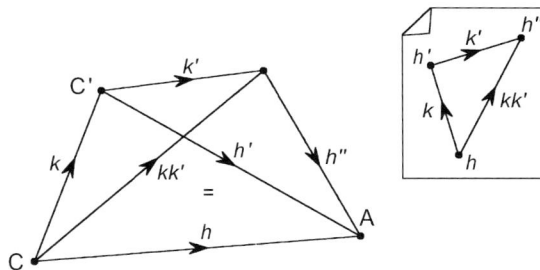

Fig. 1.8 The perception field.
The (perception) field of the object A in the category K has for objects the arrows of K arriving to A, in the figure h, h' and h''. An arrow k: $h→h'$ from h to h' is defined by an arrow k: C→C′ in K such that (h, h', k) be a commutative triangle. Its composite with the arrow k': $h'→h''$ is the arrow kk': $h→h''$. In the left figure, the elements of the field of A are seen in the category K; the field as a category is drawn on the right.

complete view of the system) only if A is a *final object*, that is, an object to which any other object is linked by one and only one link. For example, in an ecosystem an animal recognizes only its *Umwelt* (in the terminology of von Uexküll, 1956) in which he cannot see a predator which is not near enough.

We also define the *operating field of* A in K, which models the objects of the system on which A can act and what interactions it has with them, *e.g.* transmission of information or imposing a constraint. Formally, it is obtained by 'inverting the arrows' so that the operating field of A in K becomes the perception field of A in the opposite category K^{op}. However, this is a purely mathematical formulation; in the applications to systems, the opposite category has no practical meaning for the system.

Definition. The *operating field* of an object A in the category K is the category defined as follows (Fig. 1.9):

(i) Its objects are all the links f with source A;
(ii) The arrows $g: f \to f'$; from $f: A \to B$ to $f': A \to B'$ are defined by the links $g: B \to B'$ correlating f and f' in the sense that $f' = fg$;
(iii) The composite of $g: f \to f'$ with $g': f' \to f''$ is $gg': f \to f''$.

In this case also we have a base functor associating to the triangle f, f', g its base g. It is an isomorphism if and only if A is an *initial object*, that is if there exists a unique arrow from A to any object of K.

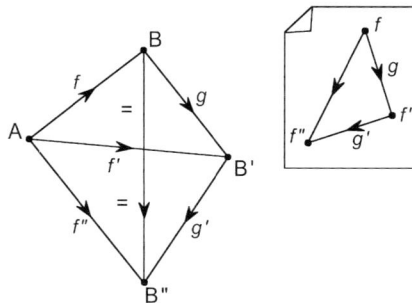

Fig. 1.9 Operating field.
For the category K depicted on the left, the operating field (depicted on the right) of A in K has for objects the links (in the figure f, f', f'') from A to other objects in K, and its links, such as $g: f \to f'$, correspond to the commutative triangles with vertex A. Thus it is defined as the perception field except that we take the arrows beginning at A instead of arriving at A. The operating field can also be defined as the perception field of A in the category opposite to K. As in Fig. 1.8, the operating field as a category is drawn on the right.

3.3. Some Examples

Let us give some examples of complex systems which will be modelled by evolutive systems. When no confusion is possible, we speak of the configuration category at a given time as if it were the system itself, and name it by the name of the components of the system; in each case the links represent the 'natural' interactions occurring between these components, and should be defined more explicitly.

(i) The objects of the *category of particles and atoms* model the elementary particles and the atoms; similarly we have the *category of molecules*, and the *category of physical objects* (*cf.* Chapter 5).
(ii) For modelling a *cell*, a configuration category will model its present components of every level: atomic, molecular, macromolecular, sub-cellular (with populations of molecules, organelles, functional complexes and so on).
(iii) For a *living organism*, we add the cellular level and the tissue or organ level.
(iv) In the category associated to a *society*, we have the level of individuals and several levels in which the objects represent the more or less complex social groups they form.

4. Construction of a Category by Generators and Relations

We have said that a category consists of a graph and a rule to compose its paths, so that functionally equivalent paths might be identified. Here we make this assertion precise by exploring more thoroughly the relationship between graphs and categories, leading to a general construction of categories: a graph generates the category of its paths (which has more arrows than the graph!); any category can be obtained as a quotient of the category of paths of its underlying graph.

4.1. The Category of Paths of a Graph

Not every graph underlies a category because it is not always possible to define an internal composition on a graph (*cf.* Fig. 1.4). However, we are going to describe how any graph generates a 'largest' category, namely the category of its paths, in which a path of the graph admits as its composite the path itself considered as a new arrow.

Let G be a graph. We recall that a *path of length n* from A to B is a sequence $(f_1, \ldots, f_n)$ of successive arrows (*cf.* Fig. 1.2); we admit paths of length 0 (no arrow) from A to A. We say that G *generates a category* K if G is a sub-graph of K and if each arrow of K is the composite in K of a path of G. A graph can generate different categories; and among them there is a largest one, defined as follows.

Definition. The *category of paths* of a graph G, denoted by P(G), has for its objects the vertices of the graph and for arrows from A to B all the paths from A to B. The composition is given by concatenation: the composite of

$$(f_1, f_2, \ldots, f_n) : A \to B \quad \text{and} \quad (g_1, g_2, \ldots, g_m) : B' \to C$$

exists if, and only if, $B = B'$, and is the path:

$$(f_1, f_2, \ldots, f_n, g_1, g_2, \ldots, g_m) : A \to C$$

The identity of A is the path of length 0 from A to A.

We identify G to a sub-graph of P(G) by identifying an arrow f of G with the path (f) of length 1. Each path of G is then its own composite in P(G), so that G generates P(G). Moreover, the category of paths has the following universal property.

Proposition. *A graph G generates the category of its paths P(G). If p is a homomorphism from G to any category K, then p extends into a unique functor from P(G) to K, which maps a path $(f_1, f_2, \ldots, f_n)$ on the composite $p(f_1)p(f_2) \ldots p(f_n)$ in K.*

This last property implies that there is a functor from P(G) to each category K which contains G as a sub-graph (we take for p the insertion of G into K); if G generates K, this functor is onto K, so that K has less arrows than P(G), several arrows of P(G) being identified by p. It explains in which sense P(G) is the largest category generated by G.

4.2. Comparison between Graphs and Categories

We have associated to a graph the category of its paths. Since a category K' is also a graph, we can form the category of its paths P(K). What is the relationship between K and P(K)? From the above proposition (applied to the identity of K), it follows that there is a functor from the category of paths P(K) to K which maps a path of K on its composite for the composition of the category K. This functor is onto K and identifies two paths if and only if they have the same composite in K. In other terms, K is the *quotient category* of P(K) by the equivalence relation in which two paths are equivalent if and only if they have the same composite in K.

This suggests the following general process to construct a category, by giving a graph G and a relation on its paths indicating which paths should have the same composite (*i.e.* for categories representing systems, which are functionally equivalent). More formally:

Proposition. *Let G be a graph, and let R be a relation on the set of its paths, formed by pairs (c, c') of paths with the same extremities. There exists a* *'universal' functor q from the category* P(G) *of paths of* G *to a category* K *such that* $q(c) = q(c')$ *if* (c, c') *is in* R.

Proof. The category K is obtained from P(G) by identifying two paths which are in the equivalence relation R' generated by R, so that K is the quotient category of K by R'. It has the same objects as G. The functor q from P(G) to K maps a path c on its equivalence class for R'. It is universal in the sense that it factors any other functor from P(G) to a category which takes the same value on two paths which are in the relation R.

In analogy with the construction of a group by generators and relations, we say that K *is a category constructed by generators and relations*, the generators being (the arrows of) G and the relations being the pairs (c, c') of paths in R to be identified.

Let us summarize the relations between categories and graphs:

(i) Given a graph G, it always generates a largest category, namely the category P(G) of its paths. This defines a one-to-one correspondence between graphs and categories of paths (also called *free categories, cf.* Section 5.3 below). G also generates a smallest category obtained as the quotient of P(G) by the relation identifying two paths with the same source and the same target. (ii) Conversely, any category K is the quotient category of the category P(K) of its paths by the equivalence relation in which two paths are equivalent if and only if they have the same composite in K.

It follows that any construction or result established for graphs can be applied to categories, by 'forgetting' the composition law. Conversely, results about categories can be extended to graphs by replacing a graph by the category of its paths. However, categories of paths are very special, so that many important categorical constructions (such as the construction of colimits, *i.e.* the 'binding process', *cf.* Chapter 2) become trivial for them; that explains why graphs are not rich enough for the applications we have in view.

4.3. *Labelled Categories*

In complex natural systems material conditions are generally imposed on the interactions between components, in particular temporal or energetic constraints; these are measured by observables which associate to each link a *weight* (a real number or vector) representing its intensity or extent in an appropriate unit system (*e.g.* propagation delay, latency, length, threshold and so on).

To model this, the configuration categories will be *labelled*, a weight being associated to each link. The weights will be taken in a monoid M (a monoid can be defined as a category with a unique object, *cf.* below, this chapter,

Section 5); most often M is a monoid of numbers (integers or real numbers) with an addition or a product operation, or a monoid of vectors with addition.

Definition. A graph G is *labelled in a monoid* M if there is given a map w associating to each arrow f of G an element $w(f)$ of M called its *weight*. A category K is labelled in M if it is labelled as a graph and if the weight $w(fg)$ of the composite of f and g in K is the composite $w(f)w(g)$ in M.

If G is labelled in M, the weights can be extended to its category of paths P(G) by taking the weight of a path to be the composite in M of the weights of its factors, so that P(G) becomes a category labelled in M. We define the *category associated to the labelled graph* (G,w) as follows: it is the category K generated by G and by the relations identifying two paths with the same extremities and the same weights. Thus K has for objects the vertices of G, and an arrow is an equivalence class of paths with the same weights. Consequently, an arrow is entirely determined by the data of its source, its target and its weight. The category K is also labelled in M.

The categories modelling natural systems are often labelled categories constructed by this process: weights are initially given on (a graph of) generators, and the category is constructed by generators and relations, as the category associated to this labelled graph. An example is the category of neurons modelling the brain of an animal at a given time t. It is associated to the graph in which the vertices model the neurons of the animal, and the arrows the synapses between them, labelled by their strengths at t. This category is defined more explicitly in Chapter 9 where it is extensively used.

4.4.   *A Concrete Example*

Let us use the above construction to construct the category K representing the different travel times between some cities of a country, knowing the possible routes between the cities and admitting that the travel time is proportional to the length of the route. First, we define the graph G formed by the routes and labelled by their lengths (in the monoid of additive integers): it has for vertices the cities and for arrows the (one-way) routes between them, the weight of a route being an integer representing its (relative) length. This graph G has seven objects A, A_1, A_2, A_3, A_4, A_5 and B, and 10 arrows representing the routes between these cities (Fig. 1.10), denoted by f_i; these arrows, followed by the corresponding lengths, are

$$(f_1, 2) : A \rightarrow A_1, (f_2, 3) : A_1 \rightarrow A_2, (f_3, 5) : A_2 \rightarrow A_3,$$
$$(f_4, 4) : A_3 \rightarrow B, (f_5, 6) : A \rightarrow A_4$$
$$(f_6, 8) : A_4 \rightarrow B, (f_7, 7) : A \rightarrow A_5, (f_8, 7) : A_5 \rightarrow B,$$
$$(f_9, 12) : A \rightarrow B, (f_{10}, 5) : A \rightarrow A_2$$

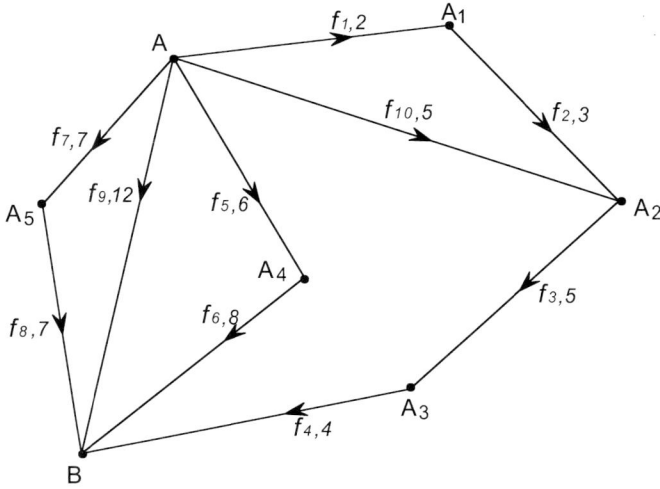

Fig. 1.10 The graph G of routes.
The vertices of G are the seven cities A, A_1, A_2, ... , A_5, B, and its arrows represent the possible routes between them. G is a labelled graph, the weight of an arrow being its (relative) length. These lengths are indicated after the name of the arrows; thus for example f_5, 6 indicates that arrow f_5 has length 6.

The category P(G) of paths of G has the same objects, but 20 arrows (Fig. 1.11), obtained by adding to G the following arrows:

$$(f_1, f_2) : A \rightarrow A_2, (f_1, f_2, f_3) : A \rightarrow A_3, (f_1, f_2, f_3, f_4) : A \rightarrow B,$$
$$(f_2, f_3) : A_1 \rightarrow A_3, (f_2, f_3, f_4) : A_1 \rightarrow B, (f_3, f_4) : A_2 \rightarrow B,$$
$$(f_3, f_6) : A_2 \rightarrow B, (f_7, f_8) : A \rightarrow B, (f_{10}, f_3) : A \rightarrow A_3,$$
$$(f_{10}, f_3, f_4) : A \rightarrow B.$$

Each path is labelled by the sum of the lengths of its factors. The composition of two successive paths is their concatenation, for instance

$$f_1(f_2, f_3) = (f_1, f_2, f_3) \text{ and } (f_1, f_2)(f_3, f_4) = (f_1, f_2, f_3, f_4)$$

In this category, we have several arrows (*i.e.* several paths of G) with the same length between two cities, such as (f_1, f_2) and f_{10} from A to A_2.

The category K of travel times (Fig. 1.12) is constructed by identifying two paths with the same length, so that

$$(f_1, f_2) \approx f_{10}, (f_1, f_2, f_3) \approx (f_{10}, f_3), (f_1, f_2, f_3, f_4) \approx (f_5, f_6) \approx (f_7, f_8)$$

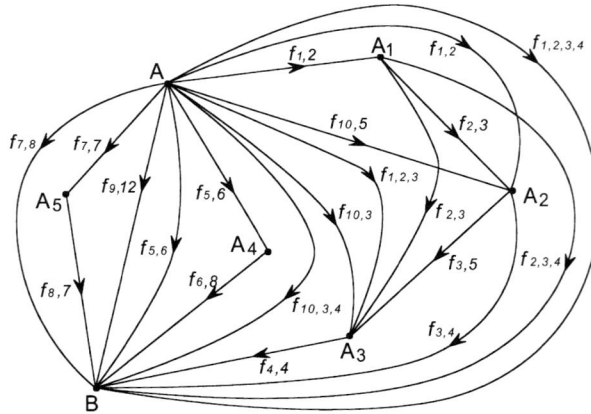

Fig. 1.11   The category of paths.
For the graph G of routes, the category P(G) of its paths has for arrows all the paths of
G (the identity paths are not drawn). A path reduced to one arrow is identified to this
arrow, so that P(G) contains G as a sub-graph. The lengths of the paths not in G are
not indicated. To simplify the notation the path (f_1, f_2) is denoted by $f_{1,2}$, the path
(f_1, f_2, f_3) by $f_{1,2,3}$ and so on.

It follows that K has the same objects, but only 15 arrows, the 5 arrows
added to the initial graph G being, with their length

$$(f_{2,3}, 8) : A_1 \to A_3, (f_{3,4}, 9) : A_2 \to B,$$
$$(f_{1,2,3}, 10) : A \to A_3, (f_{2,3,4}, 12) : A_1 \to B, (f_{1,2,3,4}, 14) : A \to B.$$

The composition is defined by

$$f_1 f_2 = f_{10}, \; f_1 f_{2,3} = f_{1,2,3}, \; f_1 f_{2,3,4} = f_{1,2,3,4}, \; f_2 f_3 = f_{2,3}, \; f_2 f_{3,4} = f_{2,3,4}, \; f_{2,3} f_4$$
$$f_3 f_4 = f_{3,4}, \; f_5 f_6 = f_{1,2,3,4}, \; f_7 f_8 = f_{1,2,3,4}, \; f_{1,2,3} f_4 = f_{1,2,3,4},$$
$$f_{10} f_3 = f_{1,2,3}, \; f_{10} f_{3,4} = f_{1,2,3,4}, \; f_{1,2,3} f_4 = f_{1,2,3,4}.$$

There are no other pairs having a composite. The length of a composite is
the sum of the lengths of its factors. In this category, $f_{1,2,3,4}$ and f_9 are two
arrows from A to B with different lengths (14 and 12).

5.   Mathematical Examples of Categories

Categories are applicable in the most varied mathematical situations: they
simultaneously generalize sets, (partially) ordered sets and groups, while
also allowing for a general theory of mathematical structures. A discussion
follows.

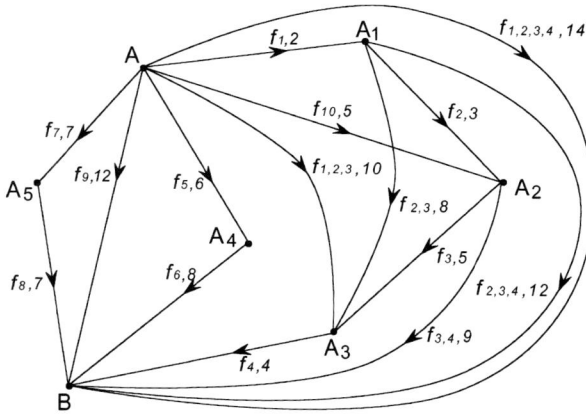

Fig. 1.12 The category of travel times.
The category K of travel times shows the possible travel times between the cities. It is obtained from the category P(G) of Fig. 1.11 by identifying two paths between two cities if they have the same length, hence the same travel time. Here the lengths of all the arrows are indicated after their names.

5.1. Categories with at Most One Arrow between Two Objects

A set E can be the set of objects of many different categories. Among these categories we can characterize those in which there is at most one arrow between two objects x and y.

(i) The smallest such category is the *discrete category* in which the sole arrows are the identities. Thus a set can be 'identified' to a discrete category.
(ii) The largest one is the *groupoid of pairs*, which has one unique arrow from x to y, namely the pair (x, y); the composite of (x, y) with (y, z) is (x, z), and the identities are the pairs (x, x); the inverse of (x, y) is (y, x).
(iii) The others are sub-categories of the above groupoid which correspond to the pre-order relations on E as defined below.

Let us recall some mathematical definitions of relations. A *relation on* E is defined as a subset R of the product $E \times E$ (hence as a set of ordered pairs of elements of E). The relation is: reflexive if (x, x) is in R for each element x of E; transitive if (x, y) and (y, z) in R imply that (x, z) is in R; symmetric if (x, y) in R implies that (y, x) is also in R; anti-symmetric if at most one of the pairs (x, y) and (y, x) is in R. A *pre-order* on E is a relation which is transitive and reflexive, an *equivalence relation* is moreover symmetric, while a *(partial) order* is a pre-order which is anti-symmetric; a set equipped with a partial order is called a *poset*.

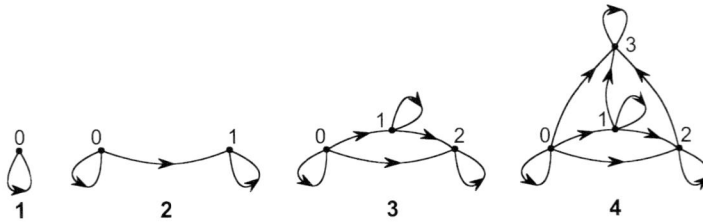

Fig. 1.13 Categories associated to integers.
The category associated to the integer n has the $n+1$ integers 0, 1, 2, …, n less than or equal to n as its objects, and the arrows determine the order between integers. Thus the category **1** has a unique object 0 and a unique arrow which is an identity. The category **2** has 0 and 1 for objects; its arrows are the single arrow from 0 to 1, and the identities of 0 and of 1. Apart from the identities, the category **3** has three arrows forming a commutative triangle, and the non-identity arrows of the category **4** form a tetrahedron.

Any relation on E defines a graph admitting E as the set of its vertices and in which there is an arrow (x, y) from x to y if and only the pair (x, y) belongs to R. It defines a category if the composite of two successive arrows exists in this graph (so that R must be transitive) and if each x has an identity (x, x) (so that R is reflexive); thus R must be a pre-order on E. In particular, if R is an equivalence relation, the corresponding category is a groupoid, each arrow (x, y) admitting (y, x) for its inverse. And if R is an order, the corresponding category has the identities as its sole isomorphisms. For example, to each integer n we associate the category defining the order of the set of integers less than n. The categories **1**, **2**, **3** and **4** thus associated to 1, 2, 3 and 4 are given in Fig. 1.13.

Beginning in Chapter 5, we take time into account by considering categories associated to the order on a (finite or infinite) subset of the positive real numbers.

5.2. *Categories with a Unique Object*

In the preceding examples, each category contained several objects with at most one arrow between two objects. Conversely, there are important examples in which there is only one object, but several closed arrows. A category with only one object is called a *monoid*. It reduces to a set on which there is given a composition law defined for every pair of elements, this law being associative and having an identity. The best known example is the *monoid of words* on an alphabet: its unique object is the null word; the arrows are all the finite sequences of letters of the alphabet, with possible repetition of a given letter. The composite of the words $(a_1,…, a_n)$ and

$(b_1, \ldots, b_m)$ is the concatenated word (which is always defined):

$$(a_1, \ldots, a_n, b_1, \ldots, b_m)$$

This monoid is widely used in applications in computer science: if the alphabet is reduced to two letters 0 and 1, the monoid of its words is formed by all the dyadic numbers.

A *group* is a groupoid with a unique object, corresponding to the unity of the group; or, equivalently a monoid in which all the 'arrows' are isomorphisms. Groups are an essential tool in most domains of mathematics, but their importance is also great in physics (examples: group of the Euclidean motions in the plane or the space, various symmetry groups, Lorentz group, quantum groups, Lie groups and others). Very recently, it has been proposed that in various applications, groups be replaced by groupoids (seen as a generalization of symmetry groups), in particular to model the network dynamic of biological systems (Stewart, 2004).

5.3. Categories of Mathematical Structures

In the definition of a graph (hence also of a category) we have assumed that the vertices form a set. This eliminates the need to tackle any foundational problems of category theory, and this condition will be fulfilled in the 'small' categories we use to model natural systems. However, this assumption is not necessary; indeed, the categories which have been most widely used are 'large' categories which do not satisfy this condition. In fact, the most usual definition of a category is a collection of objects (say A, B,...) and of sets Hom(A, B) of morphisms between them satisfying some rules. This definition is given in a theory of sets and classes (*e.g.* the Bernays–von Neumann theory [Bernays, 1937, 1941]), so that the collection of objects can be a proper class, thus allowing one to speak of *large categories*, and in particular of the large *category of sets*. This category, denoted by **Set**, has for objects all the sets and for arrows from A to B the maps from A to B. The composite of a map p from A to B with p' from B' to C is defined if and only if $B' = B$, and then it is the map from A to C which is their usual composite: it associates to an element a of A the element $p'(p(a))$ of C.

In **Set** we can illustrate how morphisms (*i.e.* maps) are privileged over objects (as said in Section 2). The 'void' set is characterized as the initial object of **Set**; that is, the unique object with one and only one morphism to every other object. A singleton is characterized (up to an isomorphism) as a final object; that is, an object with one and only one morphism coming from any other object. Moreover, the elements of a set E can be recovered as the morphisms from a singleton to E. This has led many authors, following Lawvere, to propose categories rather than sets as the foundation of mathematics. In particular, topos theory formalizes the axioms a category must satisfy to allow a development of mathematics internal to any topos instead of based on sets.

To each kind of mathematical structure there is associated a large category in which the objects are sets with such a structure, and the morphisms 'preserve' the structure. For example, the *category of groups* **Group** has for its objects the groups and for its arrows the homomorphisms between groups; there is a 'forgetful' functor from **Group** to **Set** which maps a group to its underlying set (it forgets the structure of group to keep only the underlying set). The *category of rings* has for objects the rings and for arrows the homomorphisms between rings; the *category of vector spaces* has for objects the vector spaces, for arrows the linear maps between them; the *category of topologies* has for objects the topological spaces and for arrows the continuous maps between them.

Graphs and categories are particular mathematical structures, and so they have their own associated large categories: the *category of graphs* **Graph** has for objects the graphs and for morphisms the homomorphisms between them; and the *category of categories* **Cat** has for objects the (small) categories and for morphisms the functors between them, the composition being that defined in Section 2 (above). Let us note that the structure of a category is remarkable because of its reflexivity: the category of (small) categories is itself a (large) category. There is a forgetful functor from **Cat** to **Graph** which maps a category on its underlying graph and a functor on its underlying homomorphism. This functor has an adjoint: the free object generated by a graph G is the category P(G) of paths of G; thus a category of paths is also called a *free category*.

A general theory of mathematical structures was initiated by Bourbaki (1958) in the late 1930s and early 1940s. It was influenced by discussions with thinkers and philosophers (*e.g.* Lautman, 1938; Cavailles, 1938a,b; Queneau; Levi-Strauss, 1962) who were friends of some of the first Bourbakists, and its development is parallel to that of structuralism in France. However, this theory was not really formalized; instead, category theory has provided an adequate frame for a general theory (Ehresmann, 1960, who also stresses this link with structures in the title of the book 'Catégories et Structures', Ehresmann, 1965). Lawvere used the theory of categories to present a general theory of algebraic structures in his thesis (Lawvere, 1963, 1965). The theory of sketches, initiated by Ehresmann in 1966 (reprinted in Ehresmann, 1980–82, Parts III and IV) and thoroughly developed in the 1970s and 1980s, extended this to other structures, such as categories themselves and even topologies (Burroni, 1970).

General results of category theory can be applied to the different categories of structures, thus allowing for a unified treatment; for instance, the notions of sub-structure, quotient structure, of product or sum, more generally of colimit and limit (*cf.* Chapters 2 and 4) can be generally defined. However, fine results on a specific structure (such as the study of representations of groups) necessitate a particular analysis of the structure.

Chapter 2

The Binding Problem

The objects or components of a natural system, such as a biological system, exist at different levels of complexity. For example, in an organism we have (at least) the atomic, molecular, sub-cellular, cellular, tissue and organ levels. Complexity is contextual: a cell is 'simple' at the tissue level; but 'complex' compared to its proteins. And similarly a protein admits its own internal organization, with its atoms, their spatial relations and chemical bonds, all leading to its final conformation. As Jacob writes, 'Every object that biology studies represents a system of systems; itself an element of a higher order system it sometimes obeys, following rules which cannot be deduced from its own analysis' (translated from Jacob, 1970, p. 328).

We will translate this concept of an object which is itself a system of systems to the categorical setting, and relate the properties of the overall system to those of its component systems. The idea is that a complex object has an internal organization consisting in a pattern of its more elementary components, with distinguished links between them; and the object binds together its components along these links. This is a manner to capture the sense of the three 'essential notions' expressed by E. Morin:

- the crucial idea of interaction, the real Gordian knot of hazard and necessity, since, under specific conditions, a random interaction leads to necessary effects;
- the idea of transformation, specially the transformation of separate elements into an organized whole, and conversely of an organized whole into scattered elements;
- the key idea of organization (translated from Morin, 1977, p. 80).

The model thus obtained allows us to tackle several problems: the binding problem, studied in this chapter; the hierarchy problem, studied in Chapter 3; and the emergence problem, studied in Chapter 4.

1. Patterns and Their Collective Links

We have proposed modelling natural systems by categories, wherein the objects represent the components of the system at a given time, and the arrows (called links) their interrelations. In Chapter 1, we referred to these categories as constituting the configuration or organization of the system; here, we will explore this key concept in more detail, and refine it to include the internal organization of an object of these categories.

In complex systems, we distinguish components at various levels of complexity. However, in a category, the objects themselves have no distinguishing features, and the only information we have about them comes from their links. In Chapter 1 we recalled that the idea of category theory is to encode the information within the links rather than in the objects. The question then arises: How can we use the links to recognize that a given object is complex, in the sense of having an internal organization that allows its components to operate synergistically? To answer this question, we first define the structure of such an organization as a family of interacting objects, which we call a pattern in the category. We define a pattern independently of the fact that it may or may not in turn define the internal organization of a complex object. Later we will characterize the properties a pattern must possess when it is in fact the internal organization of a more complex object.

1.1. Pattern in a Category

In Chapter 1 we employed a graph (or rather a category) to represent the configuration of a system. Since the internal organization of a complex object C of a category K can be thought of as a system itself, it seems natural to represent it by a graph G, and to relate this graph G with the category K itself. The first idea is that G is a sub-graph of K. For example, a molecule in a cell would be modelled by the sub-graph of the category modelling the cell representing its atoms and their chemical bonds. However, this representation is not always adequate. Indeed, there is the matter of the function that a component of the internal organization P of C performs, *i.e.* its role within P, or in other words, its effects upon other components of P through its links with them in this organization (for more on the notions of function and role, see Mahner and Bunge, 2001). The problem is that the same object of K can perform several different functions within P, and these have to be differentiated in the organization since they play different parts. For example, within the category K that models a country and its social groups at a given time, a professional association C, such as a scientific society, is a complex object. A member of this society is modelled by an object of K with a 'membership' link to C. However, the same member can play several roles in the association; for example, he can act as an accountant as well an editor of the association journal. In this case, the member must be counted several times in the organization of the scientific society (C), once for each of his functions in C. This would also be the case if the association were informal, with no complex object representing it. Thus, to account for the several roles an object of K may play in a pattern P, in particular if P is the internal organization of a complex object C, a component of the pattern should not reduce to an object of K, but must give an integral representation of this object and one of its (possibly multiple) roles.

Thus, a rigorous definition of a pattern P in a category K requires a further layer of abstraction. We dissociate the formal organization of the pattern, called its *sketch*, from its implementation in the system. The sketch indicates the various functions to be held in this organization, while the implementation determines which objects of the category fulfil these various functions, and how they interact. The sketch is modelled by a graph sP, whose vertices are entirely non-descript and identified only formally: they serve only as indices; its arrows determine the links between the components. Its implementation in the category associates to each vertex of the sketch (or index) an object of the category, and to each arrow of the sketch a link between the corresponding objects. In other words, the sketch allows us to dissociate the 'form' from the 'matter' which it 'informs'—much like Plato dissociated the ideal object, the pure form, from its material shadow, or implementation. More formally:

Definition. A *pattern* P in a category K is a homomorphism of a graph sP to K (*cf.* Fig. 2.1). The graph sP is called the *sketch* (or diagram scheme) of the pattern, and its vertices the *indices* of the pattern. It is generally finite. The *implementation* of P is the image in K of the sketch under P. The ordered

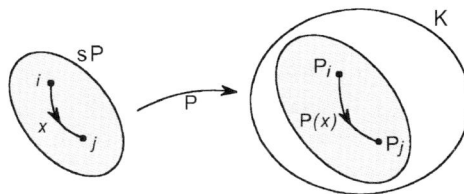

Fig. 2.1 A pattern.
A pattern P in the category K is a homomorphism of graphs P from a graph sP to K, where sP is called the *sketch* of the pattern. The *components* of the pattern are the pairs, denoted by P_i, of a vertex i of sP and its image P_i in K. The *distinguished links* of the pattern are the images in K of the arrows of the sketch sP. (In this figure, only one distinguished link is depicted.) The pattern P is entirely determined by its components and the distinguished links between them. Therefore, the sketch will often be omitted in the figures, and only the image of P (hence its implementation) in K will be drawn. (A single object of K can be associated with more than one component of the pattern; see Fig. 2.3.)

pair consisting of an index i, and its image by P is called a *component of the pattern*, generally denoted by P_i. The image $P(x)$ by P of a link x from i to j in sP is called a *distinguished link* of P from P_i to P_j.

Equivalently, a pattern P in the category K can also be defined as consisting of the following:

(i) a finite set I of indices of P and a family $(P_i)_{i \in I}$ of objects P_i of K, indexed by this set. The objects P_i taken with their index i are the *components* of the pattern.

(ii) for each pair (i, j) of indices, a set of links from P_i to P_j, called the *distinguished links* of P, from the component P_i to the component P_j.

With this definition, the sketch of the pattern can be kept in the background: it need not even be mentioned explicitly, if we do not mention the explicit origin of the index set. In the figures, sP will not be drawn if it is easily deduced from the context.

A pattern in which the objects of K associated with different indices are distinct, in other words, where $P_i = P_j$ if and only if $i = j$, reduces to a subgraph of the category, which in turn can be taken as both its sketch and its implementation; that is often the case. However, in a general pattern P, the same object of K may be associated to different indices, each one of its instances corresponding to a different function, as determined by the distinguished links of the pattern. Thus, an object that occurs once in a configuration category of a system may occur several times in the pattern modelling the organizational aspect of a complex object of the system. For example, suppose we have two different indices i and j, and nevertheless $P_i = P_j$. Let us denote this single object of K as B, so that

$$P_i = B = P_j \quad \text{with} \quad i \neq j$$

The distinguished links from P_i to another component P_k can thereby be different from those from P_j to P_k, even though they are all links in the category from B to P_k.

For example, let K be the groupoid on an alphabet: its objects are the letters and there is one arrow between any two letters. A pattern having as its sketch the 'linear' graph defining the order $1 < 2 < \cdots < n$ corresponds to a word with n letters, but the same letter can be repeated. Let us describe the pattern P corresponding to the word 'rare': it has the four indices 1, 2, 3, 4, and the letter r is associated to the indices 1 and 3, a to 2 and e to 4. In equations in K:

$$P_1 = r = P_3, \quad P_2 = a \quad \text{and} \quad P_4 = e$$

and there are three distinguished links, from $P_1 = r$ to $P_2 = a$, from $P_2 = a$ to $P_3 = r$ and from $P_3 = r$ to $P_4 = e$.

Several patterns may have the same sketch; they are said to be *analogous*. They represent different implementations (or models, in the sense of a 'working model', *i.e.* in the sense of a prototype made according to a blueprint) of the same formal organization. For instance the linear graph considered above can be implemented in any category K, and the associated analogous patterns are all the paths of length $n-1$ of K. A concrete example of analogous patterns with a more intricate sketch is given by two football teams; their sketch describes the same formal structure (forwards, defenders, goal keeper...), but the corresponding functions are fulfilled by different members, whose behaviour and interrelationships in the team are constrained by their role.

1.2.   Collective Links

The reason for introducing patterns was to characterize the internal organization of a complex object. However, even when a pattern does not represent the internal organization of an object in the category (say, the pattern represents a group of persons who cooperate informally), its components may display some unified behaviour that lends itself to study.

We have represented the interactions among the components of a natural system at or near a given time, such as a transfer of matter, energy, or information, or the imposition of constraints, by links between objects in the configuration category at this time. However, some interactions may require the cooperation of several interrelated components forming a pattern, and could not be carried out if these components acted separately. Such cooperative interactions by the components of a pattern will be modelled by what we shall term the collective links of the pattern. In other words, the behaviour of each component must be coherent with that of the components to which it is connected in the pattern, so that the constraints imposed by the distinguished links are respected. For example, the various cells of the immune system act in a coordinated way to fight against an external attack; if their actions are not coherent, the defence may fail.

Thus, in a collective link, the components of the pattern operate in synergy, their individual actions being correlated through their distinguished links. More formally:

Definition. Let P be a pattern in the category K. A *collective link* from P towards an object A of K is defined as a family $F = (f_i)_{i \varepsilon I}$ of individual links of K such that (Fig. 2.2):

(i) associated to each index i of the pattern is a link f_i from the component P_i to A;

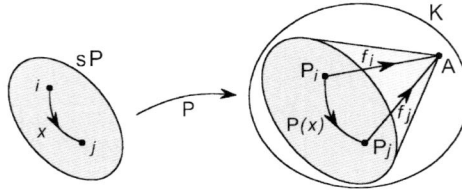

Fig. 2.2 A collective link.
A collective link from the pattern P to an object A in K consists of a link from every component of the pattern to A, these links being correlated by the distinguished links of the pattern. Thus, for each arrow x from i to j in the sketch sP of P, we have $f_i = P(x)f_j$, where $P(x)$ is the corresponding distinguished link from P_i to P_j. The collective link is schematically represented by a cone with vertex A, and whose base is the (image of the) pattern P.

(ii) for any arrow x from i to j in the sketch sP, the following correlating equations are satisfied: $f_i = P(x)f_j$.

In category theory, a collective link is also called a *cone* with basis P and vertex A; in the figures it will often be represented by such a cone. Likewise, in categorical terms, F defines a natural transformation from P towards the functor constant on A (Mac Lane, 1971).

Here we introduce a *convention*: if the set of indices is clear from the context, the notation $(f_i)_{i \in I}$ for a collective link will generally be abbreviated in (f_i); the same holds for any family.

The distinguished links of the pattern restrain the freedom of the individual links f_i of a collective link (f_i) from P to A. Indeed, a distinguished link d from P_i to P_j indicates that P_i interacts with P_j along d, and so P_i must coordinate its action f_i on A with the action f_j of P_j on A, the coordination being done along d, and thus f_i must be the composite df_j (in conformity with the correlation equations). If there is also a distinguished link from P_j to P_i, both components must reach an accord. A component P_i is 'free' only if there is no distinguished link from or to other components.

This restriction distinguishes the collective actions of the pattern from non-coordinated individual actions of its components. In the association C considered in Section 1, its different members should act coherently when they act as members of C, but they can have non-coordinated actions as individuals, when they are not acting in the name of the association. For instance, two editors of the journal of the association can have different opinions on an article submitted to the journal, but as members of the board of editors they must agree on a common decision to accept it or reject it. Thus, the larger the number of distinguished links of a pattern, the more

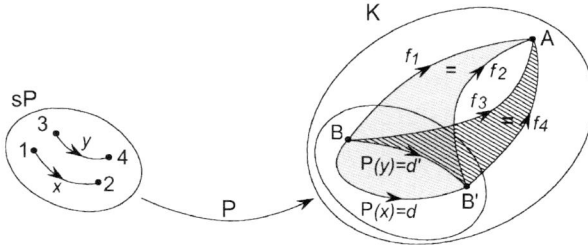

Fig. 2.3 Role of the indices.
The pattern P has four components, with indices 1, 2, 3, 4. The indices 1 and 3 have the same image B by P, while 2 and 4 have the same image B':

$$P_1 = B = P_3 \quad \text{and} \quad P_2 = B' = P_4$$

In the collective link (f_i) from P to an object A, the link f_1 from $P_1 = B$ to A is different from the link f_3 from $P_3 = B$ to A, as are the links f_2 and f_4 from B' to A. The distinguished link d from P_1 to P_2 correlates f_1 and f_2, while f_3 and f_4 are correlated by d'.

constraints are imposed on a collective link by the correlating equations, and the smaller is the number of possible collective links.

Here also, the indices play an important role. If two indices i and j correspond to the same object $P_i = P_j$ of the category, the associated individual links f_i and f_j can be the same or different. Let us give an example where they are different (Fig. 2.3): sP is the graph with four vertices 1, 2, 3, 4 and two arrows x: $1 \rightarrow 2$ and y: $3 \rightarrow 4$. If B and B' are two objects of K, one can have a pattern P having four components P_i but with

$$P_1 = B = P_3 \quad \text{and} \quad P_2 = B' = P_4$$

and with only one distinguished link $d = P(x)$ from P_1 to P_2 and only one distinguished link $d' = P(y)$ from P_3 to P_4—even though d and d' are both links from B to B' in K. In a collective link $F = (f_1, f_2, f_3, f_4)$, the two arrows f_1 and f_3 from B to A, together with the two arrows f_2 and f_4 from B' to A, must satisfy the correlation equations

$$f_1 = df_2 \quad \text{and} \quad f_3 = d'f_4$$

(but not $f_1 = d'f_2$ nor $f_3 = df_4$).

1.3. Operating Field of a Pattern

Naturally, a pattern in a category does not always represent the organization of a complex object of this category. However, in any case, a pattern P itself can be thought of as a higher level object (in a larger category), integrating the organization of its interacting components depicted by its

sketch, with its collective links thus representing the actions of this higher
object on other objects.

In Chapter 1 we have defined the operating field of an object in a category,
as a category whose objects are that original object's links to other objects.
We extend this concept to patterns, replacing the links by collective links:

Definition. The *operating field of a pattern* P in the category K is the category ΩP having for objects the collective links $F = (f_i)$ from P towards all
objects A of the category, and in which the links from F to another collective
link $F' = (f_i')$ from P to A′ are defined by the links h from A to A′ in K
which correlate F and F′, *i.e.* such as

$$f_i h = f_i' \text{ for each index } i \text{ of P.}$$

There exists a 'base' functor from the operating field ΩP of P toward the
category K, which maps F to its target A. It singles out the objects on which
the pattern acts but forgets the process by which it acts on them (*i.e.* it
forgets the links of the collective link).

A collective link models an interaction between a pattern and an object of
the category. Now an object of K can be identified to a pattern reduced to a
unique component, namely this object itself. Thus, a collective link can also
be construed as representing an interaction between two patterns, the second
one being reduced to one component. In the following chapter, we will
extend the notion of a collective link into a general notion of interaction
between two patterns, called a cluster. We will also define a category having
the patterns as its objects and the clusters between them as links.

Remark. Let P be a pattern. If we replace the sketch sP by the category of its
paths, the pattern P can be extended into a functor from this category to K,
mapping a path on the composite of the images of its factors. This gives a
pattern P^+ having the same components as P, and in which the distinguished links are the distinguished links of P and their composites. Any
collective link from P to an object A defines also a collective link from P^+ to
A (according to the associativity of the composition), so that P and P^+ have
the same operating field. Thus, if useful it is possible to replace P by P^+, *i.e.*
to consider patterns having a category for their sketch.

2. Colimit of a Pattern

The cooperation of the components of a pattern can be temporary, as in a
group of people who decide to meet to carry out a particular task. However,
if it lasts for a long period, their cooperation may be reinforced, with some
specialization of the roles of the different members. And finally the group as
such can take its own identity and become legalized as a professional

association, binding its members and requiring that they operate coherently to fulfil the functions devoted to the association.

2.1. The Mathematical Concept of a Colimit

For a category K, how do we recognize that, given a pattern P in K, there exists a particular object of K, say cP, of which P represents an internal organization, so that cP acts as a binding of P? Two conditions will be necessary and sufficient:

(i) the components of the pattern must be coherently bound to the object cP, so that the binding respects their distinguished links and
(ii) this binding must ensure that cP is functionally equivalent to the pattern operating collectively.

To explain these conditions further, let us assume that we have modelled the actions of an object in a category by its links to other objects, and those of a pattern by its collective links. Thus, the first condition translates into the existence of a particular 'binding' collective link from P to cP. The second means that the links from cP to any object A are in one-to-one correspondence with the collective links from P to A, so that each collective link is bound into a well-determined link from cP to A. In particular, the binding collective link to cP binds into the identity of cP (which is a particular action of cP on itself). This situation is well known in category theory; indeed, it means that the object cP (if it exists) is the colimit of the pattern in the category.

Definition. An object of the category K is called the *colimit* of a pattern P in K, often denoted by cP, if the two following conditions are satisfied (Fig. 2.4):

(i) there exists a collective link (c_i) from the pattern P to cP, called the *collective binding link* (or colimit-cone); c_i is called the *binding link* from P_i to cP (or simply, the binding link associated to the index i);
(ii) (*universal property*) each collective link $F = (f_i)$, from the pattern P to any object A of K, *binds* into one and only one link f from cP to A which satisfies the equations:

$$f_i = c_i f \quad \text{for each index } i.$$

This implies that the collective binding link is an initial object of the operating field of P, meaning that, for any other collective link F, there is a single arrow f from this collective binding link to F in this field.

A pattern P may have no colimit, but if the colimit of P exists it is unique (up to an isomorphism, that is any object isomorphic to cP is also a colimit

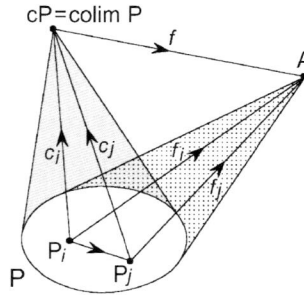

Fig. 2.4 Colimit of a pattern.
The given pattern P (its sketch is not shown) admits cP as its colimit, and (c_i) is the binding collective link from P to cP. The universal property of the colimit means that any other collective link (f_i) from P to A factors through (c_i) into a unique link f from the colimit cP to A, meaning that, for each index i, we have the equation $f_i = c_i f$. The mapping associating f to (f_i) defines a one-to-one correspondence between the collective links from P to A and the links from cP to A.

of P). On the other hand several patterns may have the same colimit (this last property will be important for our model).

Here, we provide an elementary example; a more detailed discussion occurs in Section 3. In the category K representing the social structure of a country, let C be a professional association, say a scientific society. As explained above, the pattern representing the organization of this association has for components its members indexed by their various roles in C, and its distinguished links represent their interactions in the functioning of C. This pattern admits C for colimit in K.

The general concept of a colimit has been defined by Kan (1958) under the name of inductive limit, and it has been extensively used in mathematics. In particular, many geometrical constructions reduce to the formation of colimits in the category of topological spaces. In particular:

(i) The lateral surface of a cylinder is obtained by pasting together two opposite sides of a rectangle; thus it can be defined as the colimit (in the category of topological spaces) of a pattern P which has two components, a rectangle P_1 and one of its sides P_2, and two distinguished links from P_2 to P_1, namely the insertion of this side into the rectangle and its pasting to the opposite side.
(ii) A Möbius band is constructed similarly, replacing the pasting by a pasting after torsion.
(iii) A torus is obtained by the same process as the first, applied to a cylinder and one of its bases which is pasted to the opposite basis.

In these three geometric examples, the sketch is reduced to two parallel arrows. The colimit of a pattern with such a sketch is called a *co-kernel*.
(iv) The sphere can also be constructed as a colimit, as well as any topological manifold.

2.2. The Binding Problem

The colimit of a pattern P can be seen as an actualization of the potential of its components to operate together in a coherent way. It, therefore, integrates the components into a coherent whole, rather than their behaviours being those of disparate elements. Hence, if a pattern has a colimit, it may be called a coherent assembly, and its colimit considered as an object more complex than its components, which assumes the function of the whole assembly. It follows that the colimit is a solution of the binding problem, defined as follows:

Definition. The problem of yielding an object which binds (or 'glues') the pattern P into a single object with the same coherent functional role is called the *binding problem*.

For instance, an informal organization, such as a group of people getting together to publish a journal, can be bound into a legal association, which has its own identity and operates in the name of the group.

If, in the colimit cP of a pattern P, we think that the binding link c_i associated to the index i in some way identifies the component P_i to a 'piece' of cP, then the freedom of these pieces in cP is constrained by the distinguished links of the pattern. Generally, by binding together, individuals lose some autonomy, but their cooperation may generate a collective benefit, possibly by inducing differentiation and specialization of their role.

The binding problem has been raised in many domains, and we will see applications later on. One area in which the binding problem is prevalent is in biology. Paton (1997, 2001), for example, emphasizes the importance of the gluing process for the study of functions. A molecule integrates the pattern formed by its spatial configuration, with its atoms and the chemical bonds between them. A beehive can be modelled as the population of its bees along with their chemical, biological and social interactions; some of these actions impose specialization on the bees: in particular, the actions of the queen's pheromones sterilize the worker bees. The cells of a sponge are linked by contiguity and bound together; motion of the sponge is initiated by the cells in contact with the environment, which drag the internal cells; over time these internal cells lose their loco-motor properties, which are no longer necessary, and may specialize in metabolic tasks, making the overall performance of the sponge more efficient.

Thus, the formation of the colimit of a pattern involves both local and global properties:

(i) *Locally on the structure:* the organization of the pattern is made more robust and efficient, the components being constrained to cooperate through their distinguished links, and each one contributing to the whole by the functions specific to it; for instance, the formation of the association C considered above gives it a legal status, with specification of the roles that its various members should play in it and how they should interact in order to act efficiently in C.

(ii) *Globally on the function:* the interactions of the colimit on any object of the category are characterized as the collective interactions of the pattern which they bind; the universal property means that the colimit is the object which best implements the operative functions of the pattern. For instance, the activities of the association C are exactly those its members perform collectively in the frame of the association; when these members also act collectively in another society, it will be in the name of the association.

The global property shows that the existence of a colimit imposes constraints on all the objects of the category, not only on the components of the pattern. It explains that the existence of a colimit depends in an essential way on the given category K in which the pattern is considered. If the 'same' pattern (defined by its components and their distinguished links) is considered in another category K' (*e.g.* a category of which K is a sub-category), it can have a colimit in K' but not in K. The complexification process, which will be studied in Chapter 4, consists in binding a pattern without a colimit in K, by constructing a larger category in which it acquires one. A simple example is a group of people who individually pursue a particular activity, and who decide to create a formal association.

In natural systems the binding of a pattern by formation of a colimit can be looked at from various points of view:

(i) *Informational:* improving the communication between its components through their distinguished links, in a way that restricts the freedom of the components, ensuring better cooperation. Seeing the colimit as the object which best collects information from the pattern, the formation of a colimit implements a principle of 'maximal information' (in a sense analogous to the info-max principle of Linsker, 2005).

(ii) *Functional:* the binding of the pattern into the colimit allows for the collective actions of the pattern to become more efficient (and possibly faster); this may require a differentiation of the various components, and greater internal specialization of these components by division of labour (Rashevsky, 1967, 1968).

(iii) *Morphological or developmental:* emergence of a more complex object, whose shape is an 'actualization' of the sketch of the pattern, and which integrates the various operations that the pattern can realize.

(iv) *Entropic:* formation of a macro-state admitting the pattern as (one of) its micro-states.

(v) *Newtonian or mechanist:* when the distinguished links of the pattern are weighted (*e.g.* when the category is labelled), the formation of a colimit may ensure a modification of their weights (*e.g.* a strengthening), which allows for more coherence and efficiency in the interactions.

2.3. *Decompositions of an Object*

In the preceding section, we have adopted a 'bottom-up' perspective, looking at the colimit as the solution of the binding problem for the pattern. However, the situation can also be considered 'upside-down': given a complex object C, find a pattern of which it is the colimit. Such a pattern will be called a *decomposition* of C.

In this way, the colimit C is seen as a complex object (or a hyperstructure in the sense of Baas, 1992, *cf.* Appendix) of which the pattern represents an internal organization, the components of the pattern being thought of as components of C. However, whereas the pattern determines its colimit, if it exists, in a unique way (up to an isomorphism), the reverse is not true, for the same object can have several different decompositions. For example, a single sequence of nucleic-acid codons templates for a single sequence of amino acids, but a single sequence of amino acids may be produced using any of several sequences of codons (degeneracy of the genetic code). Intuitively, the colimit forgets the structural organization of the pattern and retains only its functional, operative role, that is the collective operations it can realize; and these operations can be the same ones for other more or less different patterns. The consequences of this important property will be studied in the following chapter, for it is at the basis of the emergence process.

Taken in this direction, from a complex component C of a natural system to its decompositions into more elementary components, the various points of view on colimits indicated in Section 2.2 above correspond to the following:

(i) the distinction of the different sources of the information received by C;
(ii) the determination of functionally equivalent patterns (having the same action);
(iii) the description of ways an object may be assembled or may self-assemble;
(iv) the determination of the possible micro-states of the object, looked at as a macro-state;
(v) an analysis of the distribution of the internal forces ensuring cohesion.

3.   Integration vs. Juxtaposition

The role of the distinguished links of the pattern is essential: they impose constraints on its components which influence the manner of their binding together, and which allow the emergence of effective collective operations, transcending the individual actions of the components. Without them there would be only one amorphous collection of objects. The study of this case will allow us to measure the coherence and constraints introduced by the distinguished links, by comparing the colimit of a pattern P with the colimit (called the sum) of the pattern formed by the same components acting independently (without taking into account the distinguished links of P). An example of the difference is given by the behaviour of an unorganized crowd, and the behaviour this crowd adopts when it is directed by leaders; or the difference between a wall and a pile of bricks and mortar.

3.1.   *Comparison of the Sum to the Colimit*

A pattern without any distinguished link is reduced to a family of objects of the category. Such a pattern does not have a specific organization, since its objects are not interconnected. A collective link towards an object A is then reduced to a family of individual links from each object to A since there are no correlating equations.

Definition. The *sum* (or *co-product*) of a family (P_i) of objects of the category K is the colimit of the pattern having the P_i for its components and without any distinguished link between them (see Fig. 2.5).

More explicitly, the sum of a family of objects P_i is an object to which each P_i is linked by a link s_i satisfying the universal condition: given any family of links f_i from P_i to any object A, there exists one and only one link from the sum to A whose composite with s_i is equal to f_i for each index i.

The properties of the sum reduce to those of its components, without introducing new properties. It illustrates classical reductionism: the study of a complex object may be done through its components, so that the whole is treated as no more than the sum of its parts. It is different from the case of a complex object C, which is the colimit of a pattern P having distinguished links, for the correlating equations associated to the distinguished links restrain the behaviour of the components. The behaviour of a P_i may be very different according to its operation as a single object or as a component of C, since the colimit singles out the collective operations made possible if the various components cooperate by means of these links. As said by Aristotle: 'What is composed of something so that the whole be one is similar not to a pure juxtaposition, but to the syllabus. The syllabus is not the same as its component letters: ba is not identical to b and to a, […] it is still something

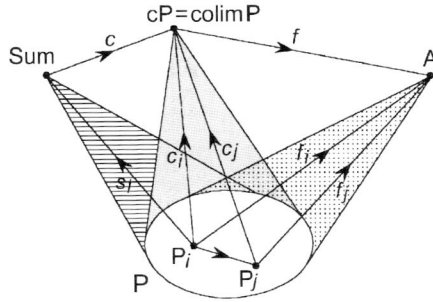

Fig. 2.5 Comparison of the sum and the colimit.
The pattern P admits both a sum (Sum) and a colimit (cP). We denote by (s_i) the collective binding link from the family (P_i) to its sum, and by (c_i) the collective binding link from the pattern P to its colimit cP. From the universal property of the sum, it follows that the family (c_i) of links from the different P_i to cP binds into a link c from Sum to cP; this c is called the comparison link from the sum to the colimit. We depict a collective link (f_i) from P to A; it binds into a link f from the colimit cP of P to A. The family (f_i) also defines a collective link from the family (P_i) to A, hence binds into a link from its sum, hence Sum, to A; this link (not depicted) is the composite cf of c and f.

else' (*Metaphysica*, Z 17, 1041b, 10). To measure this 'something', we will compare the colimit cP of a pattern with the sum of the family of its components, when both exist.

Theorem 1. *Let* P *be a pattern having a colimit* cP. *If the family* (P_i) *of its components has a sum, there is a comparison link* c *from this sum to the colimit* cP, *namely the link binding the family of the binding links* c_i *from* P_i *to the colimit* cP. *There is a one-to-one correspondence between the collective links* (f_i) *from* P *to an object* A, *and the links from the sum to* A *which factor through* c, *hence are of the form* cf *for some* f *from* cP *to* A *(Fig. 2.5).*

The comparison link c 'measures' which are the constraints imposed on the various components by their cooperation in P, and thus determines the new properties emerging for the complex object cP with respect to its components P_i. Indeed, the properties of cP are determined by the links from cP to any object A, which correspond to those links from the sum to A, which factor through the comparison link c. While all the components play a symmetrical role in the sum, in the colimit each one has its own specific contribution, depending on the constraints determined by its distinguished links to the others, so that the passage from the sum to the colimit can be interpreted as a 'symmetry-breaking', the strength of which is measured by the comparison link c. In particular if the category is labelled, the weight of the constraints is measured by the weight of c.

Intuitively, the comparison link filters the information, letting pass only that which is compatible with the internal organization of the pattern. If two components act without taking account of their distinguished links in P, this action cannot be decoded via *c*. Such individual actions cannot exist if the constraints are such that the various components lose their own identity in P. Otherwise they are possible; for instance the members of an association C (colimit of P) can collectively act in the name of the association; but they can also make a trip to a town A as separate individuals, and this will not take into account their membership in C. In other words, *c* models the information necessary to integrate the pattern into a complex object. An example of such information is provided by the formation of a tissue under the effect of cell adhesion molecules (Edelman, 1989, p. 44), which make the cells of the tissue move coherently.

The existence of the comparison link from the sum of the components of a pattern P to the colimit cP of P indicates that cP has emergent properties with respect to the components of P. Thus, a strict reductionist stance is not tenable, and should be replaced by a stance that we refer to as *operating reductionism*. This means that the properties of a complex object can be deduced from those of its components and from the correlating equations, which ensue from the distinguished links between them. This result (due to the universal property of a colimit), has deep implications since a global consequence (the behaviour of a complex object with respect to any object) is deduced from a local analysis of a pattern representing its internal organization.

3.2. *Some Concrete Examples*

The following examples illustrate the ubiquity of the concept of colimit and the roles played respectively by the sketch of the pattern (ideal form of the structure), its components (matter) and their distinguished links (constraints).

In *chemistry*, the molecular formula of a simple molecule corresponds to the sum of its atoms, while its space configuration corresponds to a colimit. Certain molecules may have the same atomic composition, but different space configurations; they are called isomers. The properties of a molecule are different from those of its atoms separately. For example, their distinguished links restrict their movement. In the case of hydrogen and oxygen, they are gases at room temperature, while water has a higher boiling point, its molecules being strongly connected by the hydrogen bonds.

The passage between molecules and macromolecules is well illustrated by the example of the haemoglobin tetramer, which absorbs oxygen, and whose role is essential in respiration. If it were only the sum of its four separated

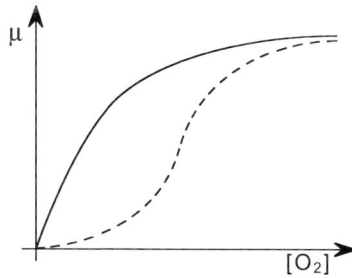

Fig. 2.6 Oxygenation of the haemoglobin tetramer.
The curves represent the oxygenation fraction μ as a function of the oxygen concentration $[O_2]$. The dotted line is for the haemoglobin molecule, while the solid line is for the case where its four constituent monomers are separated. The difference between the two curves shows the difference between a simple sum of the four units, and the colimit representing the haemoglobin molecule as a whole.

units, its oxygenation would satisfy the equation of Michael-Herni, namely the oxygenated proportion μ would be equal to the ratio $p/(k+p)$, where p is the partial pressure (dependent on the oxygen concentration) and k an equilibrium constant. The dependence of the sum upon oxygen concentration would be represented by a curve of the form given in Fig. 2.6. However, the true curve for the macromolecule of haemoglobin (colimit of its space configuration), has the form given by the dotted line of this figure. This comes from the fact that the distinguished links of the pattern are such that the receptors of three of the molecules are not accessible. In the presence of oxygen, the free receptor will bind to an oxygen molecule (by an individual link), subsequent to which the links amongst the four constituent haemoglobin monomers will impose a change of conformation, rendering the receptors of the remaining three units accessible to oxygen. It follows that the reaction starts more slowly, the delay being quantified by comparison of the two curves. The filtration effected by the comparison link between the sum and the colimit prevents direct access to the hidden receptors, whose release is controlled by their connections (distinguished links) with the monomer having the accessible receptor. Other examples of the same type are provided by allosteric proteins (for a quantitative example, see Di Cera, 1990).

Societies. An animal or human society can be modelled by the colimit of the pattern formed by the society's members and their social links, in the category whose objects are both the individuals and the different kinds of groups they form, and whose links model their social interactions. Human societies are particularly varied, from a marriage that comprises two individuals, to large international organizations such as the United Nations. Let us give some examples showing the essential role of the distinguished links and comparing the individual actions (sum) and the collective actions (co-limit) of its members.

• An individualistic society is formed by members whose social roles are not differentiated, and who have few relations between them to limit their actions. The society as a colimit functions almost as the sum of its members. Membership in the society is not onerous, but neither does it bring much benefit. The physical analogy would be with a gas.
• A democratic society is usually based on a constitution, and it relies on many reciprocal links between its members. Collective decisions (or links) must take into account the possibly conflicting interests of the various members, and balance them with those of the society as a whole. Thus, as a colimit, it is very different from a simple sum. To avoid inefficiency, a large society selects representatives whose role is to aim at the common interest, even if this interest might become manifest only later on. In this way, the society will be able to adjust to new conditions while maintaining 'home-ostasis', but possibly with some delay. The analogy with an organism is implicit in the description.
• In an authoritarian society, the links define a rigid hierarchy, with a small number of leaders that the others must follow. The collective links are imposed by the leaders, limiting the freedom of the other members. For instance in an army the links are of subordination and the colimit is personified by the chief, since the decisions which he imposes by official channels determine the actions of all the soldiers. The society might be able to act quickly and efficiently in ordinary circumstances, but it evolves with difficulty and is likely to disintegrate in the event of important external changes. The physical analogy would be with a crystal.

This analysis helps to explain the evolution of the notion of a nation (*cf.* Finkelkraut, 1987). The nation under the kings was authoritarian, with a fixed social hierarchy and the king, as a representative of God on Earth, personifying the whole nation. It was unable to withstand the change in outlook resulting from the Enlightenment. The French Revolution introduced the nation as a democratic society associating equal citizens with many reciprocal links, and subject to a common law.

3.3. *Some Mathematical Examples*

Colimits of Sets.   In the category of sets **Set** (Chapter 1), a family (P_i) of sets has for its sum their 'disjoint union', where an element common to two of the sets must be counted twice (it is thus different from the union). More precisely it is the set of all the couples (e, i) where e is any element of P_i. Let us notice in this example the role of the indices: the sum of a family with only one set E can be identified with E, but the sum of the family with two objects (P_1, P_2) which are two specimens of the same set E (in an equation: $P_2 = E = P_1$) has twice as many elements, each element e of E occurring twice, once in the couple $(e, 1)$ and once in $(e, 2)$. A pattern P in this category admits as a colimit the quotient set of the sum of its components P_i by the smallest equivalence relation which identifies (e, i) with $(d(e), j)$ for each distinguished link d from P_i to P_j, and each e in P_i.

Least Upper Bound. Let us take the category associated with a (partial) order relation $\leqslant$ on the set E (Chapter 1). Its objects are the elements e of E and there is an arrow from e to e' if and only if $e \leqslant e'$. A pattern has a collective link to an object a if and only if a is an upper bound of all its components. In this case the fact that P has distinguished links or not does not matter, because there is at most one link between two objects, so that the correlating equations associated to these links are always satisfied. Consequently, P has a colimit c if and only if the family of its objects has c for its least upper bound (lub); and in this case, c is also the sum of this family. This sum may not exist; for example in the ordered set $\mathbf{N}$ of the natural numbers, only finite families have a least upper bound (which is then their largest element). In the ordered set $\mathbf{R}$ of the real numbers, every bounded subset has a least upper bound (the order on the real line is complete).

Colimits for Graphs. In the large category of graphs **Graph** (which, we recall, has for objects the small graphs and for arrows the homomorphisms between them), a family of graphs has a sum: it is the graph whose set of vertices is the disjoint union of the sets of vertices of the various graphs, and whose arrows correspond to their arrows. A general pattern also has a colimit, which is a quotient of the sum of the family of its components. The sum of a family of categories in the category of categories **Cat** is constructed in the same way. A pattern of categories admits also a colimit in **Cat**.

In the models for natural systems that we develop, colimits will play an essential role. It is the reason why we need categories with a non-trivial composition law, instead of only graphs. Indeed, we could define colimits in graphs by replacing the graph by the (free) category of its paths (*cf.* Chapter 1), which it entirely determines, and consider colimits in this free category. However, a pattern in a free category admits a colimit only if it is of a very particular form which we have entirely characterized (Ehresmann, 1996); roughly: either the pattern is discrete and the colimit is its sum; or it is associated to a preorder with a maximum, in which case the colimit is the component of the pattern corresponding to this maximum. Such patterns are not sufficient in the applications we will develop, in particular to study neural systems.

4.   Interlude: A Transport Network

This section gives a simple example to make the abstract notions defined in this chapter more tangible and to prepare for those which will be introduced in the two chapters that follow. The idea is to model the transport network of a country divided into large areas, and in which each area is in control of part of the network and tries to make it more efficient. It leads to an explicit (and even quantifiable) construction of a colimit in a labelled category, which we have simulated in the programming language q-basic.

4.1.   *Characteristics of the Network*

The transport network is formed by various regular transport lines connecting one town to another. Some towns are not connected; if a town A is connected to another town A$'$, there can exist one or more regular lines (the direction of a line is specified). The price of a journey is independent of the specific route taken, but it depends on the means of transport and its characteristics (example: car, coach or train, time taken for the journey, comfort, frequency, number of passengers). The fare is calculated according to a rate schedule (with the rates set according to such things as comfort or convenience). In other words, rates are expressed as a 'scale' having integer values. For example, first class carriage is charged at a higher rate than second class. Thus, a line is characterized by its starting point, its destination, and its scale (carriage rate). If there is a line from A to A$'$ and another line from A$'$ to A$''$, the network has also a 'composite' line allowing travel from A to A$''$ through A$'$. The rate charged for this composite line will be at the largest of their two rates (if any part of the journey is made in first class, that rate is charged for the whole journey, even if every other part is made in second class).

The network is modelled by a category K labelled in the monoid (**N**, max) obtained by equipping the set **N** of integers with the operation 'maximum'. The category's objects represent the various towns A of the network; an arrow from A to A$'$ represents a line of the network, denoted by g: A$\rightarrow$A$'$ where g is its scale. The composite of g: A$\rightarrow$A$'$ with g': A$'\rightarrow$A$''$ is max(g, g'): A$\rightarrow$A$''$ (which corresponds to the composite line).

4.2.   *Local Area Nets, Selected Nets*

The country is divided into large areas. The network depends on a central organization which delegates to each area the power to manage its own *local area net* P; this is a sub-network connecting the various towns of the area. In general the area does not manage the lines connecting it to towns outside this area, which are managed by the central organization. However, there can exist a town A outside the area to which a certain number of inhabitants of the area must go regularly. Consider, for example, students who commute from a rural area to attend university in a city; or the movement of goods from a small community to a central trading centre. In both cases, better service (equated with more frequent connections and, perhaps, lower fares) requires the area to exploit a specific sub-network of the total network to A. This is called a *selected net* from P to A, coordinated with its local area net P, in which each town P_i of the area is connected to A by a selected line f_i: $P_i \to A$, run by the area. These selected lines are chosen so that the following constraint is satisfied: an inhabitant of P_i can, for the same rate, go to A either directly via the selected line, or indirectly by taking first any local line from P_i to another town P_j of the area, and then the selected line from P_j to A.

A local area net is modelled by a pattern P in the category K, and a selected net from P to A by a collective link (f_i: $P_i \to A$) from P to A. The scale constraint means that, for each local line h: $P_i \to P_j$ the scales must satisfy $f_i = \max(h, f_j)$.

4.3. Central Node

If the local area net P has selected nets towards various external towns A, the administration will be simplified if there is a central transit node through which its various selected nets interconnect. Such a central node will be a town C, possibly outside the area, whose site is chosen so that the following conditions are satisfied:

(i) P has a selected net towards C;
(ii) to any other selected net from P to a town A is associated a *central line* from C to A so that a traveller can go to A, for the same price, either directly by the selected line to A, or through C, by using first the selected line to C and then the central line from C to A.

In particular, the central node will simplify the organization of collective tours to any of the towns A: the members of the group living in the various towns P_i will go individually to C via the selected lines, then all together they will make the journey from C to A. It also allows people external to the area who would otherwise have to penetrate the area to get to A, to bypass the area (thus avoiding the internal traffic): they go directly to C, then use the central line from C to A.

A central node C for P will be modelled by a colimit of P in the category K. Let us give the conditions for the existence of such a colimit in this labelled category. P admits a colimit C (or central node) if there exists a selected net (c_i: $P_i \rightarrow C$) satisfying the following condition: for each selected net (f_i: $P_i \rightarrow A$), we have either:

(i) $f_i = c_i$ for each index i and there exists one and only one link f: $C \rightarrow A$ (the central line from C to A) such that $f \leqslant \min(f_i)$;

or

(ii) for each index i we have $f_i = c_i$ or $f_i = \min(f_i)$; and there is a unique link f: $C \rightarrow A$ (the central line) with $f = \min(f_i)$.

It follows that (c_i: $P_i \rightarrow C$) is the smallest selected net in the sense that for any selected net (f_i: $P_i \rightarrow A$), we have $c_i \leqslant f_i$ for each index i. These numerical conditions lead to an algorithm to determine if there exists a colimit (a central node), and in this case to compute it.

4.4.   Creation of a Central Node

Let us return to our transport network. If none of the towns to which P is connected in the total network can play the part of a central node, the area may ask the country to create such a node in an adequate place, and extend the whole network by the organization of the new lines necessary for that. However, this operation (which will extend the category K into a category K') is not always possible, at least if the scale conditions are to be strictly obeyed.

We know that, if C is a central node for P, there must exist a selected net to C in which the scales are the smallest ones. Thus, the idea will be to choose a town satisfying this condition. Two cases are possible:

(i) If there already exists a C with a selected net (c_i: $P_i \rightarrow C$) which satisfies the above condition, except that there exists no central line from C to A associated with some selected net to A, we have only to add such a line f: $C \rightarrow A$, with scale f taken as the minimum $\min(f_i)$ of the scales f_i of this net.
(ii) Otherwise it is necessary to add a new town C which will become the colimit of P in an extended category K'. For this, for each i a line is created from P_i to C, its scale c_i is taken as the minimum of the scales f_i in the various selected nets to different towns (these lines always form a selected net); and for each selected net to A, a central line f: $C \rightarrow A$ from C to A is created.

However, if some selected net does not satisfy the second condition listed above, it might not be possible to put the minimal scale on the associated central line, the result being that the travellers will have a less advantageous fare for their transport through C. Let us give such a counterexample.

• Suppose P has just 3 towns P_1, P_2, P_3 and one line of scale three from P_1 to P_2. The other lines form a selected net with scales 3, 3, 5 towards A and one with scales 4, 4, 4 towards another town A'. Then the scales of the lines to C should be 3, 3, 4, but the central line from C to A cannot be scaled in an adequate way, for its scale f should satisfy: $\max(f, 3) = 3$ and $\max(f, 4) = 5$.

The construction of the extended category K' in which P admits a colimit is a particular case of the complexification process studied in Chapter 4. And the above counterexample shows that the complexification of the category associated to a labelled graph is not always labelled.

4.5. Inter-Area Nets

This section introduces a particular case of the notions of cluster, simple link and complex link, which will be defined for the general setting in the next chapter. Until this point, we have considered a unique area and its possible management of lines from its local area net P to outside towns. Now, we will study the general transport between two areas. Various towns of another local area net Q can be connected to towns of P by lines of the total network. To increase their exchanges, the two areas can exploit jointly a sub-network of the total network, connected with their local nets. Such an inter-area net G has the following properties:

(i) Each town of Q is connected by a line of G to at least one town of P; and if it is also connected to another town of P by a line of G, then these two lines are connected by a zigzag of local lines of P.
(ii) All the composite lines formed by a local line of Q and a line of G, or by a line of G and a local line of P, also belong to G.

Such an inter-area net G gives an example of what will be called a *cluster* from Q to P (*cf.* Chapter 3). In this case, if the area P has a central node C and Q a central node D, the inter-area net G is 'integrated' into a central line from D to C; thus to go from a town of Q to C a traveller can either cross P or go to C through D. This central line is an example of what will be called a (Q, P)-*simple link*.

Alternatively, the central nodes of two areas can be connected without an inter-area net between them. An extreme case is that of two areas P and P° which have the same central node C, without having inter-area nets. This is possible if the two areas have selected nets towards the same towns, but have not agreed on coordinating their nets. In this case, if G is an inter-area net from Q to P and G' an inter-area net from P° to a local area net Q', the central nodes of Q and of Q' are connected by the composite of the central lines associated to G and G'. Thus, it is possible to go from Q to Q' via C without penetrating P or P°, even if Q and Q' are not connected by an inter-area net. This situation of a central node C common to two areas will be modelled by saying that C is a *multifold object*; and the composite line is then a (Q, Q')-*complex link* (Chapter 3).

Chapter 3

Hierarchy and Reductionism

The word 'hierarchy' has been used with very different meanings in systems theory, physics, biology, sociology and other disciplines, to denote systems whose components belong to several levels, corresponding to an internal structure with increasing complexity. Strictly speaking, a hierarchy also requires that the lower levels be subordinate or in some way inferior to the higher levels, but this aspect is typically overlooked in systems theory. Similarly, when we use the term, we do not intend this meaning. For example, a written, alphabetic language has three basic levels of complexity: the letters of the alphabet, the words and the sentences. In an organism, we distinguish the atomic, molecular, sub-cellular, cellular, tissue and organ levels. In an ecological system, there are three main levels: organism, population and ecosystem (Auger, 1989).

In this chapter, we model the concept of a hierarchy within the categorical framework, the idea being that an object of a certain level is complex with respect to lower levels if it is the colimit of a pattern representing the organization of its lower level components, where patterns and colimits are as defined in Chapter 2. It leads to the problem of reductionism, one version of which (given by Rosen, 1969) is to determine to what extent a knowledge of the system at the lowest level of the hierarchy may specify its properties at the higher level.

For that, we need to analyse how links between complex objects are formed. We distinguish 'simple links', which are mediated through their lower level components, and which relay to the complex object information already contained in these components; and 'complex links', which emerge by integration of a global structure not directly accessible via the components. This is the first question we study in this chapter.

1. P-Factors of a Link Towards a Complex Object

In a simple business enterprise (of a type perhaps disappearing in an age of computer mediated logistics and internet shopping), the procurement of merchandise might proceed as follows: a wholesale customer wants to order a product from a large company, and to do so, this customer submits a purchase order. Once received, this order is read by a worker in the marketing department, who then forwards it to her colleagues in the finance department. Here, the order is recorded and an invoice and a dispatch note made out. The finance department informs the shipping department of the order, so that the shipping staff can locate, pack and deliver the product. In other words, the order addressed to the company as such is in fact handled by employees who act as intermediaries and, directly or indirectly, transmit and process the relevant information.

The process described above is amenable to being modelled: if the company is modelled by the colimit C of its organization in the category of individuals and societies, the order of the customer becomes a link from the customer B to the company C, which is *mediated* by the various employees concerned with the order. It is this type of link from an object B to a complex object C which we study initially. Going further, if B is also a complex object, we characterize the links from B to C which bind collective links, wherein each individual link is mediated by a component of C; an example in the above would be an interaction between an association of consumers and the company.

1.1. *Links Mediated by a Pattern*

The actions of a complex object C—that is, its links to other objects—are entirely determined by the synergistic action of the components of one of its decompositions P (modelled by the collective links of P). The situation is different for the information received by C, which can be more or less independent of its lower level components. The information received by C which does depend upon at least one of these components will be modelled by P-mediated links.

More formally, let C be a complex object in the category K, and P a pattern which is a decomposition of C, so that C is the colimit cP of P. By definition of a colimit, we know that every link from the colimit to an object A binds a collective link of P, and so integrates links from the components of the pattern to A. The situation is different for a link from an object B to C which may have no relation with any component of P. However, there are some links from B to C which factor through one or several components of P, and they are obtained as follows: a link from an object B of K to one of the components P_i of P gives a link from B to C when it is composed with the binding link from P_i to the colimit. Such a composite is a link from B to C which is mediated by at least one of the components of P.

Definition. A link g from B to the colimit cP of the pattern P is said to be *mediated by* P if it is of the form $g_i c_i$, for at least one component P_i of P and

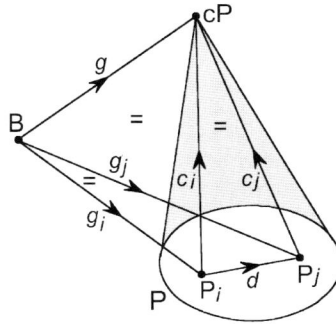

Fig. 3.1 P-factors of a link to cP.
A link g from B to the colimit cP of the pattern P is mediated by the pattern P if it factors through a link g_i from B to one of the components P_i of P. Such a link is called a P-factor of g. In this case, g also admits for a P-factor any link g_j which is correlated to g_i by a distinguished link d of the pattern (or by a zigzag of such links, *cf.* Fig. 3.3).

one link g_i from B to P_i, where c_i is the binding link to the colimit associated to the index i (see Fig. 3.1). In this case, g_i is called a P-*factor* of g.

A link to the colimit cP of P which has no P-factor represents an action of a more global nature, which does not take into account the specific organization of C given by the components of the pattern P. For example, if C is a book which is an anthology of poems, B can buy the book as such (g is not mediated), or in view of reading a particular poem, in which case g has a P-factor corresponding to this poem.

A link g from B to the colimit of P may have several more or less independent P-factors. In particular, if g admits a P-factor g_i to P_i it has also for P-factor any link correlated to g_i by a distinguished link d from P_i to P_j; that is, a link of the form $g_i d$. Indeed, the binding links c_i and c_j are correlated by d (in the binding collective link), whence the equations

$$(g_i\, d\,)c_j = g_i(dc_j) = g_i c_i = g$$

Roughly, if the links represent information transfers, B transmits the information g_i to P_i which retransmits it to cP via c_i. On the other hand, P_i also reflects it to its neighbour P_j via the distinguished link d; and in its turn P_j retransmits the information $g_i d$ so received to cP via c_j so that, by transitivity, cP also receives from P_j the composite information $(g_i\, d\,)c_j$. For the communication to be directly reliable, it is necessary that cP receives the same information g from B in the two cases, be it transmitted from P_i or P_j. Otherwise it would need some error correction techniques before being processed.

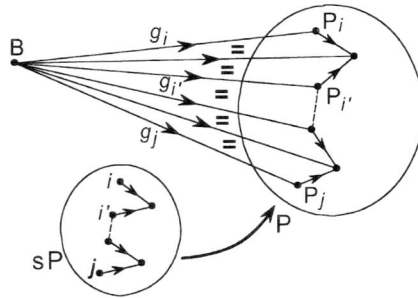

Fig. 3.2 Links correlated by a zigzag.
The zigzag of arrows from i to j in the sketch sP of the pattern P has for its image by P a zigzag of distinguished links of P. The links g_i and g_j from an object B to components P_i and P_j are correlated by this zigzag of distinguished links if there exists a sequence of commutative triangles whose bases are the distinguished links of the zigzag joining g_i to g_j. It follows that if P has a colimit cP, both g_i and g_j are P-factors of the same link g from B to cP (as indicated in Fig. 3.3).

More generally, let us define a *zigzag* in a graph as a finite sequence of arrows in which two consecutive arrows have one of their ends in common. In Fig. 3.2, we depict such a zigzag in the sketch sP of P, and the corresponding zigzag of distinguished links of P. We say that a link g_j from B to P_j is *correlated by a zigzag* of P to the link g_i from B to P_i if there is a zigzag from i to j in the sketch sP of P, and links $g_{i'}$ from B to components P_i' of P, such that the diagram (in K) of Fig. 3.2 is commutative (*i.e.* two paths on the diagram with the same ends have the same composite). The following proposition is deduced from the commutativity of this diagram.

Proposition. *If g admits a P-factor g_i, it also admits as a P-factor any link correlated with g_i by a zigzag of distinguished links of* P.

1.2. Link Binding a Perspective

In the preceding discussion, we began with a link from B to the colimit of a pattern, in order to study its P-factors. In the opposite direction, the notion of a *perspective* will characterize a family of links from B to components of a pattern P, which bind into one link from B to the colimit cP. Roughly, a perspective represents information (or constraints) transmitted by B to some (not necessarily all) components of the pattern, and which is distributed as widely as possible within the pattern, each component retransmitting that information to its neighbours via the distinguished links. It is necessary that the information that a component receives directly from B be identical to that which it receives indirectly from its various neighbours, so that all the links provide the same information at the level of the pattern as a whole.

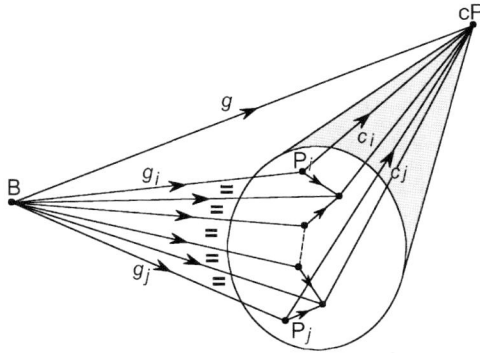

Fig. 3.3 A perspective.
Given an object B and a pattern P, a link g_i from B to a component P_i of P is called an aspect of B for P. A perspective of B for P is a maximal set of aspects of B such that two links of the set (g_i and g_j in the figure) are correlated by a zigzag of distinguished links of the pattern. Here, maximal means that any link so correlated to a link in the perspective belongs to the perspective. However, there may exist other aspects of B for P (not depicted) which do not belong to the perspective (*i.e.* they are not correlated to any link in the perspective). If the pattern P has a colimit cP, all the aspects of a perspective are P-factors of the same link g from B to cP.

Definition. For a pattern P, a *perspective* of an object B is a maximal family of links from B to some components P_i of P (called *aspects* of B for P), such that any two of them are correlated by a zigzag of distinguished links of P (see Fig. 3.3).

In this case, 'maximal' means that any link, correlated by a zigzag of distinguished links to a link of the perspective, is also in the perspective. It follows that the perspective is determined as soon as one of its aspects is known. If P has no distinguished link (*i.e.* if it reduces to a family of components), any link from B to a particular component of P is a perspective with this link for its unique aspect, and there are no other perspectives.

When the pattern P admits a colimit, the various aspects of a perspective from B act as various channels through which B transmits the same information to the colimit, these channels being coordinated by the fact that the components of the pattern are forced to act in synergy via their distinguished links. More precisely:

Proposition. *If* P *admits a colimit* cP, *all the aspects of a perspective of* B *for* P *are the* P-*factors of a unique link* g *from* B *to* cP. *In this case, we say that* g *binds the perspective. Any link from* B *to* cP *mediated by* P *binds at least one perspective for* B.

Proof. The commutativity of the diagram of Fig. 3.3, and the fact that the binding links c_i to the colimit cP are correlated by the distinguished links of P, imply the equations

$$g_i c_i = g_j c_j \quad \text{for all indices } i, j.$$

The object B may have several perspectives for P; the links from B to cP which bind them can be different or not. If two distinct perspectives are bound by the same link g, this means that the aspects of the two perspectives are P-factors of g, though this cannot be directly recognized at the level of the components of P; it is an emergent property at the level of the colimit cP.

1.3.   Examples of Perspectives

An Ecosystem.   Let us consider a group of animals of a certain species (modelled by a pattern in the category modelling the ecosystem), and one of their predators B. An animal perceives B under the aspect g_i of its odour; it emits a cry of alarm which informs its neighbours of the presence of the predator, so that they perceive B through this cry, under aspects which belong to the same perspective of B. These neighbours can themselves communicate in their turn the presence of the predator to others, and all the aspects will belong to the same perspective: namely, that formed by all the aspects relayed from B by direct or indirect communication among the animals of the group. In addition, the same predator B can also be perceived by one of the animals under the aspect f_k of its tracks, and its presence communicated to the others. This can occur without the potential prey realizing that the warning concerns the same predator; the perspective of f_k will be a different perspective from that of g_i, although it relates to the same B.

An Immune System.   In the category modelling an organism, let us con-sider the pattern P formed by the cells of the immune system with their various links, *e.g.* transmitting messages in the form of synthesized mole-cules. An antigen B will trigger the formation of complement molecules; one of these molecules will be perceived by a macrophage; in response it will synthesize interleukin-1 which will inform the B-cells and the helper T-cells of the presence of the antigen; in turn these T-cells will synthesize inter-leukin-2, which will be perceived by the receptors of killer T-cells. Each molecule of interleukin-1 and of interleukin-2, like those of the complement, represents particular aspects of the antigen for the corresponding cells; therefore the perspective of B for the immune system (P) will contain these various aspects.

The Transmission of Information.   In a category modelling a society, let us consider the pattern P modelling the contributors to a journal C. The journal may receive some information B via correspondents working together (in the same perspective). That information could also be independently received by other correspondents (in another perspective), which would make it more reliable. Problems will arise if two perspectives of the same event are contradictory. It is what often occurs in the event of war, where the belligerents seek to demoralize the adversary by propaganda. The situation when conflicting information is received will be handled later on (interplay among co-regulators, in Chapter 6).

1.4.   Field of a Pattern

By considering that a pattern P 'acts' as an object of a higher order, we have extended to patterns the concept of an operating field (Chapter 2), the objects of which are the collective links from the pattern to objects of K. If the pattern admits a colimit cP, so that it is effectively integrated into the complex object cP, the operating field of P is isomorphic to the operating field of cP in K, the isomorphism mapping a collective link to the link from cP which binds it.

We can now also extend to patterns the dual concept of the (perception) field of an object (Chapter 1), by taking the perspectives as analogues of the links 'observed' by the object. A field gives a representation of the total category such as it could be seen internally by the pattern, from only the data which are accessible to it. However, if P admits a colimit, its field may not be isomorphic to the field of this colimit.

Definition. Let P be a pattern in the category K. If h is a link from B to B', the composites of h with the links of a perspective F' of B' for P generate a perspective F of B for P; we say that F is *correlated to* F' *by* h. The *field of the pattern* P is the category defined as follows: its objects are the perspectives for P, and the links from a perspective of B to a perspective of B' are defined by the links h from B to B' correlating the two perspectives.

There exists a functor from the field of P to the category K, which associates B to a perspective of B. One can think of the objects of the field (namely, the perspectives) as obtained by identifying the aspects which provide the same information for the pattern as a whole (thus adapting the definition of a 'thing' given by Russell: 'a certain series of appearances, connected with each other by continuity and by certain causal laws' [1949, p. 111]).

If the pattern admits a colimit cP, we have a functor from the field of P to the field of cP in K, which associates to a perspective the link to cP which binds it. However, this functor is not an isomorphism; indeed, two different perspectives of an object B may bind into the same link to cP, and there might exist links to cP which are not mediated by the pattern. This is different from the case of the operating field, which is isomorphic to the operating field of the colimit; this difference is natural, for the colimit inherits only the properties the pattern has when considered as an operating agent, but not as an observer.

The field of a pattern is later used to define the notion of a *landscape* of a CR within the framework of memory evolutive systems, where it was introduced (Ehresmann and Vanbremeersch, 1989) to model the information that a certain module of the system can collect about the system, and use to exert control on the system.

2.   Interactions between Patterns: Simple Links

Looking at a pattern P of the category K as a higher order object, we have defined

• its links towards a unique object: they are the collective links of the pattern;
• the links from a unique object B to the pattern: these are the perspectives of B.

We now move to a more general analysis of the adequate links between two patterns, links which we will call *clusters*. The inter-area transport network described in Chapter 2 gives a concrete example. The clusters from a pattern Q to P model the interactions from Q to P carried out by links between their components, operating together in a manner compatible with the distinguished links. If Q and P admit colimits cQ and cP, respectively, a cluster will bind into a link from cQ to cP of a particular form: it binds a collective link from Q to cP in which all the links are mediated through P.

2.1.   Clusters

Let Q and P be two patterns of the category K. We have defined a collective link from Q to an object B′ as consisting of a family of links from the various components of Q to B′, which are correlated by the distinguished links of Q. If we replace the object B′ by the pattern P, we may similarly define a family of perspectives from the various components of Q to P, which are correlated by the distinguished links of Q. Such a 'collective link of perspectives' generates what we call a cluster. More formally:

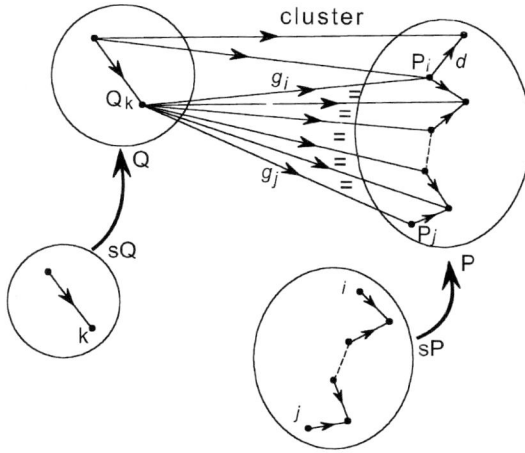

Fig. 3.4 A cluster between patterns.
Given the patterns Q and P, a cluster from Q to P is a maximal set of links from components of Q to components of P, having the following properties: Each component Q_k of Q is connected to at least one component P_i of P by a link g_i of the cluster; and if it is connected by several such links, such as g_i and g_j, these links are correlated by a zigzag of distinguished links of P (the image of the zigzag from i to j in the sketch sP of P). Moreover, the composite of a link of the cluster with a distinguished link of Q (depicted as the link to Q_k), or with a distinguished link d of P (on the right), is also in the cluster.

Definition. Given two patterns P and Q in a category, a *cluster from Q to P* is a maximal set G of links between components of these patterns satisfying the following conditions (see Fig. 3.4):

(i) For each index k of Q, the component Q_k of Q has at least one link to a component of P; and if there are several such links, they are correlated by a zigzag of distinguished links of P.
(ii) The composite of a link of the cluster with a distinguished link of P, or of a distinguished link of Q with a link of the cluster, also belongs to the cluster.

The first condition means that the links of the cluster issuing from a component Q_k of Q form a perspective of this object for P. Thus, a cluster consists of links from the components of Q to those of P, correlated by the distinguished links of both patterns, so that the information transmitted to P by each component of Q is well coordinated with the constraints imposed by the internal organization of P.

Let E be a set of links from components of Q to components of P, and E' the set obtained by adding to E the composites of distinguished links of Q with links in E. If E′ satisfies the first condition of the definition, then E *generates* a well-defined cluster, obtained by taking all the links which are correlated to a link of E′ by a zigzag of distinguished links of P.

Some Particular Clusters.   An object of the category K can be identified to the pattern having this object as its unique component; in this way, any link in K is identified to a cluster having this link as its unique element. In this case, a cluster from P to an object B′ is reduced to a collective link from P to B′; the first condition is trivially fulfilled since a pattern reduced to one component has no distinguished link.

Dually, a cluster from an object B to P is just a perspective of B for P. In particular there is the *insertion cluster* In_i from each component P_i to P, which is generated by the distinguished links of P issuing from P_i. The union of these clusters In_i generates the identity cluster id_P from P to P, consisting of all the distinguished links of P and their composites. Note that the same set also defines a cluster from O to P, where O is the 'discrete' pattern having the same components as P but without any distinguished link (hence O is reduced to a family of objects).

2.2.   Composition of Clusters

Two adjacent clusters can be composed. If G is a cluster from Q to P and G′ a cluster from P to Q′, it is easily proved that the set formed by the composites of each link in G with a link in G′ satisfies the first condition of a cluster, and thus generates a cluster from Q to Q′, denoted by GG′ and called the *composite cluster* of G and G′ (see Fig. 3.5). This composition of

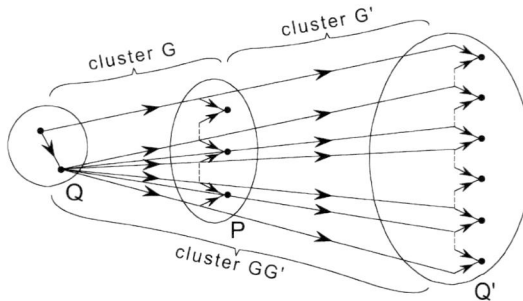

Fig. 3.5   Composition of clusters.
The cluster G from Q to P is composed with the cluster G′ from P to Q′. The composite cluster GG′ is generated by the composites of the links of G with those of G′. The figure shows why two such composites are well correlated by a zigzag of distinguished links of Q′.

adjacent clusters is associative and admits the identity clusters for identities. More precisely:

Theorem 1. *Let* K *be a category.*

(i) *There exists a category, called* **Ind**K, *whose objects are all the (small) patterns in* K; *its links are the clusters between them and its composition associates to two adjacent clusters* G *and* G' *the composite cluster* GG'. *The category* K *can be identified with a sub-category of* **Ind**K.
(ii) *Any pattern* P *of* K *considered as a pattern in* **Ind**K *admits a colimit in* **Ind**K, *which is* P *itself considered as an object of* **Ind**K, *the binding link associated to the index* i *being the insertion cluster* In_i.

Proof. To prove (ii), let us denote by P^* the pattern of **Ind**K defined by P (formally, it is the composite of P with the insertion of K into **Ind**K). The insertion clusters In_i, looked at as links from P_i to P^* in **Ind**K, form a collective link from P^* to P taken as an object P of **Ind**K. Any other collective link from P^* to an object U of **Ind**K binds into a link from P to U, namely the cluster G^* from P^* to U defined as follows:

(i) If U is in K, G^* reduces to the collective link itself, taken as a cluster from P to U;
(ii) If U is not in K, it is a pattern in K and the collective link from P^* to U corresponds to a collective link of perspectives of the components P_i of P to U; in this case G^* is the cluster that this collective link of perspectives generates.

Thus, the category **Ind**K extends the category K into a category in which any pattern of K admits a colimit. This category has been considered by many authors (Duskin, 1966; Deleanu and Hilton, 1976); these authors generally only take particular patterns, called inductive patterns (whence the name of the category), and propose a less explicit definition of its morphisms. The definition given here comes from A. C. Ehresmann (1981), where a cluster is called an *atlas*. Lastly, we note that the construction of **Ind**K is a particular case of the complexification process for a category, process which we will study in the following chapter.

2.3. Simple Links

A link between two complex objects A and C need not correspond to information or other interactions distributed amongst lower level components of the objects. However, some links—which we refer to as simple links and which we briefly introduced in Chapter 2—are of this form. A simple link from A to C is a link which is mediated by decompositions Q of A and P of C. For example, in embryology, the induction of one population of cells by another corresponds to the formation of a simple link. Such a simple link is

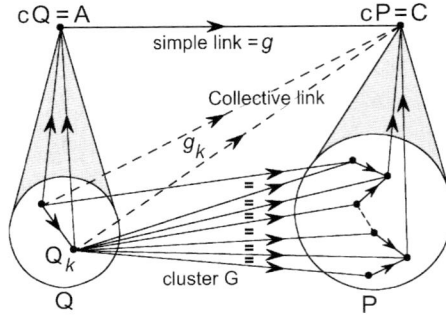

Fig. 3.6   A (Q, P)-simple link.
The cluster G from Q to P binds into the (Q, P)-simple link g from the colimit A of Q to the colimit C of P. This link is obtained as follows: the links of G issuing from a component Q_k of Q form a perspective of Q_k for P, which binds into the link g_k. The family (g_k) of the links g_k issuing from the different components Q_k of Q forms a collective link from Q to cP, which binds into the (Q, P)-simple link g.

obtained by binding a cluster G from Q to P, which thus becomes observable as a unit at the higher level, without adding any information not already available at the level of the components.

Proposition. *Let G be a cluster from a pattern Q to P. If Q has a colimit* $cQ = A$ *in the category K and P a colimit* $cP = C$, *there exists a unique link* g *from A to C binding the cluster, in the sense that the links of the cluster are P-factors of the composites of a binding link to cQ with* g (Fig. 3.6).

Proof. The links in G coming from a component Q_k of Q form a perspective for P, which binds into a link g_k from Q_k to the colimit cP. As the perspectives coming from different components of Q are correlated by the distinguished links of P, the links which bind them form a collective link from Q to cP, which binds into the link g from A to $C = cP$.

In this case, g is called the *binding of the cluster* G and the links in G are called its *factors* from P to Q.

Definition. If A is the colimit cQ of Q and C the colimit cP of P, a link from A to C is called a (Q, P)-*simple link* if it binds a cluster from Q to P. Otherwise it is said to be (Q, P)-*complex*.

In particular, the link binding a perspective of B for P is (B, P)-simple, while the link binding a collective link from P to an object B′ is (P, B′)-simple.

A (Q, P)-simple link g from A to C transmits only information already mediated through its components in Q and P, since it binds a collective link

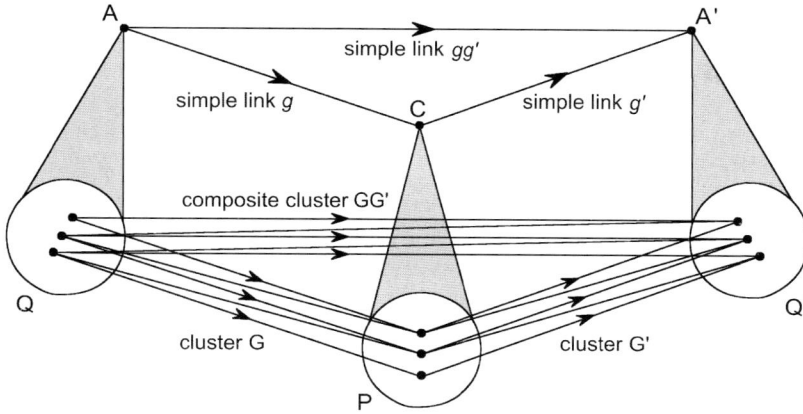

Fig. 3.7 Composition of simple links.
We have a cluster G from Q to P and a cluster G′ from P to Q′. The cluster G binds into a (Q, P)-simple link *g* from the colimit A of Q to the colimit C of P, while G′ binds into a (P, Q′)-simple link *g′* from C to the colimit A′ of Q′. The composite *gg′* from A to A′ is the link which binds the composite cluster GG′, so that it is a (Q, Q′)- simple link.

from Q, of which all the individual links are mediated by P. It follows that, if A and C have other decompositions P′ and Q′, respectively (and we know that this is a possibility), *g* may not be (Q′, P′)-simple because the information transmitted by *g* is not necessarily mediated through the components of Q′ and P′. Thus, the notion of a simple link *per se* has no meaning; it is necessary to indicate with respect to which decompositions it is considered.

A composite of simple links which bind adjacent clusters is still simple with respect to the same decompositions, because it binds the cluster obtained by composing these clusters. More formally:

Proposition. *If g is a (Q, P)-simple link from A to C binding a cluster G, and if g′ is a (P, Q′)-simple link from C to A′ binding a cluster G′, then their composite gg′ is a (Q, Q′)-simple link from A to A′, binding the composite cluster GG′* (see Fig. 3.7).

In contrast, in Section 4 of this chapter, we will see that a composite of simple links is generally not simple if it binds non-adjacent clusters.

3. Representative Sub-Patterns

We have said that two patterns may admit the same colimit. Is this still possible when one of the patterns is a sub-pattern of the other? This is the question we will study next.

3.1.   Comparison Link

Let P be a pattern of the category K. A *sub-pattern* R of P is a pattern which has for components some of the components of P (with the same indices), and for distinguished links, some of those of P. (Formally, R is a homomorphism restriction of P to a sub-graph sR of the sketch sP of P.)

   A particular sub-pattern of P is the discrete pattern O reduced to the components of P, without any distinguished link, and we have said that there exists an insertion cluster from O to P. As O is reduced to a family of objects, it admits a colimit cO if and only if cO is the sum of this family (P_i). In this case, and if P admits a colimit cP, the insertion cluster binds into a (O, P)-simple link c from cO to cP. This link is the comparison link from the sum to the colimit (defined in Chapter 2), which measures the constraints imposed on the components by the distinguished links of the pattern. This case will be extended to any sub-pattern R of P:

Proposition. *For any sub-pattern R of P, there exists an 'insertion' cluster from R to P generated by the distinguished links of P coming from components of R. If P has a colimit cP and R a colimit cR, this cluster binds into a comparison link c from cR to cP (Fig. 3.8).*

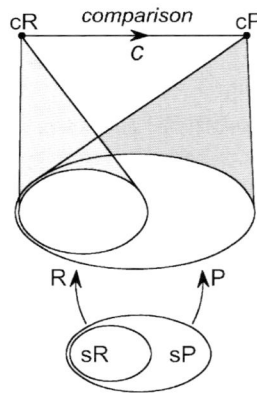

Fig. 3.8   Comparison between a pattern and a sub-pattern.
R is a sub-pattern of the pattern P, meaning that the components and distinguished links of R are some of those of P. There then exists an insertion cluster of R into P, generated by the identities of the components of R. If both R and P admit colimits cR and cP, there is a comparison link c from cR to cP that binds this insertion cluster.

The comparison link from cR to cP measures the constraints imposed by the components and links of P which are omitted in R. Let us consider two particular cases:

(i) R contains all the components of P but only some of its distinguished links; all the 'matter' is preserved, but its organization is looser (the extreme case being that of the sum). R could represent the primary structure of a protein and P its ternary structure. For a collective link from R to an object B to also define a collective link from P to B, the only condition is that its links satisfy the additional constraints imposed by the links not in R (namely, the corresponding correlating equations), and this translates into the fact that its binding from cR to B factors through the comparison link c. Thus, c represents the obstacle created by these additional constraints.

(ii) R preserves only some of the components of P, but all the distinguished links in P between them; that is to say, we have less matter but its organization is preserved. Then a collective link from R to B may not extend into a collective link from P to B, because of the missing components, and the comparison link measures the weight of the contribution of the components of P which are not components of R. For example, R could be a binding site of a protein P, with its ternary structure intact, *e.g.* the binding site allowing the recognition of an antigen by an antibody; once the antigen is bound to the binding site, the new system, consisting of the antigen–antibody complex, although retaining the binding site as a sub-system, can no longer as a whole bind to another molecule of the antigen.

3.2. *Representative Sub-Pattern*

A sub-pattern may have the same colimit as the whole pattern. For example, let P be a pattern having C for its colimit cP, and R a sub-pattern which has also C for its colimit cR. Then the comparison link is the identity of C (or an isomorphism). It means that any collective link from R to any B extends into a single collective link from P to B; or, more roughly, the collective actions of P are entirely determined by those of R: the links of P which are not in R do not impose any additional constraint and the components not in R act automatically in synergy with those of R. As an illustration, consider that the board of directors of a company determines its policy and represents it to third parties. Similarly, in biological systems, there often exist sub-systems which have redundant components, able to replace other components if those are disabled or destroyed.

Given a pattern P and a sub-pattern R which both have colimits, it is not always possible to recognize, from the patterns alone, whether they have the same colimit (this is a particular case of the problem which will be studied in the next section). A condition for this to be true is that each component of P be connected to some 'representatives' in R which act in a coordinated way via their distinguished links; that is, any collective action of the representatives commits the entire pattern. This will be achieved when there is a cluster from P to R. More formally:

Proposition. *If there is a cluster H from P to a sub-pattern R of P such that H is generated by distinguished links of P, then both P and R either do not have a colimit or have the same colimit.*

Proof. Each collective link F from R to an object B extends then into a unique collective link from P to B, namely the collective link obtained by composing the cluster H with F. Let us note that, in the category **IndK**, the cluster H is the inverse of the insertion cluster from R to P.

In the conditions of the above proposition, R is called a *representative sub-pattern* of P, and the components of R which are connected to the component P_i of P by some link in the cluster are called the *representatives of* P_i (Fig. 3.9). The usual categorical terminology for a representative sub-pattern of P is that it is *final* in P (Mac Lane, 1971).

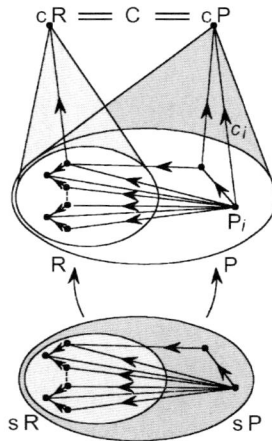

Fig. 3.9 A representative sub-pattern.
R is a sub-pattern of the pattern P. It is a representative sub-pattern of P if both P and R have the same colimit C, and if there is a cluster from P to R generated by distinguished links of P. This cluster binds into the identity of C, and its links connect a component P_i of P to components of R, called its representatives in R. Then the representatives of the component P_i of P are correlated by a zigzag of distinguished links of R.

A pattern P has a representative sub-pattern reduced to a unique component R_0 if each component of P is connected to R_0 by a single distinguished link, *i.e.* if R_0 is a final object of P. In this case R_0 is the colimit of P. For example, the structure of a society depends on the type of representatives that it admits. In an authoritarian society there are few representatives; perhaps only one, if we consider a dictatorship. In a democratic society, the representatives are elected; further, in a proportional representation system, each citizen votes for a list of candidates who (in principle at least) agree on a common program. The principle of majority rule supposes that the group R of candidates who obtain the majority of the votes are then accepted by all. However, if the opposition becomes too strong, and increasingly large number of citizens no longer accept the policy of the representatives, then R does not remain a representative sub-pattern of P. The force of discontent can be measured by the comparison link from the new colimit of R to (that of) the society.

4. Multiplicity Principle

The robustness and the adaptability of a complex natural system, say a biological system, are related to the fact that the same functions can be carried out by groups of interacting structurally different components. In our model, this is translated by saying that the patterns these groups form have the same colimit. Indeed, in the preceding discussion, we have encountered several cases in which the fact that a complex object C admits several decompositions has important consequences: dependence of the concept of simple link on the decompositions which are used; existence of non-simple links composing simple links binding non-adjacent clusters; and existence of representative sub-patterns. In the discussion that follows, we will characterize this situation and we will further show it plays an essential part in the adaptability of the system and the emergence of complex objects.

4.1. Patterns with the Same Colimit

While a pattern of linked objects in a category has at most one colimit (up to an isomorphism), the converse is not true: a complex object C can be the colimit of several decompositions into patterns, possibly without their components being connected in any specified way. In the applications, we may conceive of a decomposition of C as allotting particular values ('parameters') to certain characteristic features of the object; the existence of different decompositions gives then some plasticity to the object, by allowing for an adapted choice of parameters depending on the context.

Let P and P° be two decompositions of the same object C, so that C is simultaneously the colimit cP of P and cP° of P°; we say that P and P° are *homologous* patterns (in equation: cP = C = cP°). This situation can occur without any direct relation between the components of P and P°, so that it is impossible to recognize, at the level of their components, that both have same colimit: it is a 'global' property of the category and not a 'local' property of the patterns. For example, in a partially ordered set, two families of elements can have the same least upper bound without any of their elements being comparable in the order. On the other hand, we have seen that a sub-pattern R of P can be recognized to be representative if there exists a cluster of distinguished links from P to R. This condition is extended as follows:

Definition. Two decompositions P and P° of an object C are said to be *connected* (respectively, *strongly connected*) if there is a cluster from P to P°, or a cluster from P° to P, which binds into an isomorphism (respectively, into the identity of C). If C admits at least two decompositions P and P° which are not connected, we say that C is a *multifold object*, and the passage between P and P° is called a *complex switch*.

In other terms, C is a multifold object if it admits two decompositions with respect to which the identity of C is a complex link. Switching between such non-connected decompositions can be seen as a random fluctuation in the internal organization of C which does not modify its functionality on a higher level, where the fluctuation is not observable: different micro-states lead to the same macro-equilibrium.

As an example, take the category of particles and atoms (*cf.* Chapter 1). Its objects are the elementary particles and the atoms; an atom is the colimit of the pattern formed by its nucleus and one of its electronic orbital configurations. Quantum physics shows that each atom admits different such configurations, corresponding to different energy levels. The passage between the patterns corresponding to two of these configurations is a complex switch, and the atom is a multifold object. This example is important: we will see that it forms the basis for the evolution of more and more complex systems, from physical systems to biological organisms, up to cognitive systems. As other examples, the same amino acid can be obtained as the colimit of two different codons, while passage between two genotypes with different alleles but leading to the same phenotype is also a complex switch.

4.2.   Connected Patterns

We have defined the concepts of homologous or connected patterns when both are given as two decompositions of the same object; hence both admit this object as their colimit. Now a pattern may have no colimit in the category K, but acquire one if it is regarded as a pattern of a larger category. For instance, any pattern P of K acquires a colimit in **Ind**K, and we will give other similar constructions in the following chapter.

How may we generalize these concepts to recognize, from the patterns alone, whether they can acquire the same colimit? To answer this question, let us return to the definition of a colimit. The colimit is characterized by the fact that its links towards any other object correspond to the collective links of the pattern. Thus, for P and P° to have the same colimit, it is necessary that their collective links be in a one-to-one correspondence. To clarify this condition, we will use the operating fields of the patterns.

Let us recall (Chapter 2) that the operating field of P is the category ΩP having for objects the collective links of P, a link between two of them being defined by a link h of K correlating them. This category is provided with a base functor to K, associating B to a collective link to B (a collective link being a particular cluster, ΩP can also be defined as a sub-category of the operating field of P in **Ind**K). If G is a cluster from Q to P, it defines the functor ΩG from ΩP to ΩQ, which associates to the collective link F from P to B the collective link GF from Q to B.

Definition. Two patterns P and Q are *homologous in* K if their operating fields are isomorphic categories above K (*i.e.* there is an isomorphism between them which commutes to their base functors to K). They are *connected* if there exists a cluster G from Q to P such that the functor ΩG is an isomorphism between their operating fields, or if there exists a cluster from P to Q with the corresponding property (Fig. 3.10).

Thus, P and Q are homologous if there is a functor from ΩP to ΩQ which associates to a collective link from P to B, a collective link from Q to B, and which has an inverse from ΩQ to ΩP. It is a global property since it utilizes the synergistic behaviour of the patterns with respect to any other object of the category (via their collective links). In the stricter case where P and Q are connected, this functor is of the form ΩG for a cluster G from Q to P, or its inverse is of the form $\Omega G'$ for a cluster G' from P to Q. In this case, it can be 'locally' recognized, the isomorphism between their operating fields being instantiated at the level of their components via the composition with a cluster.

The following proposition (whose proof is a direct consequence of the definitions) shows that this more general notion of homologous patterns agrees with the special case considered in Section 4.1.

Proposition. *Given two homologous patterns (in the general sense above), one of the patterns has a colimit if and only if the other has a colimit; and in this case both have the same colimit.*

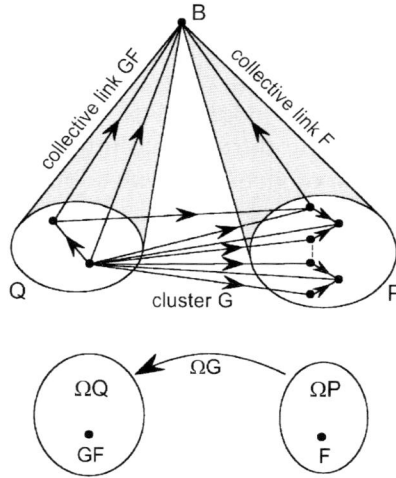

Fig. 3.10   Connected patterns.
P and Q are two patterns in the category K, and F a collective link from P to an object
B of K. The operating field ΩP of the pattern P has for objects the collective links of
this form. A cluster G from Q to P defines a functor ΩG from ΩP to ΩQ (shown at the
bottom): to the collective link F from P to B this functor associates the collective link
GF from Q to B. The patterns are connected if there exists such a cluster G for which
ΩG is an isomorphism, so that the map associating GF to F is one-to-one. This is a
stronger condition than that the patterns be homologous, which requires only that
there exists an isomorphism (of any form) between their operating fields. In the top
figure, the collective link F is shown via its links f_i in the category K; in the bottom one,
it is just an object of ΩP.

4.3.   Multiplicity Principle and Complex Links

Many characteristics of natural complex systems depend on the fact that a
complex component C acquires global properties which cannot be recognized at the level of its lower components. This is the case when C admits
different decompositions which are not 'materially' connected, so that their
homology is not observable via their components. The categories which
model them satisfy the following global property:

Definition. We say that the category K satisfies the *multiplicity principle* if it
admits at least two patterns which are homologous but not connected.

The multiplicity principle was introduced under the name of *degeneracy
principle* (see Ehresmann and Vanbremeersch, 1993), as a generalization of
the degeneracy principle considered by Edelman (1989) in a neural system.
More recently, Edelman and Gally have defined *degeneracy* as the 'ability of
elements that are structurally different to perform the same function or yield

the same output' (2001, p. 13763). They show its ubiquity in biological systems, where it 'is a prerequisite for and an inescapable product of the process of natural selection itself'. They distinguish degeneracy from redundancy, which would be the stricter case of connected patterns. Our model will justify their conjecture that complexity and degeneracy go hand in hand, by showing how emergence relies on the role of the complex links allowed by the multiplicity principle.

We saw that a composite of simple links binding adjacent clusters is still simple (with respect to the same decompositions). However, if the category satisfies the multiplicity principle, the existence of multifold objects implies that there can exist composites of simple links binding non-adjacent clusters, and such a composite is generally not simple. Indeed, let C be a multifold object admitting two decompositions P and P^o. If g is a (Q, P)-simple link from A to C binding a cluster G, and g' a (P^o, Q')-simple link from C to A' binding a cluster G', a composite gg' of these links must exist in the category K, although in general it does not bind a cluster from Q to Q'. In this latter case it is called a (Q, Q')-*complex link* from B to A (Fig. 3.11).

Let us note, however, that if P and P^o are strongly connected via a cluster H from P to P^o which binds into the identity of C, the composite cluster GH also admits g as its binding, and consequently the composite gg' binds the cluster GHG', and therefore is (Q, Q')-simple. Thus, a composite of simple

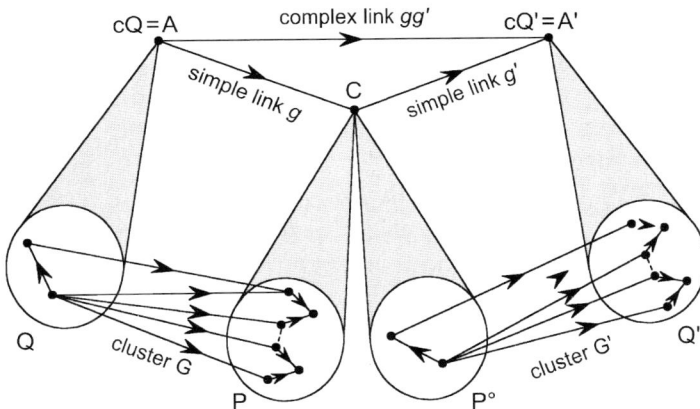

Fig. 3.11 Complex link.
A complex link is obtained by composing simple links binding non-adjacent clusters. The patterns P and P^o are homologous patterns having the same colimit C. The link g from A to C is a (Q, P)-simple link binding the cluster G, and g' is a (P^o, Q')-simple link from C to A' binding the cluster G'. If P and P^o are homologous but not connected, the composite link gg' is generally not a (Q, Q')-simple link, because it does not bind any cluster from Q to Q'.

links binding non-adjacent clusters may, in this particular case, also be simple. On the other hand, if P and P° are homologous but not connected, gg' is generally a (Q, Q')-complex link which is not mediated by links between the elementary components of A and A' in Q and Q': it does not subsume properties which are locally observable in any way at the level of these patterns, for there may exist no direct link nor even zigzag between components of Q and Q'. However, it is well rooted in this level not only locally via the clusters G and G', but also globally via the existence of a complex switch between the two decompositions P and P° of the intermediary multifold object C. Thus, such a complex link models emergent properties at the level of A and A' which are dependent on the *total* structure of the level to which pertain their components.

In the above we considered a composite of two simple links, but the result generalizes:

Definition. A (Q, Q')-*complex link* is defined as the composite of a path of simple links binding clusters between non-adjacent patterns, the intermediate objects being multifold, so that the link is not (Q, Q')-simple.

A composite of complex links is either simple or complex.

Examples.   In a category of social groups, let us consider patterns Q and Q' representing respectively a group of farmers, and a group of consumers. The farmers bring their products to the purchasing department P of a cooperative C, and they are sold to the consumers through its sales department P°. The cooperative can be identified with the colimit of both P and of P°, the group of farmers as an entity by the colimit cQ of Q, and the group of consumers as an entity by the colimit cQ' of Q'. The delivery of products by the farmers is represented by a cluster G from Q to P which binds into a (Q, P)-simple link 'delivery', and there is also a (P°, Q')-simple link 'sale' binding the cluster from P° to Q'. The composite of these two simple links is a complex link from cQ to cQ' which models the sale transfer from the farmers to the consumers via the cooperative, without any direct exchange between a farmer and a consumer. A similar example would be the link between the authors of a journal and its subscribers, via the entity 'journal' considered as the colimit of both its editorial board and its sales department.

5. Hierarchies

Now we have the tools necessary to model hierarchical systems. We have represented the configuration of a system at a given time by a category, and the complexity of an object in a category by the fact that it admits its own

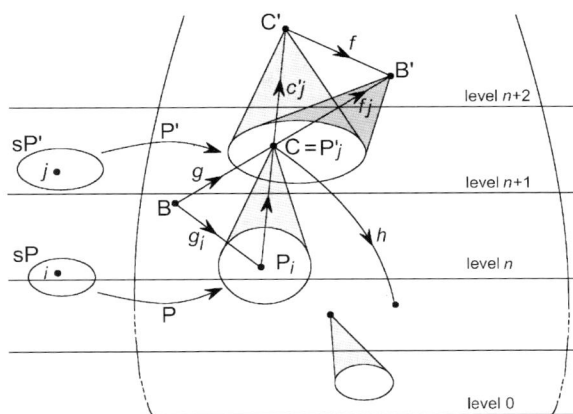

Fig. 3.12 A hierarchical category.
The objects of a hierarchical category are partitioned into several levels, represented as increasing vertically. An object C of level $n+1$ is the colimit of at least one pattern P which has components of strictly lower levels. At the same time it can be one of the components of the object C' of the higher level $n+2$ (colimit of the pattern P'). Links from C can go to any level, for instance h goes to a lower level, while f_j goes one level higher, as one of the links of a collective link from P' to B'. And links to C can come from any level; among these links, some, such as g are mediated by P, while others are not (not shown).

internal organization in the form of a pattern of which it is the colimit. As we discuss below, this leads us to model the configuration of a hierarchical system at a given time through a *hierarchical category* (Fig. 3.12).

5.1. *Hierarchical Categories*

Definition. A *hierarchical category* is a category K the objects of which are partitioned into a finite sequence of levels 0, 1, ... , N, so that any object C of the level $n+1$ is the colimit in K of at least one pattern P included in the levels $<n+1$ (*i.e.* each component P_i of P is of a level lower or the same as n).

In earlier papers (*e.g.* Ehresmann and Vanbremeersch, 1987), we had given a slightly stricter definition, requiring that any object of level $n+1$ be the colimit of at least one pattern of which all the components are of level n (and not only of a level $<n+1$). We relax this condition here because it is not essential, and it is not always satisfied in the examples, in particular in hierarchies obtained by division of levels of another hierarchy. For example, in a cell the molecular level can be subdivided into two levels, that of the small molecules and that of the macromolecules.

The level of an object can be thought of as an indication of its degree of complexity (though we shall later define a finer measure of its complexity, namely its order of complexity). No condition is imposed on the objects of level 0, which are regarded as the elementary components (sometimes called atoms). The hierarchy is of a descriptive nature, with inter-connection of levels, because, in addition to the 'horizontal' intra-level links, there are 'vertical' links from lower levels to higher levels (in particular the binding links of its lower levels components P_i to the colimit C of a pattern P), and there can exist vertical links going down from a higher level to a lower one. An object can thus receive information at the same time from objects of the same level, of lower levels, or from higher levels, and in return send messages to all the levels, in so far as the energy constraints allow it.

An object C of any intermediate level has a double aspect (like that of Janus, in the terminology of Koestler, 1965, who also speaks of holons) according to whether it is compared to higher or lower levels:

(i) Such an object admits at least one decomposition in a pattern P included in levels $<n+1$, so that it is complex with respect to its components P_i, and its actions correspond to the collective links of this pattern which it binds. Among the links from an object B to C: some are mediated by at least one of its components; others are independent of P and represent information directed to C taken as a unit, with its own identity which transcends that of its components, one not observable at the lower levels.

(ii) On the other hand, C can be one of the components of an object C' of level equal to or greater than $n+2$, and as such interact with the other components of a pattern P' decomposing C'. Compared to C', C is regarded as a 'simple' unit in the sense that one forgets the internal organization of C in components of levels $<n+1$. In this case, among the links from C to an object B', we have to distinguish: the individual links of C, corresponding to actions of C not coordinated with the other components of C'; and the joint links corresponding to the role of C as a component of C', within the frame of a common action of P'. A joint link is a link which is one of the factors f_j of a collective link from P' to B'; therefore it is the composite of a binding link from C (taken as a component of P') to the colimit C' of P', with a link f from C' to B' (Fig. 3.12).

It follows that, among the links between two complex objects, we can have simple links with respect to given decompositions of these objects, complex links obtained as composites of simple links, but also new links independent from any decomposition. In the next chapter, we will characterize 'based' hierarchical categories in which all the links can be explicitly described.

5.2.   Examples

A book: There is a hierarchical category associated with a book. The level 0 is formed by the letters of an alphabet without links between them; the objects of level 1 are words formed out of these letters, those of level 2 are sentences, those of level 3 are chapters. And the book as such is the unique object of level 4. A word is the (indexed) colimit of the (ordered) family of its letters, and there is a binding link from each one of its letters to the word. In the same way a sentence is the colimit of the ordered family of its words, with a link from each one of its words towards the sentence; the links between sentences define their order in the book. A chapter is the colimit of its sentences and the links between chapters define their order in the book. The book is the colimit of its chapters.

A cell is modelled by a hierarchical system consisting of its atoms, molecules, macromolecules and sub-cellular systems (organelles, populations of molecules, functional modules). A molecule is the colimit of its atoms, a macromolecule the colimit of its functional units, and a functional unit the colimit of its molecular composition. For example, DNA is the colimit of its different units, such as genes or non-coding stretches (introns).

A business enterprise has a hierarchical structure: The objects of level 0 represent its employees. The more complex objects represent departments, small productive units in charge of a particular task, up to the higher managerial levels (*cf.* Introduction and Fig. 6.8). The links between the members of a department represent the channels by which they exchange information and collaborate to complete their collective work. The higher levels can send orders to the lower levels, but they also depend on the results obtained by these lower levels: the work of a department building a machine is stopped if the necessary components are not produced in a sufficient quantity.

In sociology one considers hierarchies where the level 0 corresponds to individuals say the inhabitants of a country, with the links representing communication between them; the objects of higher levels are increasingly complex social groups, with their interrelations. An individual (or a group) B can communicate directly with a group C, or indirectly by the intermediary of one or more members of the group. The existence of many links which are not mediated comes from the fact that it is possible to personify groups such as families, companies, administrations, nations, international organizations and so on; for example, in France, a taxpayer will send his income tax return to the revenue department as such, but if he needs to make a correction, he can contact the taxation agent directly. However, the links are often mediated through particular members of the group who act as liaison agents, that is, agents who only relay messages, but do not act on them *per se*: secretaries filter the telephone calls of a company, while the diplomatic correspondence of a nation will be relayed by its ambassadors.

The difference between individual and joint links is well illustrated in social groups. If an individual P_i belongs to a group C, his individual links to another individual B′ represent communication on a purely personal basis, not taking account of his membership in C, while his joint links (*i.e.* his links pertaining to a collective link of the group) corresponds to his role in the group. For example, if a tax collector seeks to borrow money from a friend B′, the friend can refuse; but if he claims taxes from him as a tax collector, B′ must comply, under threat of sanctions. In the same way a property under joint ownership, with possibly unequal shares, forms a pattern whose colimit C corresponds to what is called an undivided property; if the property has a debt, the creditor B′ can require from each of the members of the society a payment proportional to his share; but if B′ is the creditor on a purely personal basis of only one of the members of the society, he can require nothing from the other members. The more numerous are the distinguished links of a pattern, the greater is the difference between individual links and joint links, the participation in the group modelled by the pattern imposing more and more constraints. In many cases, the joint links, which participate to higher level communications from the group as a whole, correspond for the individual to *meta-communication* (*i.e.* communication about communication). The interlocutor B′ will have, in his relationship with a member of a group, to disentangle, in each case, communication from possible meta-communication. Indeed, the conflation of language and meta-language can create much confusion, both being able to be more or less contradictory; and this is the source of a great number of paradoxes, as shown by Watslawick *et al.* (1967). These paradoxes seem specific to human discourse, for they rest on the use of language.

The justice system exploits the difference between individual and joint links. In a trial, the judge and the officers of the court (such as the civil prosecutor) typically speak in meta-language, since they defend the interests of the society; while the lawyers for both parts speak in ordinary language to influence the jury by appealing to their sympathy or antipathy for the defendant. And the jury is supposed to decide as representatives of the society, therefore in meta-language. In fact, however, they cannot disregard their personal feelings which, at least in an unconscious way, intervene in their judgement; whence the occurrence of more or less severe verdicts according to the emotional impact of the offence. The sentence corresponds to a constraint which the society imposes on the culprit, a constraint which would not be tolerated if it were imposed by an individual acting alone. Thus, *in extremis*, the executioner killed in the performance of his duties, whereas the law prohibits him from killing as a simple individual.

This difference between language and meta-language which leads to different moral codes for individual and society is already formalized in the Bible (and even before that in the legal codes of Hammurabi and Ur-Nammu) although this fact is little known. The *lex talionis*: 'And thine eye shall not pity; but life shall go for life, eye for eye, tooth for tooth, hand for hand, foot for foot' (Deut. 19–21) can be applied only by the society as a whole, and the judgement must be objective: 'One witness is not enough to convict a man accused of any crime or offence he may have committed. A matter must be established by the testimony of two or three witnesses' (Deut. 19–15), 'and the judges shall make diligent inquisition' (Deut. 19–18). On the other hand, for the individual, it is stipulated: 'Do not hate your brother in your heart ... Do not seek revenge or bear a grudge against one of your people, but love your neighbour as yourself' (Lev. 19:17–18). It is this commandment of love which is taken by Jesus (who quotes it explicitly), and then amplified: thus, when the crowd seeks to lynch the adulteress according to the law of Moses, he is able to use it to replace the meta-language of the society with the ordinary language of the individual, by saying: 'Let he who is without sin, cast the first stone' (John 8:7).

5.3. Ramifications of an Object and Iterated Colimits

Let us come back to the topic of a hierarchical category K. We have seen that any object C of level $n+1$ has an internal structure in the form of at least one pattern included in stricter lower levels. We now describe how it then follows that C has also a more intricate internal structure, going down to the levels less than m, for every $m<n$.

By definition, an object C of the level $n+1$ is the colimit of at least one pattern P included in levels $<n+1$ (there can be several). Similarly, each component P_i of P is itself the colimit of at least one pattern, say P^i, included in levels strictly lower than the level of P_i. We say that C is the 2-*iterated colimit of the ramification* $(P, (P^i))$ of length 2. (In the notation, the family (P^i) of patterns has the same set of indices as P.) This ramification gives an internal organization of C in two stages, namely (Fig. 3.13):

(i) the stage of its components P_i with their distinguished links in P;
(ii) the stage of the components P^i_j of the decompositions P^i of the objects P_i, with their distinguished links in these patterns; these ultimate components of the ramification will be called *micro-components*; they are of level less than or equal to $n-1$.

Thus, two stages are needed to rebuild C starting from its micro-components via the ramification: the first to bind each P^i into its colimit P_i; the second utilizing the distinguished links between the P_i (links which may

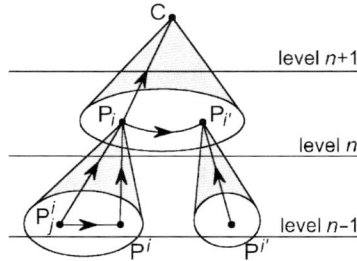

Fig. 3.13 A ramification of length 2.
The object C of level $n+1$ has a decomposition P into components P_i of levels $<n+1$.
Each of these components has in turn a decomposition P^i in components of levels $<n$.
The ramification $(P, (P^i))$ takes into account not only the micro-components P_j^i and the
components P_i, but also the distinguished links between these micro-components in the
various P^i, as well as the distinguished links of the pattern P.

not be observable at the level of the micro-components) to bind P into C.
The continuation of the same process allows us to construct increasingly
long ramifications of C, which represent the internal organization of C in
more and more stages, included in lower and lower levels.

Definition. We define a *ramification of length k* of an object C of a category
by recurrence as follows:

(i) A ramification of length 1 is a pattern admitting C for a colimit (thus it
reduces to a decomposition of C).
(ii) A ramification of length k of C consists of a pattern P having C for
colimit and, for each one of the components P_i of P, of a ramification R^i of
length $k-1$ of P_i. In this case, we also say that that C is a *k-iterated colimit* of
the ramification $(P, (R^i))$.

Let C be an object of level $n+1$ in K. For every $m<n$ we can construct by
recurrence (at least) one ramification of C whose ultimate components are of
level less than or equal to m: one takes a decomposition P of C included in
levels $\leqslant n$; then decompositions P^i of the different components P_i of P included in levels $\leqslant n-1$; then decompositions of the various components P_j^i
of P^i included in the levels $\leqslant n-2$; and so on down to the level m, whose
components are called the *ultimate* components of the ramification. If the
ramification is of length k, its ultimate components will be included in the
levels $\leqslant n-k+1$ (see Fig. 3.14). Such a ramification provides C with a kind
of *fractal structure*: the components at each intermediate stage ramify
themselves; but in addition there are correlations between their ramifications, which come from the constraints introduced by the distinguished links
between those components at each intermediate stage.

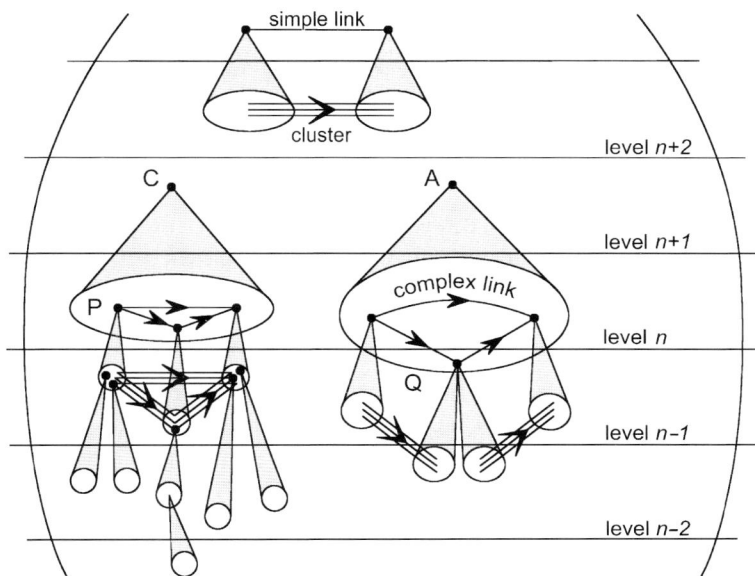

Fig. 3.14 Different ramifications.
Representation of a ramification of C of length 4 and of a ramification of A of length 2. The links between two objects with given decompositions can be simple links with respect to these decompositions (like a link binding a cluster of level $n-1$), or complex links, or independent from them. In the figure, P has only simple links, while Q has one complex link and two simple links with respect to decompositions given by lower levels of the ramifications.

As an object may have different decompositions, the above construction leads to different ramifications of C, since at each stage we have to choose a particular decomposition of the components intervening at this stage. To describe them, it is necessary to at the same time take account of the canonical vertical binding links of the components of a pattern to its colimit, and of the links formed by the distinguished links of the patterns at the various stages. For instance, C could represent the menu in a restaurant; this menu is initially described by its general composition: entrée, meat, cheese and dessert; or (another decomposition): soup, fish, fruits; then choices can be specified, such as tomatoes or ham in the entrée, chicken or sheep as meat and so on.

The number of ramifications of an object depends on the number of different choices to be made (hence the degrees of freedom), so that it increases with the length of the ramification. In the applications, this gives a great flexibility to a complex object. If we think of a decomposition as attributing particular values to some parameters, the choice of a decomposition can be seen as the filling of a slot within a frame in the sense of Minsky (1986). Then each stage gives a particular possibility to fill these slots, each one allowing for various choices, and the same process being repeated at the various intermediate stages.

Looking upwards, a ramification of length k of C makes it possible to reconstruct C in k steps, starting from its ultimate components (of level $\leqslant n-k+1$) and their links. Consequently, C seems increasingly complex when it is considered by comparison with the ultimate components of ramifications of increasing lengths, included in lower and lower levels. Yet it seems simple when compared to higher levels and forgetting its lower levels internal structures.

This concept of *complexity* thus relates to the level where the object is studied. A change of scale can be effected, forgetting all the objects of the levels strictly lower than n and the links between them; a new hierarchical category is thus obtained in which the objects of level n become its 'atoms', and which forgets all the details of their organizations at lower levels. Whence the question: does the level of an object always reflect its 'operating' complexity? It is the question which we are going to study in the next section.

6.   Complexity Order of an Object: Reductionism

Descartes gave as one of the principles of his method: 'to conduct my thoughts in such order that, by commencing with objects the simplest and easiest to know, I might ascend by little and little, and, as it were, step by step, to the knowledge of the more complex' (translated from Descartes, 1637, p. 47). This is the stance modern science has adopted through the paradigm of reductionism, which tries to reduce the study of an object to that of its lower level elementary components. For example, molecular biology would like to reduce the study of a living organism, or at least of a cell, to the level of its molecular organization. This paradigm has begun to be contested. Can the consideration of a hierarchical category give some clues to the validity of this paradigm?

6.1.   *Obstacles to a Pure Reductionism*

In a hierarchical category K, can we deduce the properties of a complex object C from that of its ultimate (atomic) components, arrived at via a ramification down to the lowest level? The analysis of the previous sections shows that there are a number of obstacles to reductionism.

First let C be of level 1, so that it is the colimit of at least one pattern P the components of which are atoms (*i.e.* of level 0). The actions of C correspond to the collective links of P; thus it is not only the atoms P_i which are necessary in the actions, but also the distinguished links between them, which impose constraints and allow for their synergistic action. So, C cannot be reduced just to its components P_i; instead, it is essential to take into account the internal organization P they form with their distinguished links. There is another problem if C is a multifold object, in the sense that it is the colimit of different patterns which are not connected (see the aforementioned multiplicity principle). The knowledge of one of its decompositions P does not determine its other decompositions, and this will have important consequences as explained below. In any case, let us note that a decomposition P of C determines its global actions, but may not trace all the lower information or constraints C receives. Indeed, P traces only those links to C which are mediated by one of its components, and we have seen that there may exist links to C not mediated by P. In particular such links can come from higher levels, and so exert some top-down influences on C.

If C is of level 2, the process indicated in the preceding section allows us to construct a ramification $(P, (P^i))$ of C of length 2, the micro-components of which are atoms. Thus, it is possible to reconstruct C in two steps from these atoms up, but this requires specifying not only their distinguished links in the various patterns P^i, but also the distinguished links of P between the colimits P_i of the patterns P^i. Is it possible to translate the constraints imposed by the distinguished links in P into constraints imposed on the atoms? In other terms, in which cases can the distinguished links of P be reduced to that of links between the atoms of C, so that the actions of C be entirely determined by the atomic level? The reduction theorem given below proves that it is possible only if the distinguished links of P are simple links binding clusters between the atoms. In this case, C is also the colimit of a large enough pattern included in the atomic level and having the ultimate (or micro-components) of the ramification as its components. If some of the distinguished links of P are complex, the reduction fails and C really has emergent properties.

More generally, let C be of level $n+1$. The above process (applied between the levels $n-1$ and $n+1$) allows us to analyse when C can be reduced to levels $<n$ by 'jumping over' level n. And the same process extends to longer ramifications, to determine cases in which a ramification of C down to a level k can be replaced by a simple decomposition of C of level k. We can then formally state the:

> *Reductionism Problem*: can a complex object C be reconstructed from atomic components up in only one step, as the colimit of a (perhaps large) pattern included in level 0?

6.2. Reduction of an Iterated Colimit

We start by examining the general problem of the reduction of an iterated colimit into a simple colimit in a given category K. Let C be an object of K which has a ramification $(P, (P^i))$ of length 2. We speak of the components of the patterns P^i (*i.e.* the ultimate components of the ramification) as the micro-components of C. In which case is it possible to replace the distinguished links of P by links between the micro-components, so that C be reduced to the simple colimit of an adequate pattern connecting its micro-components? The following theorem (Ehresmann and Vanbremeersch, 1996) shows that this is possible if each distinguished link between the components of P is 'well' mediated through micro-components.

Theorem 2 (Reduction Theorem). *Let C be the 2-iterated colimit of a ramification $(P, (P^i))$. If each distinguished link d from P_i to P_j in P is (P^i, P^j)- simple, then C is also the colimit of the following pattern R: its components are those of the various P^i and its distinguished links are those of the patterns P^i, as well as the links of the clusters G_d that the distinguished links d of P bind. Moreover, R and P are connected by the cluster generated by the binding links of the various P^i_k to P^i (Fig. 3.15). On the other hand, if certain distinguished links of P are complex links, C may not have any such decomposition.*

Proof. We suppose the sets of indices of the different P^i are disjoint, so that an index k of R is the index of just one of them, say P^i, so that $R_k = P^i_k$. Then, the binding link associated to k will be the composite of the binding link c^i_k associated to k in P^i (from P^i_k to the colimit P_i of P^i) with the binding link c_i from P_i to the colimit C of P. Any collective link H from R to an object B binds into a link h from C to B obtained as follows: for each index i of P, there is a collective link from P^i to B restriction of H, formed by the links in H corresponding to the indices of P^i; this collective link binds into a link h_i to B. It is easily proved that the various h_i form a collective link from P to B, and h is the link from the colimit C of P to B which binds it. The set formed by the binding links of the various P^i generates a cluster from R to P which binds into the identity of C, so that these two patterns are strongly connected. On the other hand, if P has complex links, a large pattern such as R above may be constructed, but it may not admit C for its colimit.

The preceding result extends to a ramification of any length: if all the links which intervene at the intermediate levels are simple with respect to the patterns of the lower levels, then the iterated colimit can be reduced to a simple colimit of a large pattern having for components all the ultimate components of the ramification. However, such a reduction may not exist if certain distinguished links are not simple.

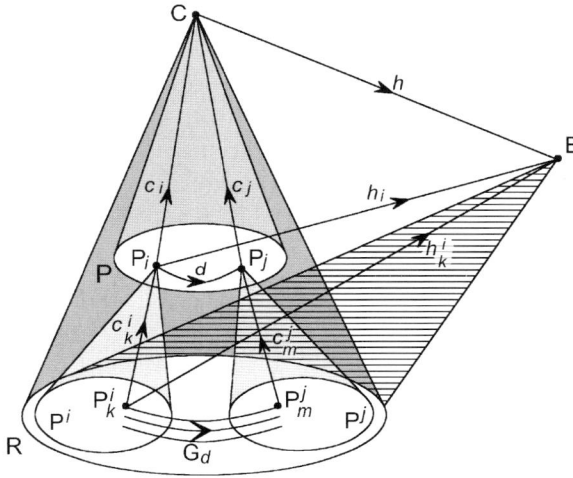

Fig. 3.15 A reducible 2-iterated colimit.
C admits the ramification $(P, (P^i))$ in which each distinguished link d of P from P_i to P_j binds a cluster G_d from P^i *to* P^j. We construct a large pattern R containing the various P^i as sub-patterns, and having also for distinguished links all the links in the various clusters G_d. This pattern R then admits C as its colimit. The binding link from a component P^i_k to C is the composite of the binding link from P^i_k to the colimit P_i of P^i with the binding link c_i from P_i to the colimit C of P. A collective link (h^i_k) from R to an object B binds into a unique link h from C to B constructed as follows: the family of links (h^i_k) for a given i forms a collective link from P^i to B binding into a link h_i from P_i to B, and these links form a collective link from P to B binding into h.

6.3. Complexity Order of an Object

Let us return to the case of a hierarchical category K. We know that, for every $m < n$, an object C of level $n + 1$ admits a ramification whose ultimate components are included in the levels $\leqslant m$, so that it allows us to reconstruct C in several steps from those levels up. From the above reduction theorem, it follows that, in certain cases, this ramification can be 'reduced', meaning that C can be constructed in only one step directly starting from its ultimate components, and from links included in the levels $\leqslant m$, and 'jumping over' the intermediate levels. Thus, the level of an object does not reflect its effective operational complexity, and to measure it, we define the *complexity order* as follows:

Definition. The *complexity order* of an object C of level $n + 1$ is defined as the smallest m such that there exists a pattern whose colimit is C and which is included in the levels $\leqslant m$. And C is said to be *q-reducible* for any q equal to or higher than its order.

By definition of a hierarchical category, any object C of the level $n+1$ is
n-reducible. In which case is it m-reducible, for an $m<n$, say for $n-1$ (the
general case being brought back to this case)? We know that C admits
always at least one two-stage internal organization given by a ramification
(P, (P^i)), whose components P$_i$ at the first stage are in the levels $\leqslant n$, and
whose ultimate (or micro-)components P^i_j are in levels $\leqslant n-1$. The reduction
theorem above gives a condition for C to be $n-1$-reducible: it is that all the
distinguished links of P are simple links binding clusters between the P^i, and
which therefore are mediated through lower levels. Indeed in this case, C is
the simple colimit of a large pattern R having the micro-components for its
components. Conversely, if this condition is not met, the result does not
extend.

Let us say that a link is *m-simple* if it binds a cluster between patterns
included in the levels $\leqslant m$. In particular any link between objects of the
levels $\leqslant m$ is m-simple. From this we deduce:

Theorem 3 (Ehresmann and Vanbremeersch, 1996). *An object C of the level
$n+1$ can be (n–1)-reducible if it admits a decomposition P of which all the
distinguished links are (n–1)-simple. If not, C is generally not (n–1)-reducible.
This result extends to lower levels, allowing a reduction to these lower levels.*

To sum up, the actions of an object C of the level $n+1$ is entirely determined by one of its decompositions of level $<n+1$; it can also be 'geometrically' reconstructed from components of level $<m$ for any $m<n$.
However, if C is not m-reducible for such an m, this construction requires
several steps, in order to unfold a ramification with complex links coordinating the components of intermediate levels. In Chapter 4, we will see how
these complex links (or rather the complex switches which enter in their
formation) make new properties emerge at each step. Let us note that, if the
'actions' of C are thus determined bottom-up, its 'behaviour' can also be
affected by top-down influences, via links coming from higher levels and not
mediated via a decomposition of C (as explained in Section 6.1, above).

6.4. Examples

In Geopolitics.   The difference between a reducible and a non-reducible
object is illustrated in the system modelling the Western society (with its
individuals and its social groups of different levels) by comparing what
would be a 'Europe of nations' with one in which the nations would fuse
into a broader entity. The latter, a 'Europe of people', would be the simple
colimit of the pattern formed by the citizens of its various nations; on the
other hand a Europe of nations would be a non-reducible 2-iterated colimit:
each nation is the colimit of the pattern formed by its citizens, and Europe

would be the colimit of the pattern whose distinguished links are the institutional links between these nations. In this situation the citizens cannot directly interact with the entity Europe, but must pass through the intermediary of the government of their own nation.

Comparison of Crystals and Quasi-Crystals. A physical example of the difference between an object of order 1 and an object of order 2 is found by comparing a crystal with a quasi-crystal. We consider the hierarchical system of the chemical units, with the atoms and their various links on level 0. Following Penrose (1992), a crystal and a quasi-crystal are modelled by objects of level 2, but the first is 0-reducible, whereas the second is not. Indeed, both are constructed in the same way in two steps beginning with the formation of aggregates of atoms assembled according to various arrangements. Once formed, these aggregates are represented in the form of objects of level 1, binding the pattern of atoms corresponding to the arrangement. We can associate with an object of level 1 (as Penrose indicates) a quantum linear superposition of different arrangements of atoms, namely its various functionally equivalent decompositions. Next, some of these aggregates will be assembled to form larger conglomerates with a precise topology, and the optimal configuration (the one having the lowest energy) will generate the (quasi-)crystal (by what Penrose calls the *quantum procedure* R). This configuration thus emerges at level 2, as the colimit of the pattern of level 1 modelling the conglomerate (passage to the optimal configuration). One can also regard it as the iterated colimit of the atoms of the various aggregates intervening in the conglomerate, and the emergence of this iterated colimit, which models the procedure R, still amounts to a unique actualization of the quantum superposition of all the atomic decompositions of the (quasi-)crystal. However, there is a difference between a crystal and a quasi-crystal, which comes from the following fact: in the case of a quasi-crystal, the second operation cannot be short-circuited, for the conglomerate assembles aggregates of differing natures; on the contrary, for a crystal, the aggregates are of a comparable nature, without the additional contribution of 'horizontal' information, so that the crystal can be directly described as the colimit of a 'large' pattern of atoms. The crystal is thus 0-reducible (its complexity order is 0), while the complexity order of the quasi-crystal is 1.

6.5. *n-Multifold Objects*

We have seen the part which the *n*-simple links play in the reducibility of the objects of a hierarchical category. The non-reducibility depends on the existence of complex links, therefore of multifold objects. In a hierarchical category, the notion of a multifold object C (and therefore the multiplicity principle) are slightly restricted by introducing some constraints on the levels in which C must have several decompositions.

Definition. We say that a hierarchical category *satisfies the multiplicity principle* if, for each n:

(i) There are objects of level $n+1$ which are *n-multifold* in the sense that they are the colimit of at least two non-connected patterns included in levels less than or equal to n.
(ii) An object of level n can belong to several patterns having different colimits at the level $n+1$.

Thus, C is n-multifold if it admits several decompositions in patterns included in the levels $\leqslant n$ which do not directly result one from the other by correlations between their components. *A fortiori,* for every $m < n$ it will have several ramifications whose ultimate components are of levels $\leqslant m$. If one thinks of C as representative of the macro-state of a system, a decomposition of C represents a micro-state of C. By analogy with the statistical definition of entropy, as proportional to the logarithm of the number of micro-states giving the same macro-state, we can posit the following:

Definition. The *n-entropy* of an n-multifold object C is defined as the number of its decompositions in non-connected patterns included in levels $\leqslant n$.

The entropy gives a measure of the flexibility of the object C, since it will be possible to pass, by a complex switch, from one to the other of its decompositions. However (contrary to Rosen, 1986), we do not consider that it determines its 'real' complexity; that we measure by its complexity order.

If K satisfies the multiplicity principle, in addition to the n-simple links which bind clusters between patterns included in levels $\leqslant n$, there will be *n-complex links* which can also be reconstructed starting from the lower levels, but in a less immediate way: for example, as composites of n-simple links which bind non-adjacent clusters. We return to this question in the following chapter, where we define the notion of a *based hierarchy*. Such links will have emergent properties which will be perceptible only at levels higher than $n-1$, for they require a total apprehension of the objects which they bind, and not only of particular decompositions of those objects. There can also be links which are not reconstructible in any specific way starting from lower levels, and which represent properties independent of these lower levels.

Chapter 4

Complexification and Emergence

In the preceding chapter, we studied the problem of reduction: given a hierarchical system, up to what point can the properties of an object or a link be reduced to those of its lower level components? This problem was of an epistemological nature. In this chapter, we will consider the constructive problem: how can a system be progressively constructed to allow the emergence of new increasingly complex objects and properties? What features of the assembly process make qualitatively new and increasingly complex objects and properties emerge? How may we describe the stepwise evolution of hierarchical systems, the emergence of matter of new types out of simpler precursors (*e.g.* Eigen and Schuster, 1979; Laszlo, 1989; Farre, 1994) and the concomitant emergence of an increasing number of levels of complexity? The challenge is no less than to understand the nature of emergence as it occurred when, starting from sub-atomic particles and their interactions, atoms were formed, then molecules, then increasingly complex material systems; the process extending to give rise to the entire evolutionary tree of living beings, from bacteria to animals endowed with consciousness and then on to societies of any nature.

All this will be modelled using a process we refer to as the *complexification of a category*, the iteration of which may lead to the emergence of a whole hierarchy of objects and links with strictly increasing complexity orders. And we will characterize the hierarchies (called *based hierarchies*) which result from such a construction. Although this process proceeds in time, here we will not approach the temporal aspects, which will be studied in the following chapters.

We begin by studying how the behaviour of a pattern of objects evolves when the system in which it is considered is modified. In particular, if a pattern has no binding, how can it acquire one within a larger system? The complexification process will answer the converse question: how may we modify a system to change the behaviour of some patterns in a specified way, in particular to force some patterns to become integrated into a higher level unit, which becomes their binding (whose precise nature may or may not have been specified from the start) in the new system?

1.   Transformation and Preservation of Colimits

In the preceding chapters, given a category K, we have studied the patterns
of linked objects in this category, and the manner by which to bind their
components together, so that the whole pattern P may be replaced by a
single more complex object (namely its colimit), which has the same behaviour as the pattern acting in synergy with respect to other objects of the
category. In concrete applications, the category plays the part of an environment for the pattern P (environment which also contains P), for instance
by representing a configuration of a natural system in which the behaviour
of P can be studied.

   We are going to examine how the pattern and its binding (if it exists)
evolve if this environment K is modified. This modification is modelled in
two different ways: one, by a functor from K to a category K' (modelling the
new environment); and the other (more abstractly) by the replacement of K
by its opposite category.

1.1.   *Image of a Pattern by a Functor*

Let p be a functor from a category K to a category K'. Let us recall that a
functor is a homomorphism between graphs which preserves the structures
of categories, namely the image of an identity is an identity, and the image of
a composite is the composite of the images. The functor p may identify some
objects or some links of K. If it does not (hence if it is one-to-one), it may
'extend' K by the adjoining of, possibly, new links and/or new objects.

   The functor p transforms a pattern P of linked objects in K into a pattern
pP in K'. pP has for components the images by p of the components of P
(with the same indices), and for distinguished links the images of the distinguished links of P. More formally:

Definition. If P is a pattern in K, with sketch sP (hence a homomorphism
from sP to K), the *image of* P by p is the pattern pP in K' obtained by
composition of the homomorphism P with p (Fig. 4.1).

   Thus, the pattern pP, image of P by p, has the same indices as P, its
component of index i is the image pP$_i$ of the component P$_i$ of P, and its
distinguished links from pP$_i$ to pP$_j$ are the images by the functor p of the
distinguished links from P$_i$ to P$_j$.

   If (f_i) is a collective link from P to an object A of K, the family of the
images $p(f_i)$ of its individual links forms a collective link from the pattern pP
to pA (the correlating equations are satisfied because the functor p preserves
the commutative triangles). Geometrically, a cone in K is transformed by
the functor p into a cone in K'. On the other hand, if P admits a colimit cP in
K, the image of this colimit is not always a colimit of its image pP in K'

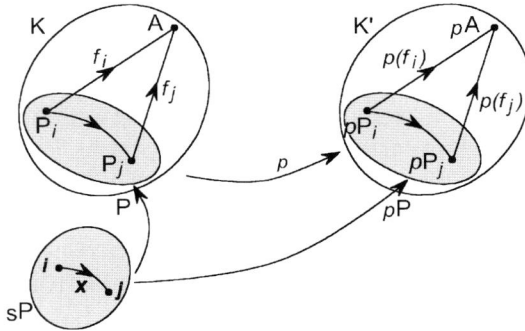

Fig. 4.1 Image of a pattern and of a collective link.
p is a functor from K to K′. A pattern P in K has for its image by p a pattern pP in K′: it has the same sketch as P; the component of index i is the image pP_i of the component P_i, and the distinguished link $p(P(x))$ is the image of the distinguished link $P(x)$ by p. If (f_i) is a collective link from P to A, the family of the images $p(f_i)$ of its links f_i forms a collective link from pP to the image pA of A.

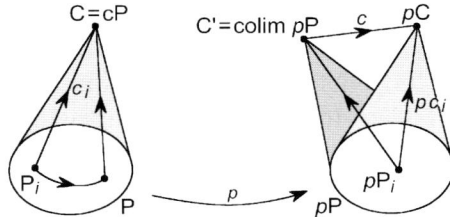

Fig. 4.2 Deformation of a colimit by a functor.
The pattern P admits C as its colimit in K. The binding collective link (c_i) from P to C has, for its image by the functor p, a collective link from the pattern pP (image of P by p), to pC. If pP admits a colimit C′ in K′, this collective link binds into a link c from the colimit C′ to pC, which 'measures' the change in the colimit imposed by the functor p.

(Fig. 4.2). It may even occur that pP has no colimit in K′. And conversely, P might have no colimit in K, whereas its image pP acquires a colimit in K′.

1.2. Replacement of a Category by a Larger One

In many applications, the category K will be a sub-category of K′, and p the insertion functor. What becomes of the patterns in K if K is so extended into the larger category K′? In particular, the complexification process defined in Section 2 will explain how to construct K′ so that a pattern without a colimit in K acquires a colimit in K′.

If p is the insertion from K to K', the image by p of a pattern P of K will have the same components and the same distinguished links, but now considered as a pattern in the larger category K'. Most often, we still denote by the same symbol P this pattern in K', specifying pattern in K or pattern in K' according to the case. Similarly, a collective link from P to A in K remains a collective link in K'. However, if C is a colimit of P in K, nothing obliges C to remain a colimit of P in K'. Indeed, the universal property of a colimit imposes constraints on *all* the collective links, which must bind into a link from the colimit. Now there may exist collective links from P to an object A' in K' which do not belong to the sub-category K, either because A' is not in K, or, even if A' is in K, because some of their individual links do not belong to K. The fact that C is the colimit of P in K does not impose any constraint on such collective links (which are partially located outside of K), and thus they may not bind into a link from C to A' in K'. In fact, all the cases are possible: P can have a colimit in K and not in K', or a colimit in K' and not in K, or different (non-isomorphic) colimits in K and K', or finally the same colimit in K and in K'. The following concrete example illustrates the different situations.

We consider a mail-order company, which operates in a certain area K and delivers its products to various customers. The company is represented by a pattern P having collective links to the regions where it operates. If the company has a depot C from which all shipments depart, the pattern P admits C for its colimit. Now, suppose the company modifies its operating field. If this field is extended to a larger area K', several cases may arise:

• If the location of the depot also makes it possible to serve the new regions, it need not be changed on that account. In this case, the colimit is preserved.
• If the depot continues to serve the regions where it used to operate, but is not adapted to the new ones, the company may replace it with a new and better adapted depot C' to which all products are sent; the products intended for the regions supplied by C will also be first sent to C', but then sent from C' to C, say by a road c, and C will deliver them to the intended customers. Then C' becomes the new colimit of P, and c models the comparison between C' and the old colimit C which is not preserved by the change.
• However, it may occur that the company does not see the utility of creating a new depot, and still uses C for the regions which it served before, while making individual deliveries to the new regions. The colimit is not preserved and P does not have a colimit in K' while preserving C as the colimit in K.
• If the company, instead of increasing its operating field, decreases it (*e.g.* for lack of customers in certain regions), the cost of the maintenance of the depot C can become too high, and it may be dismantled. This models the elimination of a colimit.

• Another case occurs when the company did not have a depot serving the area K where it operates (thus P did not have a colimit in K). It can then decide to modify its distribution network to create a central depot C′ in a region external to K. In this case C′ emerges as a colimit of the pattern P in K′, which it did not have in K. This requires that there be roads from C′ to each region served by the company.

1.3. Preservation and Deformation of Colimits

Now we come back to the case of a general functor p from K to a category K′, and a pattern P of linked objects in K. First, we suppose that P has a colimit C, and that its image pP by p has a colimit C′ in K′. We want to compare these colimits and measure how much C′ differs from the image pC of C.

Proposition. *If the pattern P has a colimit C in K and if the pattern pP (image of P by p) has a colimit C′ in K′, then there exists a comparison link c from C′ to the image pC of C (see Fig. 4.2).*

Indeed, the comparison link c binds the collective link image by p of the binding collective link from P to its colimit C.

The comparison link c 'measures' the difference between the colimit of the image pP of the pattern P, and the image of the colimit of P; that is, the deformation of the colimit during the modification of the environment described by the functor p. It is an isomorphism if and only if the image pC of the colimit C of P is a colimit of pP; in this case, we say that p *preserves the colimit of* P (Fig. 4.3).

In other terms, the global structure of P, represented by its colimit, remains stable if and only if the functor preserves this colimit, so that the comparison link c measures the barrier to stability defined by Rosen (1981)

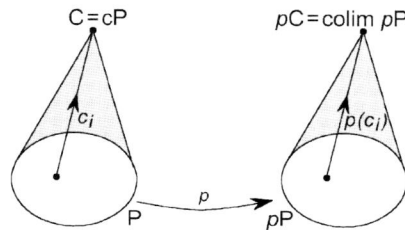

Fig. 4.3 Preservation of a colimit.
We have a functor p from K to K′, and a pattern P which has a colimit C in K. We say that this colimit is preserved by the functor if the pattern pP (image of P by p) admits the image pC of C for its colimit in K′.

as the capability of offsetting, or annihilating, a perturbation by means of an appropriate morphism (here c).

1.4. Limits

Another possible way to modify a category K is to replace it by its opposite category K^{op}: this category is the same as K, but with the arrows reversed. (This passage to the opposite is purely formal; for instance, if K models a system, K^{op} has no physical meaning; but some constructions done in K^{op} can take a meaning when re-translated into K, as we do in this section.)

A pattern P in K is identified with the opposite pattern P^{op} of K^{op}, having the same components, and the same distinguished links, but regarded as 'reversed'. By translating to the vocabulary of K what it means for P^{op} to admit a colimit in K^{op}, we obtain the notion of a *limit* of P. While the colimit (or inductive limit, in the sense of Kan, 1958) of a pattern actualizes the potential of the components of the pattern to act jointly, its limit (or *projective limit*, in the sense of Kan, 1958) can be thought of as actualizing their capacity to jointly decode and classify a message, of which each component receives only one part, the classification of the message being done by means of a link to the limit.

To define the limit directly in terms of K, let P be a pattern in the category K, and B an object of K. A message transmitted by some object B to P is modelled by the links from B to each component of the pattern, so that its content is coherently distributed amongst these components through their distinguished links. The limit will correspond to an optimal transmission. More precisely:

Definition. A *distributed link from* B *to* P is defined as a family (f_i) of individual links f_i from B to the components P_i, correlated by the distinguished links of the pattern (Fig. 4.4). We say that P has a *limit* λP, also called a *classifier* of P, if there exists an object λP and a particular distributed link (l_i) from λP to P, called the *projection distributed link*, through which any distributed link (f_i) from B to P is classified in a unique way into a link f from B to λP satisfying the equations:

$$f_i = fl_i \quad \text{for each index } i.$$

A distributed link (f_i) from B to P corresponds to a collective link from P^{op} to B in K^{op} so that it will be represented by a (now 'projective') cone with base P and vertex B, but with its generators directed from the vertex to the base (see Fig. 4.4).

If the pattern P is reduced to the family (P_i) of its components (*i.e.* if there are no distinguished links), the limit of P, if it exists, is called the *product* of the family. In the category **Set** of sets, the product of a family (P_i) of sets is

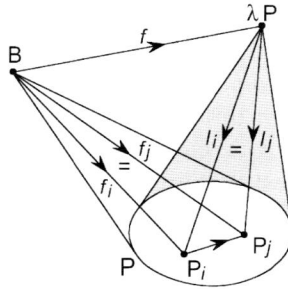

Fig. 4.4 Distributed link and limit of a pattern.
A distributed link from an object B to the pattern P consists of links f_i from B to the components P_i of P, correlated by the distinguished links of the pattern. P has a limit λP if there is a 'projection' distributed link (l_i) through which any other distributed link (f_i) from B to P factors uniquely via a link f from B to λP, so that $f_i = fl_i$ for each index i.

the usual product, that is the set of families (u_i) such that u_i belongs to P_i for each index i. If P is a pattern with this family of components, the limit of P is the subset of the product consisting of the families (u_i) such that $u_j = d(u_i)$ for each distinguished link d from P_i to P_j. In the category associated to a (partially) ordered set, a family of elements (e_i) admits a product a if and only if a is their *greatest lower bound*. In particular, in the category associated to the inclusion order on the subsets of a set E, the product of a family of subsets is their *intersection*.

If P has a limit, the objects B which may send a distributed link (f_i) to the pattern are characterized by the fact that there is a link f from B to the limit λP. Intuitively, these are the objects which have the characteristic property that the pattern, acting as a whole, can decode via (f_i), and then classify by f. In particular, λP is such an object, whence the name of 'classifier' of P, since (the identity of) λP classifies the projection distributed link from λP to P.

Earlier in Chapter 3, we defined the operating field of a pattern. The opposite notion is the *classifying field* of P: it is the category having for objects the distributed links to P, and in which the arrows from (f_i) to (g_i) are defined by the links of K correlating them, that is the links which are compatible with the transmission of the messages sent through (f_i) and (g_i). (The classifying field can also be defined as the opposite of the operating field of P^{op} in K^{op}.) If P has a limit λP, there is an isomorphism from the classifying field of P onto the perception field of λP, which associates to a distributed link (f_i) the link f which classifies it. Note that the classifying field of P should not be confused with the field of P (defined in Chapter 3) whose objects are the perspectives for P. A distributed link is a perspective only if any two components of P are connected by a zigzag of distinguished links; conversely, a perspective is a distributed link only if it has aspects arriving to all the components of P.

Two patterns are *pro-homologous* if their classifying fields are isomorphic categories over K (*i.e.* if their opposites are homologous patterns in K^op, *cf.* Chapter 3, Section 4.2). In this case, they have both no limit or the same limit. The various properties of colimits can be rephrased in terms of limits. We do this only for those properties, which we will make use of later.

In Chapter 3, we also modelled the interrelations between patterns compatible with their possible binding by clusters. The opposite notion is a *pro-cluster*, which models those interrelations between patterns which are compatible with their role as classifiers.

Definition. Let P and Q be two patterns in K. A *pro-cluster from P to Q* is a maximal set G of links of K between components of these patterns satisfying the following conditions (Fig. 4.5):

(i) Any component of Q receives at least one link from a component of P; if there are several such links, they are correlated by a zigzag of distinguished links of P.
(ii) The composite of a link of the pro-cluster with a distinguished link of Q, or of a distinguished link of P with a link of the pro-cluster, also belongs to the pro-cluster.

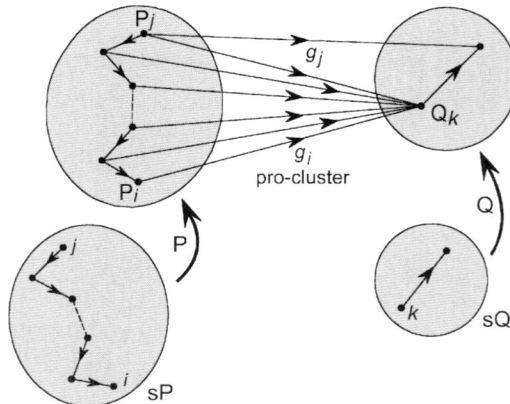

Fig. 4.5 A pro-cluster.
A pro-cluster from P to Q in a category K is a maximal set of links from components of P to components of Q such that: A component Q_k of Q is the target of at least one link g_i of the pro-cluster coming from a component P_i of P; and if there is another such link g_j they are correlated by a zigzag in P. Moreover, the pro-cluster is closed by composition with distinguished links, on the right or the left (compare with the 'opposite' notion of a cluster from Q to P, as in Fig. 3.4).

The composite of adjacent pro-clusters is defined in the same way as for clusters, and the pro-clusters between patterns in K form a category, denoted by **ProK**. If P and Q have limits λP and λQ, a pro-cluster from P to Q is 'classified' by a unique link from λP to λQ, called a (P, Q)-*pro-simple link*. However, the composite of two such pro-simple links is not necessarily pro-simple, if there exist *pro-multifold objects*, that is objects which can be the limit of two not pro-connected patterns (*i.e.* patterns whose opposites are not connected). The existence of pro-multifold objects in a category means that the opposite category satisfies the multiplicity principle.

2.   Different Types of Complexifications

The complexification process will be used to model the changes of configuration of a natural system, all of which Thom (1988) classifies as being one of four *standard changes*: birth, death, confluence and scission. For example, for a cell: endocytosis, exocytosis, synthesis and decomposition of macromolecules. Let us note that these changes correspond to internal modifications of a configuration category of the system, which models the components of the system and their interactions around a given time. This point of view is different from the models in which the changes of a system are described through the variation of some observables (increase or decrease of temperature, flip of magnetic polarity, change in volume or pressure, acceleration and so on). Here, we concentrate on the internal changes which cause this variation; for instance, an increase of temperature will be a consequence of standard changes that occur at the molecular level of the system. In terms of matter and energy, birth and death relate to exchanges with the outside, while confluence and scission relate to internal exchanges.

Given a category K which models the configuration of a system, a change of configuration is modelled by a functor p from K to the category K′ modelling the new configuration, obtained after some standard changes are applied to K. The complexification process will provide a construction of K′, which relies on two pieces of information, namely (using the terminology of Laborit, 1983): a structure-information, consisting of the configuration of the initial system K; and a circulating-information, which depends on the objectives indicating the changes to be made (enumerated in what we call an *option*), and acts like a servomechanism to implement them.

2.1.   Options on a Category

First, we must translate the standard changes into categorical operations to be effected on the category K to obtain the category K′ after the changes are realized.

(i) The *birth and death processes* depend on the relations of the system with its external environment. Birth is modelled by the addition (called *absorption*) of new elements into K', and death by the fact that some components of K are no longer in K' (*elimination*, by loss or rejection).
(ii) The *scission process* decomposes some complex objects in K so that they break up into their separate components;
(iii) The *confluence process* is modelled by the integration of a pattern in K into a complex object (possibly emerging in K'), which binds it by becoming its colimit in K'.

An *option* enumerates a list of objectives for modifying a category according to any of these standard changes.

Definition. Let K be a category. We define an *option on* K as a list Op of objectives for modifying a category by means of items of all (or some of) the following types, to be used as described accordingly:

(i) A graph U of external elements '*to be absorbed*'.
(ii) A set V of objects of K '*to be eliminated*'. Among these objects may figure the colimit C of some pattern, so that its elimination will represent a dissociation of the complex object C (the pattern remains but loses its colimit and its binding links).
(iii) A set of patterns of K '*to be bound*': if they already have a colimit in K, this colimit has to be preserved; if they have no colimit, they must acquire one which, depending on the case, might be an object which will emerge, or a specified object of K which initially is not their colimit in K.

A *mixed option* is an option specifying objectives of the three preceding types and also:

(iv) A set of patterns '*to be classified*', so that they acquire a limit (either a specified object of K, or a new object); or, if they already have a limit in K, that they preserve it.

An option having objectives of type (iv), but not (iii), is said to be a *classifying option*.
We note that an option has been called a *strategy* in our earlier articles, but this term seems to be ill adapted to some contexts (*e.g.* lower biological systems).

2.2.   Complexification

An option is a list of changes to be made to or to occur naturally in a system. The problem is to describe the category modelling the system after these changes are made, called its *complexification* with respect to the option

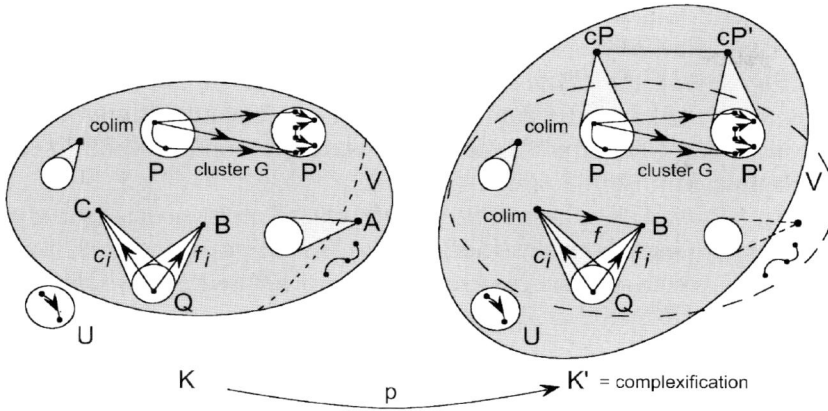

Fig. 4.6 Complexification with respect to an option.
An option Op is given in K, and K′ is the category obtained by the complexification
process, in which the changes called for by the objectives of the option are achieved.
These objectives are of the following form: The graph U external to K is 'absorbed' in
K′; the part V of K is eliminated in K′, in particular the complex object A is dissociated
into its components (which persist in K′) and the binding links are eliminated; the
patterns P and P′ that had no colimit in K acquire colimits cP and cP′ in K′, so that the
cluster G between them binds into a (P, P′)-simple link in K′; and the pattern Q admits
C as its colimit in K′, with the collective link (c_i) becoming the binding collective link to
this colimit, so that the collective link (f_i) from Q to B binds into a link *f* from C to B.
(If C had already been the colimit of Q in K, it should have been preserved.)

(Fig. 4.6). If the option is mixed (respectively, classifying), we speak of a
mixed complexification (respectively, *classifying complexification*).

For objectives involving items of the different types, the (mixed) complexification must satisfy corresponding conditions as follows:

(i) *Absorption.* The external graph U to be absorbed will become a sub-
graph of the complexification.
(ii) *Elimination.* The initial system K will be inserted in the system after
transformation, except for the set V of objects to be eliminated and the links
having at least one of their extremities in V; so that the passage from the
initial system to its complexification K′ will be modelled by a *partial functor*
p from K to K′ (partial since not defined on the eliminated elements).
(iii) *Binding.* Each pattern P to be bound is integrated into an object cP
which becomes its colimit in K′. Depending on the case, cP will be: an object
of K which was or was not the colimit of P in K, and which is 'forced' to
become its colimit in K′; or, if P had no colimit in K, cP will be a new, more
complex object, which emerges in K′. Furthermore, since two patterns P and

P° to be bound can be functionally equivalent in the initial category K—meaning that they have isomorphic operating fields in K (what we have modelled by saying that they are homologous in K, *cf.* Chapter 3)—in this case it is natural to require that they remain functionally equivalent in the complexification. This implies that they acquire the 'same' colimit in K'; and to this end, we impose the equality $cP = cP^{\circ}$.

(iv) *Classification.* To each pattern R to be classified is associated an object which becomes its limit (or classifier) in K'. Depending on the case, it will be: an object of K which was or was not the limit of R in K, and which is forced to become its limit in K'; or, if R had no limit in K, it will be a new object, which emerges in K'.

In categorical terms, the complexification K' of K with respect to the option will solve the 'universal problem' (Chapter 1): to find an optimal (or universal) solution to the problem of constructing a partial functor p from K to a category K' in which the objectives set up by the option are achieved. For the complex natural systems, which are obtained by such a process, optimal is to be understood from various points of view: material, temporal and algorithmic, as well as energetic.

2.3.   *Examples*

Consider an increase in the vocabulary of a language, a process at the base of any new knowledge: new primitive terms are added (or absorbed in the terminology of an option), new terms are defined starting from primitive terms or/and from terms of the initial vocabulary (by a binding process taking into account the relations between terms). And the process can be iterated, these new terms being then in turn used to define still more complex terms. Another example is the way in which an image, in artificial vision, is synthesized (through a binding process) starting from primitive features (*e.g.* lines, angles) and rules of association. The main difference between the two examples is the sketch of the pattern to be bound: it is 'linear' (total order) in the first case, more general in the second. This difference is similar to the one which Serres (1969) described, between a linear network corresponding to traditional dialectics, and the tabular communication networks introduced by him (and for which our formulation would give a precise operational mathematical model).

As we suggested at the beginning of this chapter, the evolution of a natural system is modelled by a sequence of complexification processes (we will return to this question later). In these cases, the emergence of a colimit of a pattern P rests:

• locally on the reinforcement of its distinguished links, which impose stronger constraints, increasing the cooperation of its components; and

• globally on the formation of a new object integrating the pattern, so that the pattern is differentiated from its context K by acquiring its own, unitary character.

Let us note that Laszlo (1989), who describes the evolution of the universe by a similar process, prefers the term 'convergence' to 'complexification', for he considers that a higher level introduces a certain simplicity compared to a lower level, since it forgets the details.

Many constructions in mathematics amount to a complexification. For example, real numbers can be constructed from the rational numbers by identifying a real number with a bounded $<$-section of $\mathbf{Q}$, that is a set S of rational numbers with no greatest element, and such that: if q_1 is in S and $q_2 < q_1$, then q_2 is also in S. This amounts to a complexification of the category defining the order on the rational numbers with respect to the option in which the $<$-sections of $\mathbf{Q}$ are to be bound (an example that we shall expand upon in Section 5 below). Other examples include the topological or differentiable manifolds, and more generally the local structures (Ehresmann, 1954), all of which are also obtained by a complexification process.

Let us note that an option on K can be interpreted as a sketch on a category (in the sense of Ehresmann, 1968) and the construction of the complexification is then a particular case of the construction of the associated *prototype* (in the sense of Bastiani(-Ehresmann) and Ehresmann, 1972).

3. First Steps of the Complexification

K is a category and Op an option given on it. The aim is to construct the complexification of K with respect to Op. In this section, we divide the construction into several steps corresponding to the successive objectives of the option, the case of each one being initially examined separately. In the following section, we will gather the results into a general construction.

3.1. *Absorption and Elimination of Elements*

Let us consider the case where the option has only objectives of the types (i) and (ii) (in the definition of an option); namely, it specifies as objectives only a graph U of external elements to be absorbed, and a set V of objects of K to be eliminated. Thus, U will have to appear in the complexification, whereas V must vanish. To absorb the elements of U, we consider the category 'sum' of K and of the category of paths of the graph U. To eliminate the objects in V, it will also be necessary to eliminate all the links admitting them as one of their extremities (since an arrow in a category must have a source and a target).

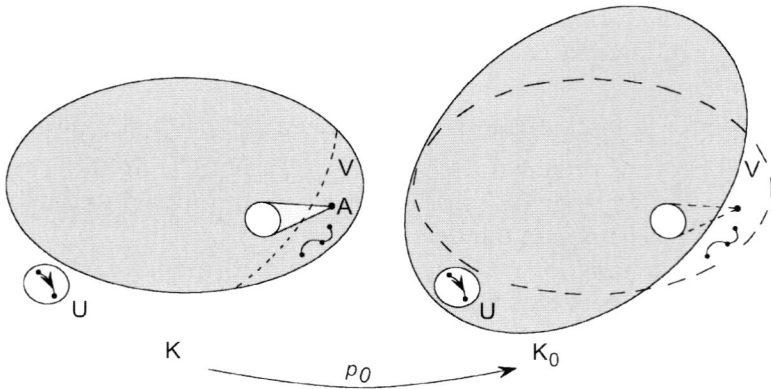

Fig. 4.7 Absorption and elimination.
K_0 is the complexification of K with respect to an option whose sole objectives are the absorption of the external graph U, and the elimination of the objects included in the part V and of all the links with at least one extremity in V. As V contains a complex object A, the elimination of this object corresponds to the dissociation of the pattern of which A is the colimit in K; the binding links to A are eliminated and the pattern has no colimit in K'.

The complexification is what remains (Fig. 4.7), namely the full sub-category K_0 of the sum having for objects the vertices of U and the objects of K not in V. In particular, if a complex object A belongs to V, a pattern which is a decomposition of A loses its colimit as well as the binding links of its components to A (Fig. 4.7). There is a partial functor 'insertion' p_0 from K to K_0, defined on the full sub-category of K whose objects are not in V.

3.2.   *Simple Adjoining of Colimits*

Here, we suppose that the option specifies, as its sole objective, a set of patterns P without a colimit in K, to be bound into emergent objects cP in K' (objective of type (iii)). We have seen that there exists a category, namely **Ind**K, which contains K and in which each pattern P in K acquires a colimit (Chapter 3). The first idea is to take for the complexification the full sub-category K' of **Ind**K having for objects the patterns P that the option requires be bound. (We recall that the objects of **Ind**K are the patterns in K and its links the clusters between these patterns.) More precisely, this category contains the category K as a sub-category (by identifying an object of K to a pattern reduced to one component), and has a new object denoted by cP, for each pattern P to be bound. The links come from clusters between the objects of K, between patterns to be bound and/or between objects of K and patterns to be bound. As a result of the properties of **Ind**K, the new

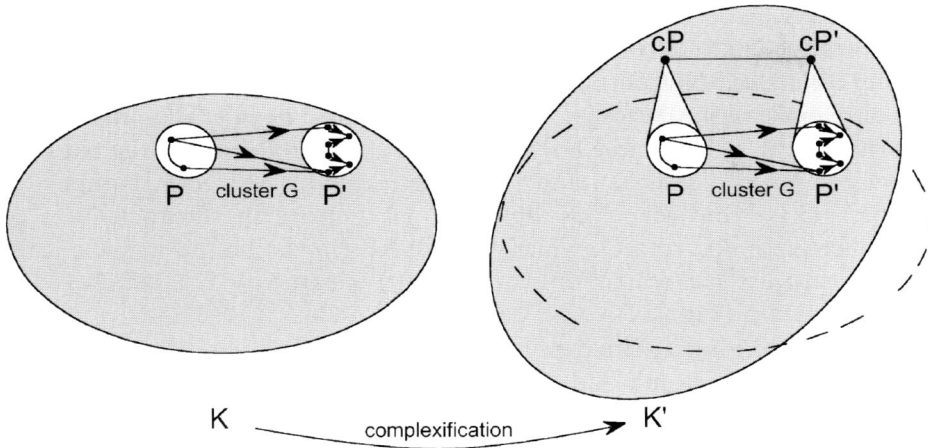

Fig. 4.8 Binding of patterns.
K′ is the complexification of K with respect to an option whose only objectives are to bind certain patterns lacking a colimit in K. It contains K as a full sub-category, and, for each pattern to be bound, K′ also contains a new object which becomes the colimit of the pattern in K′; in the figure, cP and cP′ are the colimits which bind P and P′. The links are simple links binding clusters of K.

object cP becomes the colimit of the pattern P considered as a pattern in K′, and a cluster between two patterns P and P′ to be bound is a (P, P′)-simple link from P to P′. Thus, there are only simple links, binding clusters of the sub-category K (Fig. 4.8).

However, this construction raises two problems: firstly, if a pattern has a colimit in K, this colimit is not necessarily preserved (by the insertion) in K′, so that the construction does not extend to the case where the option requires that some given colimits in K be preserved in K′. Secondly, two patterns to be bound can be functionally equivalent patterns in K (in the strict sense of being homologous) and not acquire the same colimit in K′. This is not satisfactory in the applications. We treat the first problem below and the second problem in the next section.

3.3. 'Forcing' of Colimits

Here, we consider an option on a category K_1 in which the only objective is to bind a set of patterns Q satisfying either of the following two conditions:

(i) If the patterns have a colimit in K_1, their colimit must be preserved in the complexification K_2.
(ii) If the patterns do not have a colimit in K_1, we require that a specified object of K_1 becomes their colimit in the complexification K_2.

More precisely, for each one of these patterns Q to be bound, there must be a collective link (c_i) from Q to an object C, which remains or becomes the binding collective link to the colimit. In this case no new objects are added in the complexification, the objects of K_2 being those of K_1. The links are at least those of K_1, plus simple links binding the collective links of the patterns Q towards objects B of K_1; and, in order to have a category, composites of such links. However, these links can participate in the formation of new collective links from Q, which we will have to be bound in their turn. And the process is iterated until it leads to a category K_2, which is the complexification (Fig. 4.9).

We develop this idea in an explicit construction (adapting the construction of Bastiani(-Ehresmann) and Ehresmann, 1972). To avoid a transfinite construction, we suppose that the patterns to be bound have only a finite number of components, which is generally the case in concrete applications.

First Step.   Let Q be one of the patterns to be bound and (c_i) the collective link from Q to an object C which is to be forced to become its colimit, so

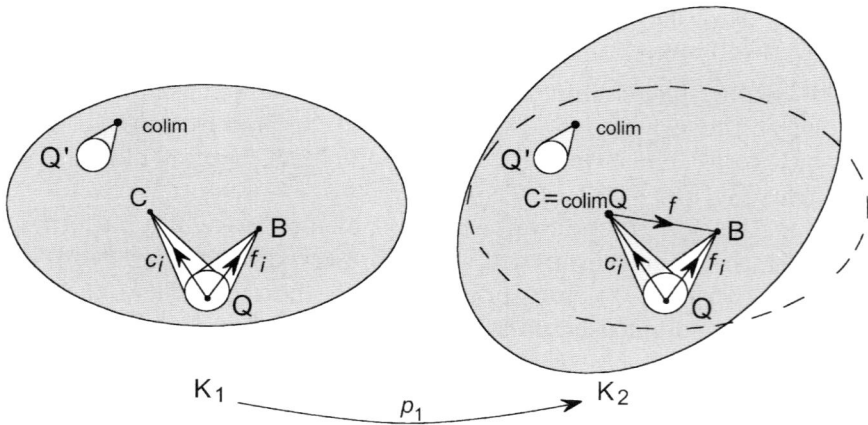

Fig. 4.9 Forcing of colimits.
K_2 is the complexification of the category K_1 when the only objective of the option is to 'force' some given patterns to admit a specified object of K_1 (which may or may not be their colimit in K_1) as a colimit in K_2, with a specified binding collective link; in the figure Q is a particular pattern forced to admit C as its colimit in K_2, with (c_i) as the binding collective link; and Q' is a pattern which must preserve its colimit and binding collective link. No objects are added. By definition of a colimit, in K_2 each collective link (f_i) from Q to an object B must be bound in a link f from C to B, and such a link is added in K_2 (since it does not exist in K_1). As the construction can introduce new collective links from Q, it must be iterated to also bind them. Finally, we obtain the functor p_2 from K_1 to K_2.

that $C = cQ$ in the complexification. To any collective link (f_i) from Q to an object B, we associate a link f from $C = cQ$ to B which will bind it; for that we impose the equations:

$$c_i f = f_i \quad \text{for each index } i \text{ of Q} \tag{1}$$

If there already exists such a link in K_1, it is chosen for f and if there exist several such links, they are identified. Otherwise, such an f is added. A category L_1 is constructed by addition of the composites of these links with those in K_1. Formally, this category admits for generators the graph obtained by adding the above arrows f to K_1, and for relations those given by the composition in K_1 and the equations (1) for each Q. There is a (not necessarily one-to-one) functor q_1 from K_1 to L_1.

Induction. Because of the composites added in L_1, a pattern image by q_1 of a Q to be bound into C may not have C for colimit in L. The same construction is iterated starting from L_1, and then by induction. It leads to an increasing sequence of categories L_m connected by functors q_m from L_{m-1} to L_m. All these categories have the same objects as K_1.

Let K_2 be the category union of the L_m (*i.e.* the colimit in the category of categories **Cat** of the sequence of functors q_m). There is a functor p_2 from K_1 to K_2. It is proved that the image by p_2 of the pattern Q to be bound into C effectively admits C for its colimit in K_2. This comes from the fact that Q has only a finite number of components, so that a collective link from Q to any B has only a finite number of links f_i, which are necessarily contained in one of categories L_m, and then their collective link is bound in L_{m+1}.

4. Construction of the Complexification

We will join together the constructions made in the preceding section for the various objectives of an option. Moreover, we require that, if two patterns to be bound are functionally equivalent in K, they remain so in K'. Two patterns are *functionally equivalent* if there is a one-to-one correspondence between their collective links, and this has been defined strictly by the notion of homologous patterns. Thus, the supplementary condition becomes:

(SC) *If two patterns to be bound are homologous in K, they must acquire the same colimit in the complexification.*

Let us consider an option Op on the category K, which specifies a graph U to be absorbed, a set V of objects of K to be eliminated and a set of patterns to be bound. We are going to construct a category K' fulfilling its objectives and satisfying SC. Then we will show in which sense this construction is optimal, so that K' is really the complexification of K with respect to Op. The construction will be done in several steps.

4.1.   Objects of the Complexification

We start by constructing the category K_0 in which the graph U has been absorbed and the objects of V eliminated (as in Section 3.1), and by constructing the partial functor p_0 insertion of K into it. The objects of the complexification K' will be the objects of K_0 as well as the objects cP defined as follows: To each pattern P to be bound, we assign an object cP, which will become the colimit of (the image of) P in K'. This object will be selected so that the objects associated to two homologous patterns P and P^o in K are equal (or at least isomorphic). For that:

(i) If P has a colimit in K, which the option demands be preserved, then we take this colimit for cP.

(ii) If P does not have a colimit in K, but if P or a homologous pattern is forced by the option to accept a specified object C of K for its colimit, then $cP = C$.

(iii) In the other cases where P has no colimit in K, we add a new object cP, but the same cP is selected for all the patterns P^o homologous to P, so that $cP = cP^o$. This object cP can be seen as a higher order object which emerges by integration of the pattern into a more complex unit, one which assumes its own identity.

4.2.   Construction of the Links

First, we construct a category K_1 which has for objects the objects defined above. Its links are defined as follows:

Simple links. To each cluster G from P to P' in K_0 where P and P' are objects of K_0 or patterns to be bound, there is associated a (P, P')-simple link g binding it. More precisely (Fig. 4.10):

(i) If P and P' have colimits in K, the link g already exists in this category (from the properties of colimits). In the particular case where P and P' are reduced to one component, so that G is reduced to only one arrow of K_0, the simple link is identified with this arrow. Thus, K_0 is identified with a sub-category of K_1.

(ii) For any pattern P to be bound, and for any of its components P_i, the insertion cluster from P_i to P binds into a link c_i from P_i to cP, and these c_i form a collective link c_P from (the image of) P to cP; it will become (or, if P already had a colimit, remain) the collective binding link in K_1.

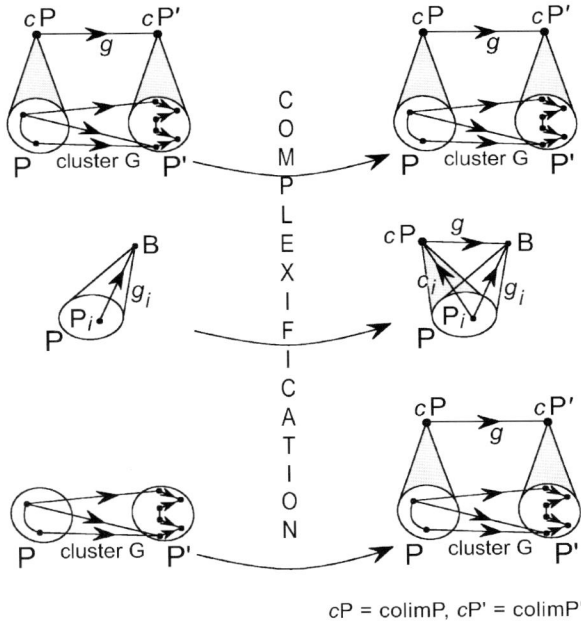

Fig. 4.10 Construction of the simple links.
One of the objectives of an option is to bind the patterns P and P'. These patterns acquire colimits cP and cP' in the complexification. Three different kinds of simple links in the complexification are represented successively: a (P, P')-simple link g in K which is preserved in the complexification; a link g (initially existing in K, or added in the complexification) binding a collective link (g_i) from P to B; an added (P, P')-simple link g binding a cluster G from P to P' when these patterns initially had no colimit.

(iii) If P had no colimit in K and if P' is reduced to one component B, the cluster G is a collective link (g_i) from cP to B, and one adds a link g binding it which satisfies the equations:

$$c_i g = g_i \quad \text{for any index } i \text{ of P.}$$

(iv) More generally, if P and/or P' do not have colimits in K, we add a link g from cP to cP' so that there is equality of the composite clusters:

$$Gc_{P'} = c_P g.$$

Complex Links. The simple links thus defined may not form a category if there are two homologous patterns P and P^o which are forced to have the same colimit cP. Indeed, in this case, there can exist a (Q, P)-simple link arriving at cP and a (P^o, P')-simple link issuing from cP which have no composite.

A composite will have to be abstractly added to them. Thus, composites of paths of simple links binding non-adjacent clusters are added. Formally, K_1 is the category defined by generators and relations (Chapter 1) the generators being the simple links, and the relations coming from the composition in K_0 and the above equations. It admits K_0 as a sub-category.

4.3.   End of the Construction

In the category K_1 it is not certain that the patterns P to be bound really have cP for a colimit. Indeed, we have bound only the collective links formed of links in K, but there can exist new collective links in K_1 formed of added simple or complex links. To force P to admit cP as its colimit, we apply to K_1 and to these patterns the induction construction of the 'forcing' (described in Section 3.3). It leads to the complexification K' and a functor from K_1 to K' which, composed with the insertion from K_0 to K_1, gives a functor from K_0 to K', hence a partial functor p from K to K'. This functor can possibly identify two arrows of K with the same extremities, if one objective of the option is that some patterns without a colimit in K acquire for a colimit a specified object of K. This will generally not be the case in the applications to natural systems, and in that circumstance we will be able to identify K_0 with a sub-category of the complexification.

  If the option requires the binding of several patterns which are homologous, we have imposed the condition that they acquire the same colimit cP; in other words, cP integrates not only one pattern, but a set of homologous patterns, and thus it is a multifold object, so that the complexification K' satisfies the multiplicity principle (Chapter 3). In particular, if K already satisfies this multiplicity principle and if the option requires that colimits in K be preserved, K' will also satisfy the multiplicity principle.

5.   Properties of the Complexification

Here, we give the main theorems which result from the preceding constructions, and which will be used extensively later on.

5.1.   Complexification Theorem

Theorem 1 (Complexification Theorem). *Given a category* K *and an option* Op *on it, a partial functor p from* K *to a category* K' *can be constructed so as to satisfy the following conditions:*

(i) *The objectives of* Op *are realized in* K'.
(ii) *If two patterns that* Op *requires be bound are homologous in* K, *they have the same colimit in* K'.

(iii) *The functor p satisfies the 'universal' property: if q is another partial functor from K to a category K* which satisfies the conditions (i) and (ii), it factorizes in a unique way as the composite of p with a functor q' from K' to K* which preserves the colimits cP of the patterns P that are to be bound.*
   Moreover, if K satisfies the multiplicity principle, so does K'.

   K' is called the *complexification of* K *with respect to* Op. It is defined up to an isomorphism.
   This theorem specifies the sense in which the objectives of the option are carried out in an optimal way in the complexification. The interest of the preceding construction is that it makes it possible to describe explicitly what are the links between the new objects cP emerging in the complexification, and therefore how these new objects interact in a more or less complex manner.
   Note that a similar theorem is obtained if condition (ii) is removed, so that two homologous patterns do not necessarily acquire the same colimit. The construction is similar, the only difference being that the cP and cP^o associated to homologous patterns without a colimit in K may now be different. The category thus obtained is called the *broad complexification* of K for Op. It can be proved (Bastiani(-Ehresmann) and Ehresmann, 1972) that the (not broad) complexification is the quotient category of this broad complexification by the relation of identifying cP and cP^o, if P and P^o are homologous in K. In fact, the links of the broad complexification may be defined in only one step by taking as links from cP to cP' the functors from the operating field of P' to that of P. (This construction is to be compared to that made in Benabou, 1968, in the framework of his 'virtual colimits'.) However, then the distinction between the simple links directly deducible from those in K, and the emerging complex links, is thus blurred, all the links being constructed in the same way, while they are neatly separated in the construction given above.

5.2. *Examples*

In the example of the transport network of Chapter 2, the creation of central nodes corresponds to a complexification K' of the category K associated with the network. However, whereas K is labelled in (**N**, max), we saw that K' may not be so labelled. The complexification of the category associated to a labelled graph (as was the category in this example) is not necessarily labelled.
   Let us take for K the category associated to the order on the rational numbers, the option having just one objective of type (iii), namely to bind each bounded-above set P of rational numbers. Hence, they each acquire a least upper bound. P and P' are homologous if and only if they have same upper bounds, so that P is always homologous to the set $P^>$ obtained by

adjoining to P all the rational numbers less than at least one element in P, and it must have the same colimit in the complexification (condition (ii) of the complexification theorem). Thus, the complexification will add only one object $cP^>$ for each set of the form $P^>$. Moreover, there will be a link from $cP^>$ to $cQ^>$ if and only if $P^>$ is contained in $Q^>$, this link binding the insertion cluster from $P^>$ to $Q^>$. Therefore, the complexification is the category associated to the order on the real numbers.

Let us extend the above result to (partially) ordered sets in general. Take for K (the category associated with) an arbitrary poset E, the option having for objective to bind each bounded above subset of E (which amounts to adjoining least upper bounds). It has been shown (Ehresmann, 1981) that the complexification of K for this option corresponds to the *MacNeille completion* of E (Banaschewski and Bruns, 1967). Its elements are the *Dedekind cuts* of E, namely the subsets D such that D contains any subset of E having the same upper bounds as D. Indeed, every subset P of E is contained in a smaller Dedekind cut D; this D is homologous to P in K, and therefore must admit the same colimit. On the other hand, the broad complexification of K then corresponds to the *universal completion* of E (Herrlich, 1979): its elements are the subsets P of E which contain: with an element of P all its lower bounds in E, and with a subset of P its least upper bound if it exists. It is larger than the completion of MacNeille because two such subsets can generate the same Dedekind cut.

5.3. *Mixed Complexification*

Complexification models the process whereby a natural system undergoes the four principle changes we described above: birth, death, confluence and scission. Systems of higher complexity, such as cognitive systems, are able to undergo another type of change, namely by 'classifying' objects according to some of their attributes. In other words, they form new objects representing the characteristics common to a class of objects of a certain type. In the categorical setting, we have modelled the classification process using the concept of a limit. As a limit in a category is a colimit in the opposite category, by applying the preceding constructions to the opposite category we obtain classification theorems. More precisely:

Theorem 2 (Mixed Complexification Theorem). *Given a category* K *and a mixed option* Op^m *on it, we construct a partial functor* p' *from* K *to a category* K^m *with the following properties:*

(i) *The objectives of the option are realized in* K^m.
(ii) *If two patterns to be bound are homologous in* K, *they acquire the same colimit in* K^m.

(iii) *The functor p′ satisfies the 'universal' condition: any other functor q from K to a category* K* *satisfying the conditions (i) and (ii) decomposes in a unique way as the composite of p′ with a functor q^m from* K^m *to* K* *which preserves the colimits of the patterns to be bound and the limits of the patterns to be classified.*

K^m is called the *mixed complexification* of K with respect to the mixed option.

Proof. *Construction of* K^m. Let us break up the mixed option Op^m into an option Op which has the same objectives except for the patterns to be classified, and into a classifying option Op^c which has the latter for its sole objectives.

• First, we consider the case where the option Op^m is reduced to Op^c, so that it is a classifying option; we translate it into a (non-mixed) option on the category opposite to K, in which the patterns to be classified become patterns to be bound. The opposite of the broad complexification of K^op for this option is the mixed complexification of K for Op^m.
• In the general case, the construction of K^m is done by recurrence (*cf.* Bastiani(-Ehresmann) and Ehresmann, 1972). The construction begins with the complexification K′ of K with respect to Op, followed by the complexification of K′ with respect to Op^c. The process is re-iterated by induction: complexification with respect to Op, then with respect to Op^c. It leads to a sequence of increasing categories whose union is the mixed complexification K^m of K with respect to Op^m.

Let us notice that in a mixed complexification there is emergence of complex links of a new nature, these links being composites of links binding collective links or classifying distributed links, themselves formed of such complex links.

6. Successive Complexifications: Based Hierarchies

The complexification process models the changes of a natural system between two moments. It results in the emergence of complex objects, namely those which bind a pattern initially without a colimit. Thus, it would seem that the iteration of this process during the evolution of the system could lead to the emergence of a hierarchy of increasingly complex objects. Is this really the case, or is there only a quantitative difference bearing on the number of objects which emerge during successive changes, instead of a qualitative increase in their order of complexity? As an application, does the evolution process lead to the creation of new forms, or is everything contained in germ *ab initio* and could emerge in only one step?

In terms of complexifications, the question becomes: can we replace a sequence of complexifications by a unique one and get the same final result? In this sequence, the objectives are imposed progressively, the option imposed at each step taking account of the results of the changes already achieved in the preceding steps; is it possible to obtain the same final result by imposing objectives simultaneously in a unique step, some- how anticipating the intermediate changes? To answer this *emergence problem*, we analyse the situation created by a sequence of complexification processes. If the initial category satisfies the multiplicity principle, it will lead to a particular kind of hierarchy, called a based hierarchy, with increasingly complex objects, for which we will characterize how new properties emerge at each level, dependent on the whole structure of the lower levels.

6.1.   Sequence of Complexifications

First, we consider two successive complexifications of the category K. Let K′ be the category constructed above by complexification of K with respect to the option Op. We give an option Op′ on the complexification K′ and construct the corresponding complexification K″ of K′. We suppose that the sub-categories of K and K′ generated by elements which are not eliminated are identified with sub-categories of K″. Does there exist an option Op″ on K such that the corresponding complexification of K (see Fig. 4.11) is the same as (or isomorphic to) K″?

In Op″, the elements to be absorbed would be those specified by Op and Op′. The same is true for the elements of K to be eliminated (though Op′ could specify that some objects of K′ not coming from K be eliminated). The real problem is for the patterns to be bound. If an object emerges in K″ as the colimit cP of a pattern P in K′ that Op′ requires be bound, can it also be obtained as the colimit of a pattern of K (in which case Op″ will require that this pattern be bound)? Naturally, this is possible if P is completely in K. The case which raises problems is when some components P_i of P do not come from K, and therefore have themselves emerged in K′ as colimits of patterns P^i that Op requires be bound. In this case, cP should be the 2-iterated colimit of the ramification $(P, (P^i))$ (the components of P which come from K are regarded as the colimit of the pattern reduced to one component). The reduction theorem for an iterated colimit (Chapter 3) shows that two cases have to be distinguished:

(i) If all the distinguished links of P are simple links binding clusters in K between the patterns P^i, the colimit cP of P can be constructed as the colimit of a large pattern R in K, whose components are those of the various P^i, in which case it will be obtained in one step if R is to be bound by Op″.

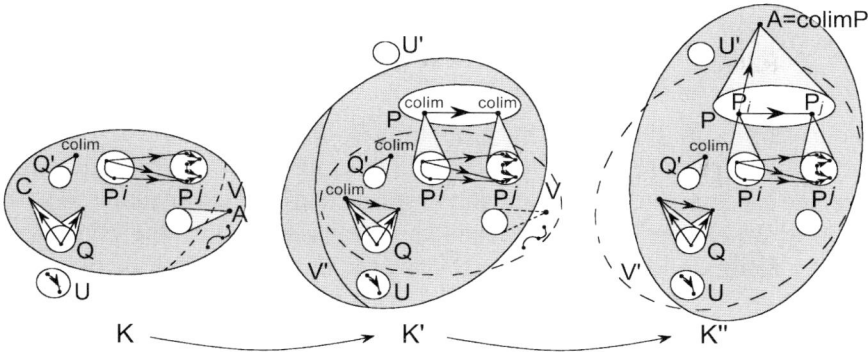

Fig. 4.11 Two successive complexifications.
The category K admits a first complexification K′ with respect to an option Op; among
its objectives: to absorb U, to eliminate V, to bind P^i and P^j, to force C to become the
colimit of Q and to preserve the colimit of the pattern Q′. The category K′ admits K″ as
a complexification with respect to an option Op′ on K′; among its objectives: to absorb
U′, to eliminate V′ and to bind P. Under what circumstances can there be an option on
K for which K″ appears as a one-step complexification of K? Only when all the
distinguished links of the patterns in K′ which Op′ requires be bound in K″ are simple
links binding clusters of K; it could be the case for P in the figure. Otherwise, if some of
the distinguished links of one of the patterns to bind are complex links, the colimit A of
P emerging in K″ will not be the colimit of any large pattern of K (see text).

(ii) However, such a reduction is not possible if P admits some distinguished
links which are complex, in the sense that they do not bind a cluster of K,
e.g. if they are composites of simple links binding non-adjacent clusters of
K. In this case the construction of cP necessitates two steps, and there is
no option Op″ on K allowing the emergence of cP in a unique step: The
complexity order of the colimit of P is then 2 with respect to K.

This shows that the construction of the second complexification K″ can
effectively require two successive steps and cannot be reduced to a unique
complexification if there are complex links in K′. There exist such complex
links if K′ satisfies the multiplicity principle, and we have proved that this
is the case in particular as soon as K satisfies the multiplicity principle
(*cf.* the complexification theorem).

The preceding result extends to a longer sequence of complexifications: a
sequence of complexifications of a category K which satisfies the multiplicity
principle cannot be reduced to a unique complexification of K with respect
to an option with more objectives, and leads to the emergence of objects of
increasing complexity. From this we derive

Theorem 3 (Iterated Complexification Theorem). *If the category K satisfies
the multiplicity principle, a sequence of complexifications of K cannot be*

replaced by a single complexification of K; *it leads to the emergence of a hierarchy of objects with strictly increasing complexity, necessitating a construction having more and more steps.*

For example, in the category modelling a cell, the synthesis of a poly-peptide chain requires three successive complexifications of the atomic (*i.e.* in this case amino acid) level: the first one leads to its primary structure, the second one forms the basic repeating configuration of the chain (wavy ribbon or helix) and the third one consists of the folding into the final conformation, in other words the formation of the tertiary structure. Let us note however that, in the hierarchical system associated with the cell, these three complexifications of the atomic level are done inside the macromolecular level.

The above theorem is essential, for it uncovers the condition at the root of the emergence of objects of increasing complexity order: it is the possibility for two patterns to be functionally equivalent (or, more precisely, homo-logous) without being connected (multiplicity principle). For a natural sys-tem satisfying this multiplicity principle, it means that two patterns of linked components may have independently the same operational behaviour or actions, without direct communications between their components to coor-dinate these actions. This gives a great plasticity to the system, since it allows switches between these patterns without affecting the global behaviour of the system. If the patterns are, or later become, decompositions of the same complex component, this component may be 'activated' via one decompo-sition in some contexts, via the other in other contexts, allowing for some adaptability to the context. Or it can switch from one decomposition to the other, allowing to take in account two different aspects of the situation more or less simultaneously; it is the case in particular when a multifold object plays the part of an intermediary object in the formation of a complex link; in natural systems, the switch could be internally directed, or the result of a probabilistic event, or else.

6.2. *Based Hierarchies*

A sequence of complexifications of a category K leads to the formation of a hierarchical category in which the objects of K and the objects to be absorbed are of level 0, those emerging in the first complexification are of level 1, those emerging in the second are of level 2 and so on. This hierarchy is of a particular type: not only is each object constructed in several steps starting from level 0 by a ramification (this is true in any hierarchical category), but the links can also be constructed in steps from level 0 up. We will say that it is a *based hierarchy*, a notion that we are going to define more precisely.

Let H be a hierarchical category, and H_n the full sub-category of H whose objects are all the objects of H of level less than or equal to n, and whose links are all the links between them. The category H is the union of the increasing sequence of these H_n. In Chapter 3, we defined the n-simple links as being links, which bind clusters between patterns included in the levels up to n (thus between patterns included in H_n).

Definition. A link of H is called an *n-complex link* if it is not n-simple but belongs to the smallest set containing: the n-simple links; with two links their composite; and with any cluster formed of n-simple links, the link binding the cluster, if it exists in H. The hierarchy is said to be *based on* H_0 if any link of H_{n+1} is either n-simple or n-complex.

To say that a link of H is n-complex means that it can be reconstructed in steps starting from the levels up to n (taken with its objects and the links between them, therefore starting from H_n), using only the operations of composition of links and binding of clusters. It is n-simple if it can be obtained by binding a single cluster contained in H_n. The complexity of a link in H thus depends on the level n where its complexity is observed, and reflects the number of steps required to reconstruct it starting from this level n and the lower ones. Thus, an n-complex link is at the same time m-simple for any m such that it belongs to H_m; indeed, observed on such a level m, there is no need to construct it—the way it depends on lower levels is forgotten.

The following proposition gives a characterization of a based hierarchy.

Proposition. *A hierarchical category* H *is a based hierarchy if and only if each link is* 0-*simple or* 0-*complex.*

Proof. First, let us prove that, in any hierarchical category, an n-complex link is m-simple or m-complex for every $m > n$. Indeed, an n-simple link is *a fortiori m*-simple. The class L_m that contains the m-simple links and is closed by composition or binding of a cluster thus contains the n-simple links, as well as those deduced from them by these operations. Therefore, it contains L_n. In other words, any n-complex link is m-simple or m-complex for $m > n$. It results from this that if any link of H is 0-simple or 0-complex, it is *a fortiori n*-simple or n-complex for any n, so that H is then based.

Conversely, let us suppose that H is based. We then have to prove that an n-simple or n-complex link is also $(n-1)$-simple or $(n-1)$-complex, *i.e.* belongs to L_{n-1}. By induction it will follow that any link is 0-simple or 0-complex. Indeed, if h is n-simple, it binds a cluster of H_n and, by definition of a based hierarchy, each link of this cluster is either $(n-1)$-simple or $(n-1)$-complex, and therefore belongs to L_{n-1}. This class being closed by the operation of binding a cluster, h also belongs to it. Thus, L_{n-1} contains the n-simple links, and as it is closed by composition and binding of clusters, it contains L_n. Thus, any n-complex link is $(n-1)$-simple or $(n-1)$-complex.

Construction of Based Hierarchies.   In a based hierarchy H, the operations used to gradually construct a link from the level 0 up, namely the composition of links and the binding of clusters, are exactly those carried out in the process of complexification, so that H results from H_0 by a sequence of complexifications. More precisely:

Theorem 4. *If* H *is a based hierarchy, for any n there is an option* Op_n *on* H_n *such that* H_{n+1} *is identified with the complexification of* H_n *with respect to* Op_n *(the objectives of* Op_n *being solely the binding of patterns). Conversely, the hierarchy associated with a sequence of complexifications of a category* K *is based.*

   Thus, a based hierarchy can be stepwise constructed from its lowest level up by successive complexifications.

6.3.   Emergent Properties in a Based Hierarchy

Let us consider the based hierarchy H obtained by a sequence of complexifications of a category K; the objects and links of H are obtained by steps (corresponding to the successive complexifications) starting from K. We suppose that K satisfies the multiplicity principle; in this case, we know that H also satisfies this principle, so that there exist complex links.
   By definition, an n-simple link g from A to C is the binding of a cluster from Q to P, for decompositions P of C and Q of A included in the levels up to n. Thus, it is entirely determined 'locally' by the links between the components of A and C of this cluster, which it just integrates in a unit of a higher level, without adding anything new. However, the situation is different for an n-complex link: though explicitly constructed from objects and links of lower levels, it reflects their global structure, which emerges at the higher levels.

Theorem 5. *In a based hierarchy, an n-complex link of level* $n+1$ *from* A *to* A$'$ *is not 'locally' deduced from links between the components of* A *and* A$'$ *of the levels up to n, but it takes account of the whole structure of these levels. It represents properties emerging at the level* $n+1$, *which project this whole structure into the higher levels* (see Fig. 4.12).

Proof. First, we prove the theorem for an n-complex link gg' from A to A$'$ obtained as the composite of two n-simple links, say a (Q, P)-simple link

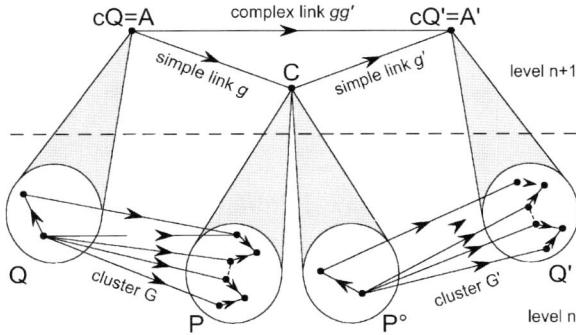

Fig. 4.12 Emergent complex link.
C is a multifold object of level $n+1$, A and A' are objects of this same level binding patterns Q and Q'. The link gg' is an n-complex link from A to A' composed of two n-simple links g and g'. This link represents properties emerging at level $n+1$ from the whole structure of the lower levels, properties which are not localized in the components of Q and Q'.

g from A to C and a (P°, Q')-simple link g' from C to A', where C is a multifold object admitting both P and P° as decompositions. The properties of this link depend not only on the 'local' properties of the two clusters in the levels up to n that g and g' bind, but also on the fact that the patterns P and P° have the same colimit C. This last condition requires that both patterns be homologous in H_n, meaning that they operate in a similar way with respect to the various objects of levels up to n. This condition thus takes into account the entire structure of H_n. It follows that the properties of the n-complex link gg' are not direct translations of local properties of links between the components of A and A' (as is the case for an n-simple link), but 'emerge' at its own higher level from the total structure of the levels up to n. The same would be *a fortiori* true for an n-complex link binding a cluster of n-complex links.

From the preceding theorem, we deduce:

Corollary. *An object of level $n+2$ which is not n-reducible has properties which are not reducible to those of its micro-components of levels up to n, but are contingent on the total structure of these levels.*

Proof. Let D be an object of level $n+2$. It can be constructed from the levels up to n as the 2-iterated colimit of a ramification (R, (R^i)). Let us suppose that some of the distinguished links of R are n-complex; for instance, one such link is the n-complex gg' considered in the proof of Theorem 5. Then, D is not n-reducible and it follows from the preceding proof that these links impose on D constraints that take account of the total structure of the levels up to *n*, and not only of the local properties of the micro-components of the ramification. Therefore, D has emergent properties compared to its micro-components, although they are deducible from the total structure of the levels up to *n*.

7.   Discussion of the Emergence Problem

The complexification process has been introduced to model the progressive changes that occur during the evolution of a complex natural system, and, in the following chapters, it will be used to model transitions between its successive configurations. As explained above, if the initial system satisfies the multiplicity principle, a sequence of such complexifications leads to the emergence of a based hierarchy with increasingly complex objects and links.

7.1.   Emergentist Reductionism

In a based hierarchy H, there is an operational reduction of a level to its lower levels, since the objects as well as the links of H can be reconstructed starting from its lower levels up (and even starting from the level 0). However, Theorem 5 above proves that this construction requires several steps, with the emergence of new properties at each step, this emergence relying on the holistic properties of the preceding levels. The construction is of a 'morphological' nature, and not an algorithmic one, since it takes account of the 'form' of the various lower levels structures. Intuitively, the non-linear language of a based hierarchy is entirely decoded via the primitive terms, namely the atoms of level 0 and their links, and the 'syntax' which indicates how to bind them to gradually construct the objects and links of the higher levels in several steps (by the intermediary of successive complexifications). However, the construction of each level leads to the emergence of new properties at that level, which however depend on the total structure of the preceding levels. Thus, from a philosophical point of view, we can speak of an *emergentist reductionism*, thereby giving mathematical expression to the concept introduced by Bunge (1979). Let us compare this idea of emergentist reductionism with other philosophies.

For the atomists, following Democritus, there would be a basic level of indivisible units. Any object would be obtained by successive bindings starting from these atoms; and any link between the objects would be mediated by this basic level. The advances of physics have pushed the basic level further and further down: it is now known that what we still refer to as 'atoms' are in fact formed of still more elementary particles; and that these particles are formed of quarks, which are perhaps not the ultimate atom, in Democritus' sense.

A strict reductionism would demand that each object is reducible to the lower level in only one step (*i.e.* is of complexity order 0), and that any link is 0-simple. That is generally not the case in a hierarchy, even a based hierarchy. Such a reductionism *pur et dur* (pure and hard) forgets the essential role played by the links between objects and denies the concept of emergence.

On the contrary, a holistic approach begins with the most complex level and deduces the lower levels from it, for 'the more perfect generates the less perfect' (Bateson, 1985, clearly explains the difference). In this case, the levels are often numbered in the opposite direction (with 0 for the most complex one). To compare with our hierarchies, for objects of level $n+1$ to be able to generate lower levels, they would have to have only one ramification down to the level 0, so that there would be no multifold objects. Thus, the emergence problem is again bypassed.

The emergentist reductionism defined above in a based hierarchy, which is of an epistemological nature, is intermediate between the two preceding approaches. At each level it extends the assertion of Rosen that 'every object in the big category can be regarded as a limit of elements in the small one' (1985a, p. 196)—with 'limit' translated in the categorical sense of 'colimit'— by also determining the appropriate links. Indeed, it makes it possible to construct the higher levels starting from the lower levels, provided that at each level not only the objects but also the links between them are taken into account, and that the construction be done in several steps. At each step there will be emergence of new properties, emanating from the holistic structure obtained at the preceding steps, as explained in Theorem 5.

7.2. The Emergence Problem

The emergence problem arises when trying to explain such phenomena as the evolution of the material universe; or of living beings, from the simplest ones to higher animals with higher order cognitive processes up to consciousness; or of society and culture. This problem has been thoroughly discussed by a number of authors (*e.g.* Farre, 1994; Salthe, 1985; Bunge, 1979, 2003), but without any consensus on a general process. As complex natural systems evolve through successive complexification processes, the preceding results indicate such a process:

Theorem 6 (Emergence Process). *The root of emergence is the existence of multifold objects (modelled by the multiplicity principle). If it is satisfied, the emerging objects and links are explicitly constructed through sequences of complexification processes.*

Roughly, an object is multifold if it can oscillate between two decompositions into more elementary components that are not locally connected. This gives it the freedom to interact with other objects under these different configurations. Thus, the multiplicity principle (as discussed in Chapter 3) formalizes what Edelman calls degeneracy. The model we have developed (and which we extend in Part B with the introduction of the memory evolutive systems) answers the question of Edelman and Gally; namely, how degenerate systems become linked and synchronized across levels, a problem they consider 'a major challenge in modern evolutionary biology' (2001, p. 13767).

The based hierarchies can also be compared to Bohm's theory of implicate orders (Bohm, 1983), which has influenced their development. Indeed, the complexification process could be a model of 'structuration' in Bohm's terminology, for adding or suppressing a colimit means exactly 'to create or dissolve what are now called structures' (Bohm, 1983, p. 120). Thus, what he calls a structure in its essentially dynamic nature could be modelled by a based hierarchy, the construction of which, through successive complexifications, corresponds to the unfolding of a specific order (which may be much more complicated than a fractal order), the successive steps representing 'sequences of moments that "skip" intervening spaces' (*op. cit.*, p. 211). Each level implicitly contains folded orders depending on the various possible options on it, and which may become manifest at its own level once one option is selected. The *degree of implication* of an explicit order corresponding to an object of a certain level is measured by its complexity order (in the sense of Chapter 3). Thus, based hierarchies could progress in the direction suggested by Bohm, to develop 'more general sorts of mathematization that may prove to be relevant' (*op. cit.*, p. 165).

7.3.   *Causality Attributions*

Let us recall that Aristotle distinguished four causes: material, efficient, formal and final, the last one being properly restricted to organisms or to nature as such. Consideration of the material and formal causes is generally abandoned in physics (the relevant ideas having been switched from the causation family to the family of concepts concerning state, circumstance, agent and the like), and the efficient cause is seen as an event involving a transfer of energy that produces the effect. However, Rosen (1985a) comes back to Aristotle, and he suggests that biological systems could be distinguished from what he calls simple physical systems by their behaviour with respect to these four causes. This distinction can be made more precise if we model a complex natural system, such as a biological or social system, by a based hierarchy, and Rosen's idea gives another insight into these complex systems. We develop this more thoroughly in Chapter 7, in the framework of memory evolutive systems, but there the analysis will rely on the following remarks.

How do the four Aristotelian causes come into play in a complexification process? The change introduced by one complexification, in particular the emergence of new objects, will have for its material cause the system in its initial configuration, for its formal cause the chosen option on it and for its efficient cause the implementation of the corresponding complexification process. There is no final cause as long as only the formal aspect is taken into account; a final cause would be concerned with the reasons that led to the choice of the objectives of the option. Up to this point, there is no difference from 'simple' physical systems.

However, if we now consider the change introduced by a sequence of complexifications, the causes will be more difficult to analyse. Can we say that the material cause is the system in its initial configuration, and identify a formal and efficient cause? This is true for the first complexification, but this complexification modifies the configuration of the system by introducing new objects and links, which must be taken into account in the choice of an option on it and in its implementation, leading to a second complexification; and so on up to the last complexification. And the iterated complexification theorem above proves that, when the multiplicity principle is satisfied, it is not possible to replace the sequence of complexifications by a single one, with respect to an option collecting their various objectives. In terms of causality, it follows that *the material, formal and efficient causes have to be updated at each step*: the system in its initial configuration is no longer the material cause of its final configuration after the sequence of complexifications, and we cannot identify either a global formal cause or a global efficient cause leading from the initial configuration of the system to its final one. Thus, the material, formal and efficient causes are intermingled together and with their effects in the global transition from the initial to the final configuration.

At a 'meta-level', we could say that the multiplicity principle appears as the formal cause of the emergence of complexity, while its material cause is 'hidden' in the lowest level, from where it must be unfolded (as in the implicate order of Bohm, 1983) by the efficient cause, which is the iterated complexification process as such, which, in a physical system, necessitates a transfer of energy. To sum up:

Theorem 7 (A Characteristic of Complexity). *In a sequence of complexifications of a category* K *which satisfies the multiplicity principle, there is emergence of objects of increasing complexity; the material, formal and efficient causes of the global transition from* K *to the final category are intermingled together and with their effects, and cannot be untangled.*

This characteristic feature of complex natural systems explains why their study cannot be brought back to that of the classical models used in physics, such as dynamical systems, except as a valid approximation on a particular level, and for a more or less short period, as will be explained in Chapter 7.

PART B. MEMORY EVOLUTIVE SYSTEMS

Chapter 5 Evolutive Systems

1 Mechanical Systems vs. Living Systems
    1.1 Mechanical Systems
    1.2 Biological and Social Systems
    1.3 Characteristics of the Proposed Model
2 Characteristics of an Evolutive System
    2.1 Time Scale
    2.2 Configuration Category at Time t
    2.3 Transition from t to t'
3 Evolutive Systems
    3.1 Definition of an Evolutive System
    3.2 Components of an Evolutive System
    3.3 Boundary Problems
4 Hierarchical Evolutive Systems and Some Examples
    4.1 Hierarchical Evolutive Systems
    4.2 The Quantum Evolutive System and the Cosmic Evolutive System
    4.3 Hierarchical Evolutive Systems Modelling Natural Systems
5 Stability Span and Temporal Indices
    5.1 Stability Span
    5.2 Complex Identity
    5.3 Other Spans
    5.4 Propagation Delays
6 Complement: Fibration Associated to an Evolutive System
    6.1 Fibration Associated to an Evolutive System
    6.2 Particular Cases
    6.3 The Large Category of Evolutive Systems

Chapter 6   Internal Regulation and Memory Evolutive Systems

1   Regulatory Organs in Autonomous Systems
   1.1   General Behaviour
   1.2   The Co-Regulators
   1.3   Meaning of Information
2   Memory and Learning
   2.1   Several Types of Memory
   2.2   Different Properties of Memory
   2.3   Formation and Development of the Memory
3   Structure of Memory Evolutive Systems
   3.1   Definition of a Memory Evolutive System
   3.2   Function of a Co-Regulator
   3.3   Propagation Delays and Time Lags
4   Local Dynamics of a Memory Evolutive System
   4.1   Phase 1: Construction of the Landscape (Decoding)
   4.2   Phase 2: Selection of Objectives
   4.3   Phase 3: Commands (Encoding) of the Procedure, and Evaluation
   4.4   Structural and Temporal Constraints of a Co-Regulator
5   Global Dynamics of a Memory Evolutive System
   5.1   Conflict between Procedures
   5.2   Interplay Among the Procedures
6   Some Biological Examples
   6.1   Regulation of a Cell
   6.2   Gene Transcription in Prokaryotes
   6.3   Innate Immune System
   6.4   Behaviour of a Tissue
7   Examples at the Level of Societies and Ecosystems
   7.1   Changes in an Ecosystem
   7.2   Organization of a Business
   7.3   Publication of a Journal

Chapter 7   Robustness, Plasticity and Aging

1   Fractures and Dyschrony
   1.1   Different Causes of Dysfunction
   1.2   Temporal Causes of Fractures
   1.3   Dyschrony

2 Dialectics between Heterogeneous Co-Regulators
   2.1 *Heterogeneous Co-Regulators*
   2.2 *Several Cases*
   2.3 *Dialectics between Co-Regulators*
3 Comparison with Simple Systems
   3.1 *Classical Analytic Models*
   3.2 *Comparison of Time Scales*
   3.3 *Mechanisms vs. Organisms*
4 Some Philosophical Remarks
   4.1 *On the Problem of Final Cause*
   4.2 *More on Causality in Memory Evolutive Systems*
   4.3 *Role of Time*
5 Replication with Repair of DNA
   5.1 *Biological Background*
   5.2 *The Memory Evolutive System Model*
6 A Theory of Aging
   6.1 *Characteristics of Aging*
   6.2 *Theories of Aging at the Macromolecular Level*
   6.3 *Level of Infra-Cellular Structures*
   6.4 *Cellular Level*
   6.5 *Higher Levels (Tissues, Organs, Large Systems)*

Chapter 8   Memory and Learning

1 Formation of Records
   1.1 *Storage and Recall*
   1.2 *Formation of a Partial Record*
   1.3 *Formation and Recall of a Record*
   1.4 *Flexibility of Records*
2 Development of the Memory
   2.1 *Interactions between Records*
   2.2 *Complex Records*
   2.3 *Examples*
3 Procedural Memory
   3.1 *Effectors Associated to a Procedure*
   3.2 *Basic Procedures*
   3.3 *Construction of the Procedural Memory*

4   Functioning of the Procedural Memory
    4.1   Activator Links
    4.2   Recall of a Procedure
    4.3   Generalization of a Procedure
5   Selection of Admissible Procedures
    5.1   Admissible Procedures
    5.2   Procedure Associated to an Option
    5.3   Selection of a Procedure by a Co-Regulator
6   Operative Procedure and Evaluation
    6.1   Interplay among the Procedures
    6.2   Formation of New Procedures
    6.3   Evaluation Process and Storage in Memory
7   Semantic Memory
    7.1   How Are Records Classified?
    7.2   Pragmatic Classification with Respect to a Particular Attribute
    7.3   Formation of an E-Concept
    7.4   Links between E-Concepts
    7.5   Semantic Memory
8   Some Epistemological Remarks
    8.1   The Knowledge of the System
    8.2   Acquisition of Knowledge by a Society
    8.3   The Role of the Interplay among Procedures
    8.4   Hidden Reality

Chapter 5

Evolutive Systems

Up to now we have not spoken about *time* as such, though it is implicitly present in the stepwise construction of a based hierarchy, the construction of each step taking some duration. However, time will play an essential part in what follows. Indeed, a complex natural system, such as a biological or social system, is an open system and its composition and organization change in time: the components of an organism are renewed unceasingly; in a society, some members leave, others arrive, relations are formed between them or are broken, new groups are formed, others dissociate. To study the dynamics of such a system, it is not enough to describe its configuration at a certain instant, *i.e.* its components and the interactions between them in progress at that moment. It is also necessary to describe the changes between its successive configurations, as it interacts with the surroundings, and as internal processes occur.

In particular, it will have to be explained how the complex system itself, and its sub-systems, can preserve their identity (*genidentity* in the terminology of Carnap, 1928) over time in spite of the progressive changes of their own components. For example, how may we identify, and consider as an individual, a particular cell of an organism at different instants of its life, even though it may differentiate, divide and otherwise change radically?

The evolutive systems defined in this chapter give a model for these systems, a model which will be completed in Chapter 6.

1. Mechanical Systems vs. Living Systems

The complex natural systems, in particular those studied in biology or in sociology, are of a very different nature than the mechanical systems studied by the physicist, and will have to be modelled differently.

1.1. Mechanical Systems

In Newtonian physics, a mechanical system is generally described by the successive values of the system's dynamical variables, in particular the positions and momentum components of its points or of its extended parts.

These variables are taken as the coordinates of a fixed space, called the phase space (or the *state space*), which represents all the possible states of the system, the 'state' of the system at time t being represented by a point in this space. This state varies with time, and its trajectory in the phase space can be computed using differential equations (*e.g.* Hamilton's or Lagrange's equations) depending upon the forces and dynamical laws to which the system is subjected.

Such systems are regarded as closed, that is without exchange of matter nor energy between the system and its external environment (but there can be interchanges of matter or energy between its sub-systems). It is possible that some components be replaced (in the course of years, every component of a bicycle wheel, from tyre and tube to rim to spokes to hubs to bearings to grease, may get gradually replaced), but this replacement is made with similar components and does not modify the analytic description. Thus, the phase space remains fixed.

1.2. Biological and Social Systems

The situation is very different for a complex natural system, in particular for a biological or a social system. For example, a very simple animal, able to move to find food and to eat it, can survive only thanks to the exchanges it has with its environment, be they primarily material or energetic (food, air, light and so on) or informational. It must move to where it finds food, either by chance or using sensors, and then eat it. The food which it will absorb at a certain time t will become part of its body, since it will be transformed internally by the digestive organs into molecules usable for its metabolic needs; and waste will be removed outside. Even its form can change, temporarily (consider a bacterium which incorporates an external element by endocytosis) or permanently (*e.g.* metamorphosis of a larva). The neural system of a slightly more advanced animal will be able to remember significant objects and form better adapted behaviours.

Complex natural systems have internal organizations (formed by their components and their interactions) which change in the course of time, the variation depending both on material and energetic exchanges with the environment and on internal processes; the system is said to be *open*. The traditional physical models do not apply any more. The laws used to determine the motion of a ball are not adapted to the study of the inter-actions between players of the two teams in a match of football! How may we model an open system? How may we recognize that it preserves its 'identity', in spite of changes which can lead to a complete renewal of its components at the end of a long enough period, a possible re-organization of them and the formation of new components?

1.3. Characteristics of the Proposed Model

The concept of an evolutive system defined in the next section models the evolution of a complex system. It does not describe both the system and its external environment, but works from an internal point of view, describing the successive configurations of the system and the transitions between them. These transitions result essentially from the four standard changes, birth, death, scission and confluence, which we have modelled by the complexification process (Chapter 4). Elements coming from outside appear as new components which do not have a former configuration; those eliminated (destroyed or rejected) disappear in later configurations; some new more complex components can be internally generated over time. The changes are described according to their internal consequences, by the fact that they modify the system composition, and by the fact that they modify the interactions between certain components, *e.g.* the transfers of energy (via chemical or metabolic reactions) or information. For example, a muscular movement of an animal will be triggered by a neuronal signal, which, as with all transfers of information, also involves some transfer of energy.

To show the difference from the traditional models, let us consider the case of the motion of a ball. The ball will be described here by its various atoms and molecules, with their spatial relationships and chemical bonds, and the changes will be modelled only in so far as they modify the atomic and molecular interactions: if the ball has a uniform movement, it does not have internal consequences; but if it undergoes an acceleration, this will have implications for the forces experienced by the individual components. The model may take account of the case in which the ball is inflated and later go flat, while classical models for closed systems would not apply easily. Naturally, this example is somewhat rudimentary, in comparison with complex natural systems, which are characterized by an internal control on the dynamics as we explain now.

A main characteristic of complex natural systems is their *autonomy*. They are *self-organized* in the sense that their changes are to some extent self-regulated by a network of sub-systems which act as internal regulatory organs (we call them co-regulators); these co-regulators operate under the system constraints, be they internal or external, and with the help of an internal memory; the dynamics of the whole system is modulated by the cooperative or conflicting interactions between these co-regulators. For example, when an animal eats, it starts a whole chain of internal reactions to break up the food and digest it. This self-organization will be one of the characteristics of the memory evolutive systems, to be introduced in Chapter 6. In a memory evolutive system we will consider simultaneously an evolutive system modelling the global system, an evolutive sub-system modelling its memory, and various evolutive sub-systems modelling its different co-regulators.

2.   Characteristics of an Evolutive System

In the preceding chapters we have associated to a natural system, say a cell, a category modelling its configuration at a certain time t, formed by its components of all kinds (*e.g.* atoms, molecules, macromolecules, organelles for a cell), and the interactions between them around t. To take account of the changes over time, this static representation must be upgraded to a dynamic one. For that, we will use, not a single category, which would impose some stability, but a family of categories indexed by time, which model the successive configurations of the system, with functors, called *transitions*, between them describing how they change. Such a model is what we call an evolutive system, which we describe in more detail in succeeding sections.

2.1.   Time Scale

At the basis of an evolutive system we have a *time scale*. We conceive of time primarily as the measure of internal change (following Saint Augustine). We adhere to Bergson (1889) who makes a distinction between the continuous and homogeneous notion of time used by the physicists, and a heterogeneous internal notion of time as 'pure duration', which he describes as a 'succession without reciprocal exteriority' (Bergson, 1889, p. 72). In a memory evolutive system both notions will be used: the first one in the description of the complete system by its successive configurations; the other in the description of the stepwise actions of its various internal regulatory organs, the co-regulators, which (as stated above) are sub-systems controlling its dynamics.

   The time scale of an evolutive system modelling a complex natural system will generally be continuous. In particular in a memory evolutive system, the time scale T of the whole system corresponds to ordinary homogeneous and continuous time, but the time scale of each one of its co-regulators will be regarded as discrete. Indeed, this time scale corresponds to the sequence of instants of T where a new step begins for the co-regulator, a step representing the duration (measured on T) necessary to carry out one particular operation at the level of the co-regulator. The steps can be of similar duration (*e.g.* the cycles of biological rhythms, the seasons, annual plans and so on) or depend strongly on the context (*e.g.* some arbitrary sequence of steps in some goal-seeking behaviour). Thus, each co-regulator is an evolutive system (an evolutive sub-system of the whole system) which has its own discrete time scale, extracted from T; for instance, the molecular phenomena proceed at a faster rate than the replication of a cell.

Thus, the time scale of the complete system will be represented by a finite interval of non-negative real numbers, from the birth of the system up to its disappearance; and the time scale of each of the co-regulators by a particular finite sequence of points of this interval. Let us insist on the fact that the discrete time scales of the co-regulators relate to their internal functioning, whereas the continuous time of the overall system allows to integrate and formally compare these independent discrete time scales (as in Bachelard, 1950, p. 92), in particular measuring and comparing the lengths of their successive steps.

Since an evolutive system must be capable of modelling a whole system, as well as the various co-regulators of a memory evolutive system, we define the time scale T of a general evolutive system as being either an interval of the non-negative real numbers, or a finite sequence of them. What matters is not the topology of T but its order (induced by the order on the real numbers), and we will associate to T the category defining this order, which has for objects the elements of T, and an arrow from t to t' if and only if $t \leqslant t'$.

2.2. *Configuration Category at Time t*

To define an evolutive system, we first give its time scale T as above. Then, to every instant t of T, we associate a category K_t which models the configuration of the system in the neighbourhood of t, as given by its internal composition and organization. K_t is called the *configuration category at time t*. It is a structural or relational concept, not a spatio-temporal one (nearer to Leibniz than to Newton).

- The objects A_t of K_t represent the *configurations at time t* of the components of the system which exist at t. These components may have a previous history and still exist at t; or they may be new, introduced at t by absorption of external elements (endocytosis for a cell), or by binding together patterns of already existing components (protein synthesis).
- The problem is more difficult for the links which must represent the interactions between these components (*e.g.* attachment of a protein to a receptor). Since these interactions are not instantaneous; what we should represent are the interactions around t. If the time scale T is continuous, the links represent not just events occurring at t, but, in the terminology of Whitehead (1925), 'germs' of interactions in small intervals of time around t.
- The law of composition for the category K_t allows sequences of interactions which have the same total effect to be identified; thus it leads to a functional classification of paths of links, while making it possible to

characterize the various equivalent ways (from the temporal and/or energetic point of view) information can be transmitted between two components.

2.3.   Transition from t to t'

To model the dynamics of the system, it is not sufficient to know the categories K_t representing its configurations at the different instants t of the time scale T. It is also necessary to connect them and describe how the configuration at a later instant t' of T unfolds from the configuration at t via changes of any nature. Consequently, we would like to:

(i) determine if a constituent (object or link) which existed at t still exists at t', and in this case describe its new configuration and
(ii) identify any new constituents at t'.

In an evolutive system, the change of configuration from t to a later time t' is modelled by a partial functor $k(t, t')$ from K_t to $K_{t'}$ called the *transition from t to t'* (Fig. 5.1). The idea is that this transition makes it possible to identify what a component at t has become at t', as one would point, in two successive photographs of a landscape, to the respective sites of its various elements. It also makes it possible to determine which components were removed permanently (by destruction, rejection in the environment, decomposition), which temporarily (short absence, inactivation by latency), and which have newly appeared (by absorption of an external element or formation of a new object or complex link).

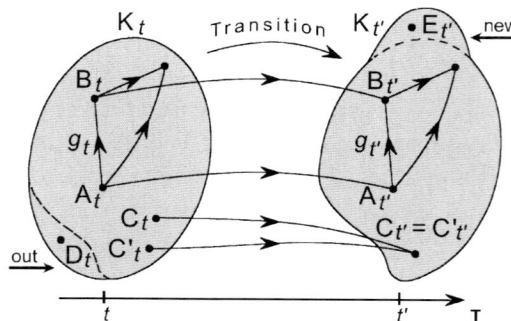

Fig. 5.1   Transition between two configurations.
The evolutive system **K** admits T as its time scale and K_t for its configuration category at t. The transition (partial) functor from K_t to the configuration category $K_{t'}$ at a later time t' models the change of configuration from t to t': the link g_t from A_t to B_t has become a link $g_{t'}$ from $A_{t'}$ to $B_{t'}$, an object D_t has been lost, and a new object $E_{t'}$ has appeared. The functor is not one-to-one since it maps the two distinct objects C_t and C'_t on the same object, which becomes the configuration at t' of both C_t and $C_{t'}$.

For example, the evolutive system associated to a country has for its components at t the people residing there, with their various social links and the groups they form at this date. An inhabitant will have disappeared at t' if he is dead or if he has left for abroad, in which case he might come back later; new people will enter the country, and new social groups may form. Here we speak of residence at a given time and not citizenship (Marquis, 1996).

Similarly, for the links: the transition functor either

(i) associates to a link g_t from A_t to B_t a link from $A_{t'}$ to $B_{t'}$, denoted $g_{t'}$, which represents its new configuration at t'; or
(ii) is not defined on g_t. In particular this is the case if at least one of the components A_t or B_t has disappeared at t'; but a link can disappear without either of its extremities disappearing (think of divorce).

To expand on this, we recall that to say that a functor is partial means that it is only defined on a sub-category of K_t. The objects D_t of K_t which do not belong to this sub-category (*i.e.* on which the functor is not defined) represent the components which disappear between t and t'. If an object A_t of K_t is in the sub-category, its image by the functor represents its new configuration at t', and this will generally be denoted by $A_{t'}$ (keeping the same letter A with a different time index). Two different objects C_t and C_t' at t can 'fuse' to have one overall configuration $C_{t'} = C_{t'}'$ at t' (*e.g.* two companies may merge; see Section 3). The new objects $E_{t'}$ at t' are distinguished by the fact that they are not the image of any object of K_t.

When the time scale is discrete, the transition from t to the next instant $t + 1$ often results from a complexification process. This process was introduced precisely to describe the standard changes of a natural system, namely absorption or elimination of elements, and dissociation or binding. In this case, the sequence of transitions since the initial date corresponds to a sequence of complexifications which can lead to a based hierarchy.

3.   Evolutive Systems

We have now all the ingredients to give a formal definition of an evolutive system, and to determine how it maintains temporarily stable components in spite of their progressive transformation.

3.1.   *Definition of an Evolutive System*

In the preceding section, we introduced the concept of an evolutive system. To give a formal definition, we impose a constraint to ensure that successive transitions add up to the net transition over the complete period, so that the later configurations of an object A_t of K_t are well defined. For that, we ask that, if A_t has a new configuration at an instant $t'' > t$, this configuration can be computed either directly by the transition functor from t to t'', or indirectly by dividing the period from t to t'' into several shorter periods, and composing the associated transition functors, provided that A_t did not disappear temporarily during one of these periods; and the same for the links.

Definition. An *evolutive system* (or ES) **K** consists of the following (see Fig. 5.2):

(i) A *time scale* T, which is an interval or a finite subset of the non-negative real numbers.
(ii) For each instant t of T, a category K_t called the *configuration category at t*. These categories are disjoint.
(iii) For each instant $t' > t$, a partial functor $k(t, t')$ from K_t to $K_{t'}$, called the *transition from t to t'*. These transitions satisfy the following transitivity condition (TC), given $t < t' < t''$ in T:

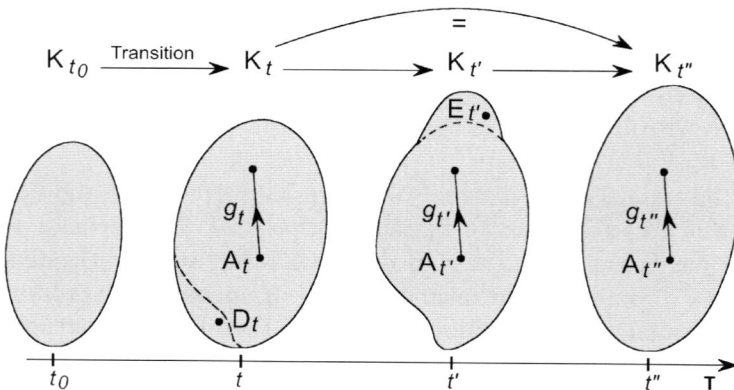

Fig. 5.2 An evolutive system.
The figure shows several successive configurations of an evolutive system **K**, beginning from an initial configuration at time t_0. The transitions between the successive configurations are partial functors between configuration categories. The transition from t to t' maps an object A_t of K_t to its new configuration $A_{t'}$ (if it exists) in $K_{t'}$, and the same for the links. The transition from t to t'' is obtained by composing the transition from t to t' with the transition from t' to t'', so that the new configuration $A_{t''}$ of A_t can be defined either directly, as the image of A_t by this transition, or in two stages, the first being the transition from t to t' mapping A_t on $A_{t'}$, and the second being the transition from t' to t'' mapping $A_{t'}$ to $A_{t''}$ ('transitivity' of the transitions). D_t has been lost at t', and $E_{t'}$ appears at t'.

(TC) If the object A_t has $A_{t'}$ for its new configuration at t', and if $A_{t'}$ has a new configuration $A_{t''}$ at $_{t''}$, then $A_{t''}$ is also the image of A_t by the transition from t to t''. Conversely, if B_t transitions to a configuration $B_{t'}$ at t' and to a configuration $B_{t''}$ at t'', then $B_{t'}$ must transition to a configuration at t'', and this configuration is $B_{t''}$. Similarly for the links.

Let us give some more definitions. The time scale T has a greatest lower bound in the ordered set of the real numbers; it is called the *initial date* (birth) of the evolutive system. If T is bounded above, it has also a least upper bound, called the *final date* (death or disappearance) of the evolutive system. Lastly, the objects and the links of the categories K_t are called the *constituents* of the evolutive system.

An *evolutive sub-system of* **K** is an evolutive system **L** such that its time scale S is contained in T, its configuration category L_s is a sub-category of K_s for each instant s of S, and its transitions are restrictions of those of **K**. More generally, we define interactions between evolutive systems as follows. If **K'** is an evolutive system with a time scale T' containing T, an *evolutive functor* **p** from **K** to **K'** is a family of functors (p_t) indexed by T, where p_t, for each t in T, is a functor from K_t to K'_t, called the *configuration of* **p** *at* t; these functors must be compatible with the transitions in both evolutive system, meaning that they satisfy the following family of equations:

$$p_t k(t, t') = k'(t, t') p_{t'} \quad \text{for each } t \text{ and } t' \text{ in T.}$$

The evolutive systems with evolutive functors between them form a large category (*cf.* Section 6).

3.2. Components of an Evolutive System

In complex natural systems modelled by evolutive systems, the components can vary over time. Nevertheless these systems have components which retain more or less persistent identities during their own lifetimes (which can be shorter than that of the system). For example, the biologist thinks of a tissue or an organ, such as the skin, as having an enduring existence within the lifetime of the organism, in spite of its internal changes over time, and in spite of the fact that, over the lifetime of the organism, there may be con-siderable turnover in the tissue or the organ's component cell population. And, at a lower level, he also thinks of a cell of a tissue as having an identity of its own, though the cell has a shorter life than the tissue, and may renew all its components during its lifetime.

In an evolutive system, the objects of the different configuration categories are all distinct, and all we know to relate objects at different times is how they are connected by the transitions. To construct the evolutive system modelling a natural system, we take for these objects the transient

configurations of the components of the system (meaning a component existing at t, indexed by this t), and the transitions connect the successive
configurations of a component. Conversely, given an evolutive system **K**
(with no other indication), how can we recognize persistent components of
the system through their successive changes, so that, when **K** models a
natural system, we recover the phenomenological components of the natural
system? Roughly, a component A of **K** will be a family (A_t) of objects of
configuration categories, composed of an object A_i which appears at a given
time i, and all its existing images by the transition functors. For example, in
the evolutive system modelling an organism, a cell will be modelled by the
family of its successive configurations at the various instants t of its life,
which delineate its history.

   An example which shows that we must be cautious with the formal definition is the case of the evolutive system associated to a country. A resident
A is represented by the family of configurations A_t associated to the instants
t where he resides in the country. These configurations correspond to each
other via the corresponding transitions, and they can be ordered by their
dates t. As with the case of a cell, there is an initial date, namely the citizen's
birth-date or the date where he first came to the country; and a last date,
that of either his death or his definitive departure from the country. However, between these two extreme dates, there can be greater or lesser intervals
during which he does not reside in the country, and so is not a component of
the evolutive system.

   In this example, we have no problem because we have identified the
component beforehand and have just to look at its different configurations. However, how can we recognize that some family of successive
configurations in an evolutive system corresponds to the same component,
when the only data are the configuration categories (which are disjoint by
definition) and the transitions between them? The idea is that the family of
successive configurations of a component forms a maximal set of objects of
the various configuration categories that correspond to each other via the
transitions.

   Given a constituent (object or link) of K_t, its image by any transition is
called a *later configuration* of the constituent, while any constituent of
which it is the image by a transition is called an *earlier configuration*. For
each $t' > t$, a constituent has at most one later configuration, but for any
instant $s < t$, it can have one, more than one (the transitions may not be
one-to-one), or no earlier configuration. Let us denote by $|K|$ the set of all
the objects of **K** (*i.e.* the objects of its different configuration categories).
The order relation *'earlier than or simultaneous with'* (denoted $\leqslant$) is
defined on $|K|$ by:

   $A_t \leqslant A_{t'}$ if and only if either $t = t'$, or A_t is an earlier configuration of $A_{t'}$.

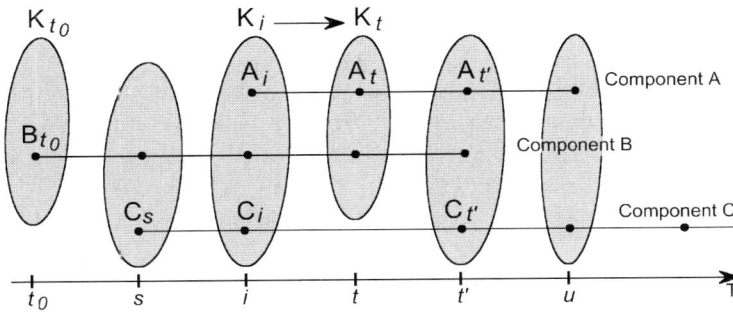

Fig. 5.3 Components of an evolutive system.
A component consists of a maximal family of objects of successive configurations corresponding to each other through transitions over time. The component A has the time i for its initial date, and the time u for its final date. The component B has the same initial date t_0 as the system, and its final date is $t' < u$. The component C has s for its initial date and has temporarily disappeared between times i and t'.

Definition. A *component* of an evolutive system $\mathbf{K}$ is defined as a family $A = (A_t)$ of objects of $\mathbf{K}$, indexed by a subset T_A of its time scale T, and satisfying the following conditions (Fig. 5.3):

(i) For each t in T_A, the object A_t, called the *configuration of* A *at* t, is an object of K_t.
(ii) T_A has a first element i, and the configuration A_i, called the *initial date* of A (or 'birth-date'), has no former configuration.
(iii) A is formed by all the later configurations of A_i.

T_A represents the period of time during which the component exists, *i.e.* its lifetime. Its length may be infinite. Otherwise it has an upper bound, and its least upper bound in the real numbers corresponds to the final date (or death) of the component A; this date can belong to T_A or not. The component can be born later than the evolutive system (this occurs when its initial date is strictly greater than the initial date of the evolutive system), disappear earlier, and between these two dates there may exist instants of temporary disappearance of the component, corresponding to dates between its birth and its death which belong to T but not to T_A. At any instant t of T_A, a component A has a single configuration A_t. The configurations of A form a completely ordered subset of $|\mathbf{K}|$, with respect to the order 'earlier than or simultaneous with', which has A_i for its smallest element. The category associated with this order will be called the *category associated with component* A.
Similarly, a link of the evolutive system is defined as being a family $g = (g_t)$ of links g_t between components A_t and B_t of the configuration

categories, formed by an initial link and all its later configurations. Since the corresponding A_t are configurations (not necessarily all of them) of a single component A, and the B_t are configurations of a single component B, we say that *g is a link of the evolutive system from* A *to* B. Such a link can model either a persistent or an intermittent interaction between these components, for example the successive phases of activation of a synapse.

A pattern in the evolutive system is defined in a similar way, as being a family of patterns in the configuration categories, formed by an initial pattern and all its later configurations; and the colimit of this pattern, if it exists, will be a component C whose successive configurations are colimits of the corresponding configurations of the pattern.

If no ambiguity is possible over the instant t which is considered, a configuration A_t of a component at t will often be simply denoted by A, without specifying the indexing time t, and the same for a link or for a pattern. For example, a business can be modelled by an evolutive system. Its components represent the personnel (from the managerial staff to the workmen), the resources necessary for the activities (installations, machines, supplies and so on), as well as the various departments and divisions. Each component has its own lifespan. It is the same for the links representing their functional interactions, which can be hierarchical relations between people, communications in an office, transport of material, interactions between products or connections between the different sections of the organization.

3.3.   *Boundary Problems*

A complex natural system is modelled by the evolutive system formed by its successive configurations and the transitions between them. In this case, a component as defined above models the history or the trajectory of what is generally called a component of the system, for example an atom, a molecule, a cell, an organ, an animal, a social group or whatever. The point is that it is an invariant which persists in spite of changes through its different configurations.

However, there are cases where the invariance, or lack thereof, is not clear. A good example is a population of organisms of a single species. A population should be a component of the evolutive system modelling an ecosystem; however, the specialists are not able to agree on when speciation occurs (*cf.* any of a great number of articles on the subject, such as may be found in the journal *Biology and Philosophy*). For example, if some individuals of a mono-specific population are reproductively isolated, and their descendants differentiate under the effect of different ecological conditions, so that they give rise to a new species, to which species do these pioneers and their early descendants belong?

Let us see how we can represent this situation in our notion of a component. We have said that a component A disappears at the instant u if u is the least upper bound of T_A. If u belongs to T_A (this is always the case when T_A is finite), the configuration A_u is the final configuration of A. If u does not belong to T_A what is generally called the component will have a sequence of configurations at instants tending towards u, but without a configuration at u: the precise time of its death is not reached. A component can also combine with another component, this being accompanied or not by the death of one and/or the other. Whence the following cases (depicted in Fig. 5.4):

(i) Two components can have common configurations (it will be said that they mix). Indeed, transitions are not required to exhibit a one-to-one correspondence, so that two objects C_t and C'_t may have the same configuration $C_{t'}$ at $t' > t$. They still correspond to two different components C and C′ (since a component has a unique configuration at each instant where it exists), but C and C′ have the same configurations for each instant after the instant where they mix, in particular at t'. An example is given by the mixture of two immiscible liquids, such as water and oil, where each one is still identifiable within the mixture; or by a merger of two companies, where each one still operates with some autonomy within the merged company.

(ii) A component A may 'absorb' a component A′ at t' so that A′ dies at t' (it has no later configuration), but in such a way that the own components of A′ become components of A. For instance, in an ecosystem, A may represent a predator which eats its prey A′ at t'.

(iii) Two components B and B′ may *fuse* at t' and form a new component S which is born at t' and subsumes both of them. This is the case when a sum $S_{t'}$ of B and B′ emerges, while B and B′ die at t'. An example would be the mixing of two miscible liquids, such as water and wine, after which the two component liquids can no longer be individually identified; or the creation of an entirely new company through the fusion of two old companies.

(iv) Two components E and E′ may *unite* at t' to form a more complex component M, while they also individually continue to exist. After t' there then exist three components E, E′ and M; for example after a marriage, the husband and the wife keep their identity, but their union also takes its own identity.

The preceding problems relate to the final date of a component. With the adopted definition, the initial date does not pose a problem since we have supposed that a component always has an initial configuration which has no former configuration. However, the example of a species recalled above shows that this condition can be too restrictive, and we are going to slightly generalize the definition. It then becomes more technical, but is seldom necessary in the concrete examples which we will study.

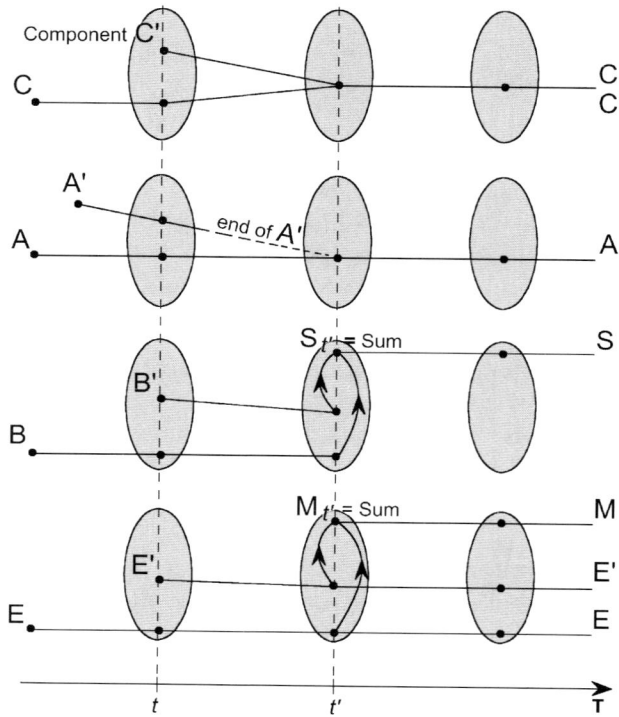

Fig. 5.4   Different interconnections between components.
Two components of an evolutive system may interact over time. Several cases are shown.

(i) The components C and C' are independent up to t', at which time they mix, but they retain their own identities in the merger.
(ii) A and A' are also independent up to t', but then A absorbs A', so that only A remains afterwards.
(iii) Both B and B' coexist until t', when their sum S is formed; they then disappear while S persists.
(iv) E and E' bind at t' in $M_{t'}$, but after that, E, E' and their sum M continue to exist.

From the given definition, it results easily that a component A is a maximal family of objects of the configuration categories containing with an object A_t all its later configurations. However, if T is not discrete, there can exist maximal families $B = (B_t)$ which have no initial configuration, whence the more general definition:

Definition. A maximal family $B = (B_t)$ of objects B_t of an evolutive system **K** which contains, with an object B_t of K_t all its later configurations, is also called a *component* of the evolutive system.

Such a family can be defined as forming a maximal totally ordered subset of the set $|K|$ of objects of **K**, ordered by the order 'earlier than or simultaneous with'. If the set T_B of instants at which such a component B exists does not have a least element, it has a greatest lower bound j in the ordered set of real numbers, and the object B_t has a sequence of earlier configurations B_s at instants $s > j$ tending toward j, though it has no earlier configuration at j.

4. Hierarchical Evolutive Systems and Some Examples

The evolutive systems which will be considered later have further properties which they inherit from those of their configuration categories. These properties are suggested by the applications to modelling natural systems.

4.1. *Hierarchical Evolutive Systems*

Definition. An evolutive system is called a *hierarchical evolutive system* (or HES) if all its configuration categories are hierarchical (Chapter 3) and if its transitions preserve levels. It is a *based hierarchical evolutive system* if these categories are based hierarchies (Chapter 4). The evolutive system satisfies the *multiplicity principle* if its configuration categories satisfy the multiplicity principle (Chapter 3).

For example, the evolutive system modelling a business is a hierarchical evolutive system: a machine decomposes into its various parts; a warehouse into the products which it contains; while a workshop represents the colimit of the pattern formed by the people who work there at a given moment; and is itself but one of the components of the production department.

In a hierarchical evolutive system, a component A has all its configurations A_t of the same level, since they result from one another by transitions, and the transitions respect the levels. This unique level will be called the *level of the component* A. Over time, the number of levels may increase, with formation of more and more complex components. This will be the case when transitions correspond to functors from a category to one of its complexifications, and the multiplicity principle is satisfied, so that there exist multifold components. In particular, a based hierarchical system can be constructed by the unfolding over time of a sequence of complexifications of the initial configuration category (or of its lower levels if it has several levels). From the Iterated Complexification Theorem (Chapter 4), if this category satisfies the multiplicity principle, its successive complexifications also satisfy the multiplicity principle, and lead to the formation of components of increasing complexity orders. Most complex natural systems are obtained in this manner. Below, we provide some examples.

4.2.   *The Quantum Evolutive System and the Cosmic Evolutive System*

Cosmic evolution has been divided into three phases: the expansive phase at the moment of the big bang; the constructive phase in which the nuclear particles are formed from the virtual particles of the quantum vacuum; and the evolutionary phase, which begins with the emergence of atoms and in which more and more complex matter evolves, up to the most complex natural systems (Farre, 1997).

The constructive phase leads to the formation of what we model by an evolutive system, called the *quantum evolutive system* (QES). It has for components the elementary particles and the atoms; the links correspond to their specific interactions, deduced from the fundamental forces; thus the configuration category at t is the corresponding category of particles and atoms. In Chapter 3, we presented an atom as a multifold object of this category, so that the QES satisfies the multiplicity principle. As recalled above, it follows that successive complexifications of the QES also satisfy the multiplicity principle, and thus lead to the emergence of a hierarchy of objects of increasing complexity order, namely the more and more complex material objects.

These complexifications lead to a based hierarchical evolutive system, called the *cosmic evolutive system*, which models the universe obtained during the evolutionary phase. Its components are material objects of any complexity, as well as living beings and social groups. Its transitions correspond to complexification processes, so that it is obtained by unfolding the QES by a sequence of complexifications over time. At each step, the emergence of an object C as a colimit of a pattern relies on the entanglement of the energy fields acting on the pattern, which extract energy from the components of the pattern and invest it to bind the pattern into the higher object C. By closure, C becomes a higher component of the system, which takes on its own identity (in the sense of Section 5 hereafter).

4.3.   *Hierarchical Evolutive Systems Modelling Natural Systems*

Complex natural systems such as biological, cognitive or social systems, are based hierarchical evolutive systems obtained by unfolding over time a sequence of complexifications starting from an evolutive sub-system of the cosmic evolutive system (and *a fortiori* from a sub-system of the QES). Since they ultimately are based on the quantum level, which satisfies the multiplicity principle (as stated above), these hierarchical evolutive systems also satisfy the multiplicity principle and they have multifold components of increasing complexity order. Let us analyse how the properties of these components, though entering the realm of classical physics, are inherited from (and in some way reflect) quantum physical laws.

Multifold 'macro' components have several non-connected decompositions. Much as with atoms, the complex switches between these decompositions will generally correspond to energy transformations that produce random fluctuations of the internal organization, without changing the higher level functions: several micro-states lead to the same macro-state. When these multifold components intervene as intermediary objects in the formation of a complex link (obtained as the composite of simple links binding non-adjacent clusters), they figure simultaneously with two non-connected decompositions, so that they operate as a kind of superposition of these decompositions, generalizing to higher levels the role of quantum superpositions. Moreover, as explained in Chapter 4, the properties of such higher level components emerge from an integration of the total structure of the lower levels, which confers on them a kind of *non-localization*, translating the non-localization of quantum physics to higher levels.

We do not mean that multifold macro objects 'possess' quantum properties. Their properties are classical, but they ultimately depend on quantum properties from which they emerge through complexification processes. Thus, we differ from authors who consider that the laws of quantum physics remain valid as such for macro objects, in spite of the problem of *decoherence*. In our view, the passage from a level to a higher level through a complexification process is at the root of the emergence of new properties, even if they adapt, extend and generalize quantum properties to macro objects as explained above. For instance, we differ from the interpretation of Aerts *et al.* (2000), who speak of quantum-mechanical properties, in particular for cognitive interactions (a case which will be developed in Part C). In fact their examples are easily translated into our setting by saying that the objects they consider are multifold, this condition ensuring that they have the described properties.

5. Stability Span and Temporal Indices

Here we consider a hierarchical evolutive system, so that its components have a well-defined complexity level. If A is a component of level $n + 1$, its configuration A_t, in the hierarchical category K_t, has an internal organization included in the levels less than or equal to n. Is this internal organization transformed into an internal organization of the later configurations of A, so that the component A itself keeps a stable internal organization of its components of lower levels? This would be the case if the system had an invariant structure, with fixed components, as in simple material systems: a machine remains made up of the same functional parts, the only possible changes replacing a component by a functionally similar one.

The situation is different in the complex natural systems which we model. In this case, the internal organization of a component may vary; however, it preserves a certain local continuity and changes only gradually, so that some kind of *complex identity* emerges. For example, the components of a cell are slowly renewed over time without affecting its identity; and an association can continue to exist in spite of the departure of certain members, the admission of new ones, the replacement of part of its personnel and even possibly some changes in the governing statutes.

To model the complex identity of a component over time, we are going to associate to a complex component three indicators which measure the rhythm of change of its lower level organization. The first one measures its *stability span*, the two others, its variations.

5.1. Stability Span

Let A be a component of level $n+1$ in a hierarchical evolutive system **K**. The configuration category $\mathbf{K}_t$ being hierarchical, the configuration A_t of A at t is the colimit of at least one pattern P_t included in the levels less than or equal to n. In the course of time, the components and distinguished links of this pattern change. Until which moment do the transformations of A and this pattern remain correlated?

For an instant $t' > t$, we denote by $P_{t'}$ the new configuration of P_t constituted by the images under the transition from t to t' of the components and links of P_t which have not disappeared. This pattern $P_{t'}$ may or may not admit $A_{t'}$ for its colimit. Indeed, between t and t', a number of components of P_t may have disappeared or been replaced, or simply have broken their binding links to A; conversely, new objects may have been connected to A. However, in a stable enough system, there exists a more or less long period $e(t)$ during which there will be a decomposition Q_t of A_t whose successive configurations remain decompositions of A, in the sense that they admit the corresponding configuration of A for their colimit (Fig. 5.5). Roughly, A admits an internal organization Q which is 'locally' preserved from t to $t+e(t)$.

Definition. The *stability span of* A *at* t is defined as the greatest real number $e(t)$ such that, in the category $\mathbf{K}_t$, there exists a pattern Q_t included in the levels less than or equal to n, which satisfies the following two conditions:

(i) Q_t admits A_t for its colimit in $\mathbf{K}_t$;
(ii) for any instant t' of the time scale between t and $t+e(t)$ (non-inclusive), its new configuration $Q_{t'}$ admits $A_{t'}$ for its colimit.

Intuitively, the stability span of A at t is the longest duration during which there remains a core of the components of A at t that suffice to maintain its organization.

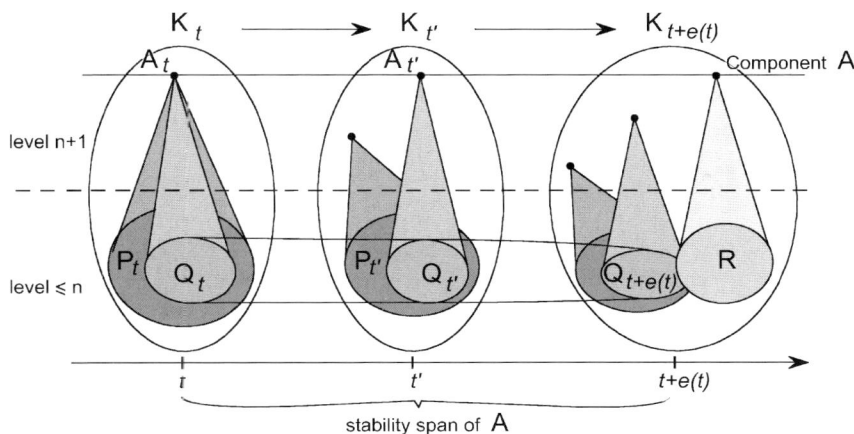

Fig. 5.5 Stability span.
At t the component A of level $n+1$ is the colimit of the pattern P_t in the levels $\leqslant n$, and also the colimit of another such pattern Q_t. The pattern P_t then varies independently of A such that A is no longer its colimit at t'; but A remains the colimit of the successive configurations of Q_t, up to $t+e(t)$ non-inclusive. The stability span $e(t)$ measures the longest period (beginning at t) during which there exists such a pattern Q_t, the successive configurations of which admit the successive configurations of A as a colimit.

The stability span depends on the level, and generally increases with the level: while the stability span of a metabolite is of the order of a minute, it can reach several days at the cellular level, and years at the organ level. (However, this is not always the case: the stability span of a proton is near-infinite.) The variation of the stability span in the course of time tells us about the pace of change. The longer the span is, the less there occur acquisition, modification, disappearance or replacement of components. Thus, during stability periods, this span is long, whereas it is shorter during periods of development or decline.

For instance, at the macromolecular level, for a certain population of proteins, the span decreases if they are denatured more rapidly, or if the DNA coding for them undergoes too many mutations. In a business, the stability span of a workshop corresponds to a period during which there is relatively little turnover of personnel, so that the workshop preserves a stable enough composition. For a stock of goods, the stability span corresponds to a period during which a majority of the goods are preserved.

5.2. *Complex Identity*

A nation acquires its own identity as a collective because it keeps its permanence in spite of the fact that some citizens die, others are born, strangers come to reside, its laws and even its constitution are progressively modified. This identity may extend for centuries, a duration which is not comparable with its stability span, which is related to the period required for the renewal of half of its population. This is made possible by the continual 'sliding' of the stability span. We can say that the nation as such keeps its *complex identity*. It is this general notion which we are going to define next.

Let A be a component of level $n+1$ of a hierarchical evolutive system. The stability span reflects a certain invariance of its internal organization over the short term, but this organization can vary considerably over the long term, say between t and a time s very distant. However, this variation is accomplished gradually; as the stability span at each date extends over a sufficient period, there is a partial covering of successive stability spans, over which successive decompositions are intertwined.

Indeed, by definition of the stability span at t, there exists a pattern included in the levels strictly less than $n+1$ which remains a decomposition Q of A during this stability span, from t to $t^* = t+e(t)$ non-inclusive. Let t' be an instant of the time scale between t and t^*; the stability span at t' extends up to t'^*. If $t^* < t'^*$ there is only a (more or less extended) sub-pattern Q' of Q which remains linked to A from t^* to t'^*. However, we may find a pattern P which contains the sub-pattern Q', and which remains a decomposition of A during the whole stability span from t' to t'^*. Thus, during the life of A there is a sequence of more or less gliding patterns, each remaining a decomposition of A temporarily, during a stability span.

More formally, we can construct a sequence of instants (t_m), $m = 1, 2, \ldots, u$; and of patterns (P^m) included in the levels strictly less than $n+1$, satisfying the following conditions (Fig. 5.6):

(i) $t_1 = t$, and $t_u^* = s$; for each $m < u$, $t_{m+1} < t_m + e(t_m) = t_m^*$;
(ii) P^m is a decomposition of A at t_m and remains so for its later configurations up to $t_m + e(t_m)$ non-inclusive.
(iii) P^{m+1} admits a common representative sub-pattern with the new configuration of P^m at t_m.

These conditions agree with the notion of identity in the terminology of Rosen (1985a), who explains that we keep track of a single system, even though the characteristics used to define it may be changing more or less radically in time.

In particular, let us suppose that A has emerged at time i as the binding of a pattern P (colimit added through a complexification process). Up to $i+e(i)$, the evolutions of A and of P remain correlated, but then they can deviate. Thus, A will take its own identity which 'transcends' that of P, while 'keeping its identity in spite of change' (Hegel, 1807, p. 199), this provided that the change is gradual enough. In this case we say that A *acquires an*

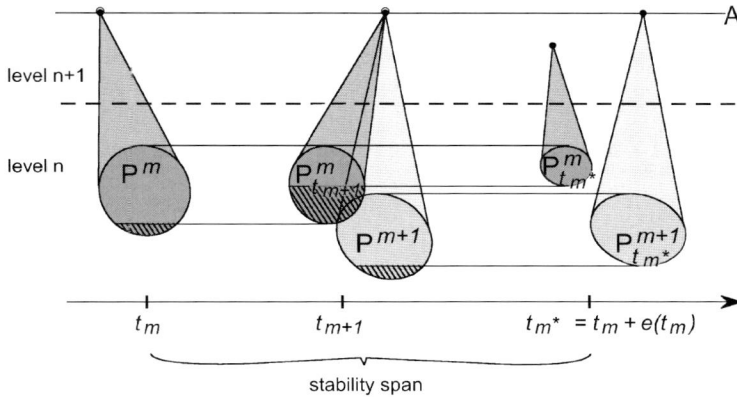

Fig. 5.6 Complex identity.
Component A of level $n+1$ has an internal organization at lower levels, which is progressively modified: A admits a sequence (P^m) of decompositions, each one valid over one stability span: The decomposition P^m starts at t_m and its successive configurations (which have lost some components as indicated at the bottom) remain bound to A up to $t_m^* = t_m + e(t_m)$ non-inclusive. The decomposition P^{m+1} starts at a time t_{m+1} strictly between t_m and t_m^*, and its successive configurations remain bound to A up to t_{m+1}^* (not depicted) non-inclusive. At t_{m+1} it has a common sub-pattern with the configuration of P^m at this time.

n-complex identity, its successive configurations being colimits of patterns included in the levels less than or equal to n, each one gradually coming forth from the preceding one, with existence at every moment of a decomposition preserved temporarily during a stability span, by the process described above. A maintains its identity as long as its stability span does not decrease too much; when this span tends towards 0, A will disappear.

For example, if the letters of a sentence are randomly changed at the rate of one per second, for a few seconds (its stability span), the difference is sufficiently small to not alter it in practice, and the sentence will continue to retain its meaning. However, this stability span is short, and at a certain moment the sequence is no longer understandable.

5.3. Other Spans

The stability span of a component A depends primarily on the rate of renewal of the internal organization of A by disappearance of its components or addition of new ones. Two other temporal indicators are introduced to specify the influence of these changes.

The *renewal span* will measure the speed with which the components are renewed. In the evolutive system modelling a business, the renewal span of a stock of goods of a certain type is inversely proportional to its rate of depletion, and it will be advantageous that it be short enough to prevent the goods from becoming obsolete. On the other hand, it is better that the renewal span of the staff be long, to ensure proper continuity of operations and to benefit from the experience gained by the employees.

Definition. Let A be a component of level $n+1$. Its *renewal span at t*, denoted by $r(t)$, is the smallest period during which the following condition is satisfied: the configuration of A at $t+r(t)$ is the colimit of at least one pattern included in the levels less than or equal to n, made of components which were not part of A at t, and which are introduced into the system between t and $t+r(t)$.

The stability span measures the period during which A preserves an almost constant composition and organization. The *continuity span* will measure a less strong stability, of a functional type, taking into account not the individuality of the components, but their function: it is not affected if components disappear in so far as new components replace them function-ally (*e.g.* an electron leaves an atom, but is replaced by another one).

Definition. Let A be a component of level $n+1$. The *continuity span of* A *at* t is the greatest period $c(t)$ during which there exists in K_t a decomposition of A into a pattern P_t included in the levels less than or equal to n, satisfying the following condition: For each t' between t and $t+c(t)$ (non-inclusive), the configuration of A is the colimit of a pattern $R_{t'}$ made of components which are either new configurations at t' of components of P_t, or new components that functionally replace them.

The continuity span is usually longer than the stability span, though they are equal if there is neither loss nor renewal of the components.
For a stock of goods, the continuity span measures the period during which its flow remains constant, removals being exactly compensated by arrivals; it shortens if the flow becomes irregular. For a business, the continuity span might indicate the time during which its organizational flowchart is preserved (with functions possibly performed by new people). The business could alternatively be modelled by an evolutive system in which we take as components not the employees as individuals, but only their role in the business (this model being thus more like the model depicted by the flowchart). In this case there would be no difference between the stability span and the continuity span.
The three spans are partially intertwined.

• A decrease of the continuity span $c(t)$ with an increase of the renewal span $r(t)$ implies a reduction of the stability span $e(t)$, and represents a progressive diminution of the number of components remaining in A, since the lost elements are not replaced quickly: this is what occurs during periods of decline.

• A simultaneous decrease of $e(t)$ and $r(t)$ implies a strong rate of change of the components; it indicates a great instability if $c(t)$ also decreases. It can however be advantageous when $c(t)$ remains constant. For example, the 'just in time' management of supplies (Hall, 1989) amounts to maintaining such a fast turnover of supplies that their stocks can be reduced to the minimum.

• During expansion periods, $r(t)$ will decrease and $e(t)$ will remain constant. During steady periods, the changes measured by $r(t)$ and $c(t)$ more or less compensate each other, so that these spans, as well as $e(t)$, remain almost constant.

The different situations are represented in the diagrams of Fig. 5.7 which could be compared with the cycle of the commercial life of a product (Pourcel, 1986).

In Chapter 7, we give a theory of the aging of an organism through a 'cascade of re-synchronisations' (proposed originally in Ehresmann and Vanbremeersch, 1993). It indicates as one of the characteristics of senescence the fact that the stability spans of higher and higher level components decrease, down to 0 where death occurs. This deterioration of the components may have multiple causes, based on external disturbances or internal deficiencies.

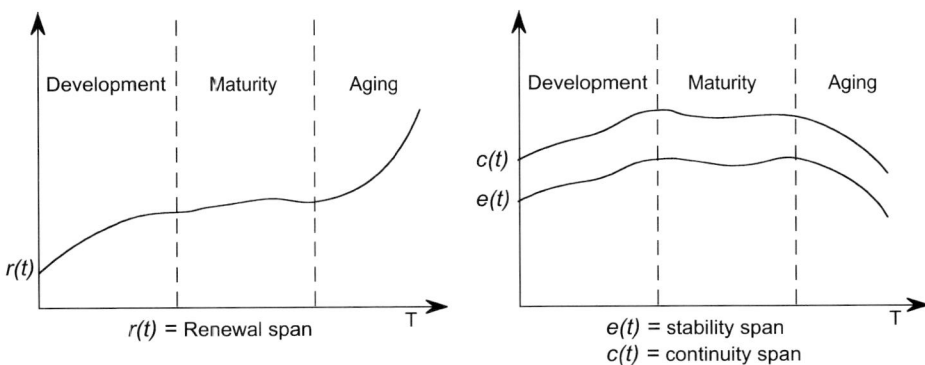

Fig. 5.7 Variations of the different spans.
The renewal span $r(t)$ of a component A is short during the development phase, when modifications are fast, more or less constant during maturity and increasing during aging. The stability span $e(t)$ and the continuity span $c(t)$ increase during development and decrease during aging, with $c(t)$ remaining greater than $e(t)$.

5.4. *Propagation Delays*

In a natural system, time is involved not just as a parameter indexing successive changes. The interactions between components are subject to various temporal constraints: delays necessary for the propagation of signals, activation delays, latency periods and so on. To take them into account in an evolutive system, we introduce another temporal marker, the *propagation delay* of a link, which represents the time necessary for the transfer of energy or the transmission of information between the components that the link connects.

More precisely, we generally suppose that the configuration categories of the evolutive system are labelled in the additive (or sometimes multiplicative) monoid of the non-negative real numbers. This means (Chapter 1) that to each configuration g_t of a link g of the evolutive system, a positive real, called its *propagation delay*, is associated, in such a way that the propagation delay of the composite of a path is the sum (or the product) of the propagation delays of its factors. The propagation delay of the link g may change with time. Consequently, when considered as a property of the evolutive system, the propagation delay is a function of time, associating to each successive time value the propagation delay for the configuration of the link in the corresponding configuration category.

In complex natural systems, the propagation delay generally increases with the complexity level of the linked components: the time taken by an interaction between two molecules is negligible compared to the time required to transmit a signal between two cells, the latter process requiring a whole succession of molecular operations; in a business, information is generally transmitted more quickly between two people of the same department than from department to department. The formation of a colimit may decrease the propagation delays of the distinguished links of the pattern which it binds together. For example, in the category of neurons, the propagation delay of a synaptic path corresponds to the time taken for the transmission of a nerve impulse through it; when a pattern acquires a colimit in a complexification, it is transformed into a synchronous assembly of neurons, so that its components act in synergy (*cf.* Chapter 9).

Propagation delays play an important role in the dynamics of systems, and modifying them can increase the robustness and/or the efficiency of the system. For example, in a business, one way to improve the output is to decrease the propagation delays at all levels: *e.g.* by acceleration of the transfers of products between warehouses and workshops by rational centralization of the installations, reduction of waste time resulting from technical or human problems, faster circulation of information between the various departments.

6. Complement: Fibration Associated to an Evolutive System

In an evolutive system, it seems that we have separated the structural data, namely the organization of the system around a given time, represented by its configuration, from the dynamics describing its changes in the course of time, modelled by the transitions. The following mathematical construction allows them to be combined. As it is rather technical, it will not be used in later chapters.

6.1. Fibration Associated to an Evolutive System

Let **K** be an evolutive system. Its components and its links between them (defined in Section 3) form a graph, but this graph is 'static', without indication of the periods during which the components or the links exist. And even in systems which have fixed components, no information is provided about their dynamics. However, the structure and the dynamics of the evolutive system can be integrated in a large quasi-category **FK** which models and subsumes all the data. (A *quasi-category* satisfies all the axioms of a category except that the composite of a path is not always defined.)

Definition. If **K** is an evolutive system, the *fibration* **FK** *associated to* **K** is a quasi-category which has for its set of objects the set $|K|$ of all the objects of **K**, and which is generated by its following sub-categories (Fig. 5.8):

(i) The configuration categories K_t for each t; their links are called *vertical links*. K_t is also called the *fibre at t*.
(ii) The category associated to the order 'earlier than or simultaneous with' on $|K|$, the links of which are called *horizontal links*.

   To say that these sub-categories generate **FK** means that all its other links, called *transverse links*, are composites of horizontal and vertical links. A transverse link from A_t to $B_{t'}$ models an action of A on B which extends at least from t to t', and has an effect $g_{t'}$ at t'. It is obtained by composing the horizontal link from A_t to $A_{t'}$ with a vertical link $g_{t'}$ from $A_{t'}$ to $B_{t'}$ and will be denoted by $g_{t'}: A_t \to B_{t'}$. Then a unified notation for all the links is obtained by denoting the horizontal link from A_t to $A_{t'}$ by id: $A_t \to A_{t'}$ (where id is the identity of $A_{t'}$). With this notation, the composition of **FK** is defined as follows:

Definition. The composite of $g_{t'}: A_t \to B_{t'}$ with $g'_{t''}: B_{t'} \to C_{t''}$ is defined if and only if g has a new configuration $g_{t''}$ at t'', and this composite is $g_{t''}g_{t'}: A_t \to C_{t''}$.

   From this we derive the following proposition:

Proposition. *The graph having for its vertices the objects of the evolutive system and for its arrows the vertical, horizontal and transverse links between*

FK

K_{t_0} K_t $K_{t'}$ $K_{t''}$

vertical

g_t

transverse

horizontal

A_t

$C_{t''}$

$B_{t'}$ $B_{t''}$ $g'_{t''}$

$g_{t'}$ $g_{t''}$

$A_{t'}$ $A_{t''}$

time t_0 t t' t''

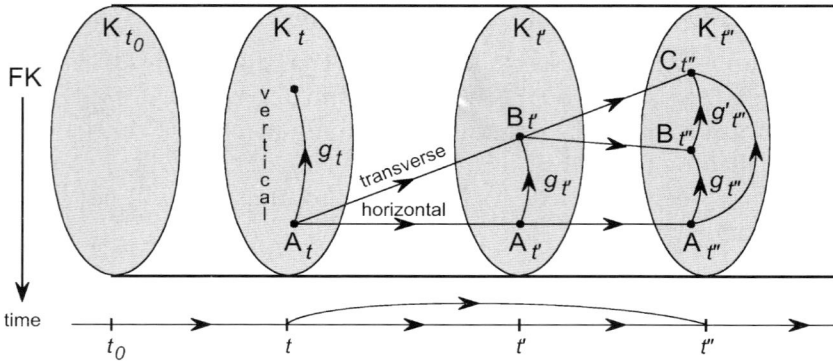

Fig. 5.8   Fibration associated to an evolutive system.
The fibration associated to **K** is the quasi-category **FK** whose objects are all the objects of the various configuration categories of the evolutive system, and whose links are:

(i) the links (such as g_t) of these categories, called 'vertical' links;
(ii) the 'horizontal' links from an object A_t to its later configurations; and
(iii) 'transverse' links from A_t to $B_{t'}$ obtained as the composite of the horizontal link from A_t to $A_{t'}$ with a vertical link $g_{t'}$ from $A_{t'}$ to $B_{t'}$.

The composite of this transverse link with the transverse link from $B_{t'}$ to $C_{t''}$ with $g'_{t''}$ as its vertical part is the transverse link from A_t to $C_{t''}$ having for its vertical part the composite of the new configuration $g_{t''}$ of $g_{t'}$ with $g_{t''}$. There is a functor from **FK** to the category 'time' associated to the order on the time scale of the evolutive system.

them is a quasi-category **FK** *for the partial composition law defined above. There exists a functor 'base' from* **FK** *to the category* T° *associated to the order on the time scale* T *of* **K**.

The functor from **FK** to the category T° maps the link $g_{t'} : A_t \to B_{t'}$ to the arrow in T° from t to t'. In particular, the image of a category K_t is reduced to t (whence the name of vertical for its links), while the image of the horizontal link from A_t to $A_{t'}$ is the arrow from t to t'.

Using **FK**, the components of the evolutive system can be defined as follows:

• A *local section* of the base functor is a partial functor α from T° to **FK** which associates, to the times t for which it is defined, an object αt in K_t, and to the arrow from t to t' the horizontal link from αt to $\alpha t'$.
• We define an order on the set of local sections: α is less than α' if α is a restriction of α' (as a map).
• Then a *component* of **K** corresponds to a maximal local section for this order.

6.2. *Particular Cases*

The quasi-category **FK** is a *category* (*i.e.* every path has a composite) if there is no disappearance of links between components without disappearance of at least one of these components. In other words, it is a category if the evolutive system satisfies the following *conservation principle*: If two components A and B existing at t still exist at t', any link existing between them at t has a new configuration at t' (divorce is not possible!).

When the links have a propagation delay, **FK** contains a sub-graph reflecting the time that an interaction between two components takes. It contains all the vertical and horizontal links, but only those transverse links $g_{t'}: A_t \rightarrow B_{t'}$ such that the propagation delay of $g_{t'}$ is less than $t'-t$, so that the interaction in progress at t' had already started at t and still persists at t'.

An evolutive system in which there is no disappearance of components can formally be described as a functor from the category T° defining the order on T towards the category of categories **Cat**. In this case, **FK** is the fibration associated to this functor (*cf.* Grothendieck, 1961; or Ehresmann, 1965, who calls it a crossed product).

6.3. *The Large Category of Evolutive Systems*

We have defined evolutive functors between evolutive systems. Such functors extend into functors between the associated fibrations. If p is an evolutive functor from **K** to **K'**, it extends in the functor **Fp** from **FK** to **FK'** such that

$$\mathbf{F}p(g_{t'} : A_t \rightarrow B_{t'}) = p_{t'}(g_{t'}) : p_t(A_t) \rightarrow p_{t'}(B_{t'}).$$

The evolutive systems and the evolutive functors between them form the (large) *category of* evolutive system. From general categorical results it follows that limits and colimits of patterns exist in this category.

Chapter 6

Internal Regulation and Memory Evolutive Systems

In the preceding chapters, we introduced hierarchical evolutive systems, which give a model for hierarchical natural systems. We explained how these model systems can be made to evolve by iteration of the complexification process with respect to an option. Such an option is a list of objectives, whose realization may lead to the emergence of increasingly complex components and interactions. The problem is: how are these objectives chosen in the real world?

Here, we study this problem of choice in autonomous hierarchical natural systems, such as biological or social systems. The concept of a memory evolutive system alluded to in the preceding chapter will give a morpho-dynamic model for such systems. In this model, a net of internal regulatory organs of various levels of complexity, called co-regulators, partially controls the dynamics by participating in the selection of the objectives and in the control of the procedure which carries them out.

Following the convention of Chapter 1, we adopt a vocabulary, which seems to confer intentions to the co-regulators (they 'act', 'observe', 'communicate', 'select objectives', 'send commands' and so on). However, (with the exception of a few cases), the operations they perform are not deliberate, but a consequence of the organization of the system and of its normal functioning. Each co-regulator acts according to its own time scale and in its own frame of reference (its landscape; see below). The communication between the various co-regulators is often indirect and partial, their objectives may conflict, and an equilibration process must develop to maintain the integrity of the system as a whole, with a concomitant risk of damage (or fracture) for some of the individual co-regulators. The system's autonomy is strengthened by a capacity to learn how to recognize specific features of the environment, and to develop and remember best-adapted responses.

In this chapter, we give an overall picture of a memory evolutive system and of its co-regulators; a more detailed analysis is given in the two following chapters.

1.   Regulatory Organs in Autonomous Systems

A complex hierarchical natural system autonomously maintains its identity over time; it evolves and adapts to changes in its environment, in spite of the multiplicity of its complexity levels, each one with its own structural and temporal constraints. If we think of a biological system, from its atoms or even its sub-atomic particles to the molecules, macromolecules, organelles, cells, tissues and organs, they all collaborate to ensure the preservation of the organism.

1.1.   General Behaviour

How is it possible that an autonomous system of this sort has both robustness and plasticity, along with fault-tolerance, so that its behaviour remains coherent despite its more or less variable environment? For that, it needs to possess a variety of internal mechanisms that function at its different levels and cooperate to regulate its overall activity. It is also very useful for it to have a 'memory' from which it may recall inner processes and past experiences (for further explanation, see Section 2, below). In this way, it can avoid re-analysing a situation it already recognizes, and so give a faster and better adapted response. This is different from cybernetic systems, which operate by feedback and without a memory. As Kitano (2002) writes:

> Unlike complex systems of simple elements, in which functions emerge from the properties of the networks they form rather than from any specific element, functions in a biological system rely on a combination of the network and the specific elements involved … Robust systems maintain their configuration and functions against external and internal perturbations, and robustness is an essential feature of biological systems … even damage to their very structure produces only minor alterations of their behaviour. Such properties are achieved through feedback, modularity, redundancy and structural stability. (p. 206)

Each group of components of the system connected by a specific activity will function with a regulating organ, one which takes account of both external and internal requirements and constraints on its level, and of the more or less partial information received from the memory and from other modules and other levels. For example, the receptors on the membrane of a cell receive and transmit information between the external medium and the cytoplasm, and via second messengers, exert a control on the metabolic activity of the cell.

The connections are not only multiple and complex, but they are also specific and used for the selection of adapted objectives and procedures to achieve them. It is necessary that these choices be coherent, and respect the global constraints of the system, so that the system may maintain sufficient homeostasis; and furthermore, that the system adapts to the changes of its environment and possibly progresses, either in size or in quality.

There is great variability in the composition of the various levels; they are susceptible to multiple material constraints, and their interactions may require more or less time; in the case of stability spans (which measure the turnover, *cf.* Chapter 5), it will take incomparably longer to renew the cells of the body than the proteins of a cell. Consequently, there cannot exist an all-pervasive regulating mechanism, a process present everywhere, at all moments and at all levels, covering all the specific rhythms. An internal observer (in the sense of Matsuno, 1989), or an external one, will be able to form only a partial image of the system, relating to a particular level of description, and a specific time scale.

1.2.   The Co-Regulators

Because there is no central regulating mechanism, it is necessary that the system have a network of partial regulatory organs, the co-regulators, as said at the start of this chapter (and alluded to in Chapter 5).

A co-regulator is a sub-system whose components, called its agents, act together to perform a specific function. This function is defined by the local operations or 'procedures' (in earlier works we called them 'strategies', but the term is not neutral enough) that the co-regulator may engage in (we will say selects), in response to the information it receives, to achieve well-adapted results. In a nervous system, an example of a co-regulator would be a specialized brain area, such as that treating colour or shape. Each co-regulator operates with its own rhythm. Thus, 'the temporality is not unilateral nor uni-modal, but plural, polyphonic and polysemic' (translated from Draï, 1979, p. 42), and it is one of the characteristics which ensures the plasticity of self-organized systems, and distinguishes them from simple mechanical systems.

The co-regulators operate in parallel, both horizontally, in time, and vertically, in the hierarchical structure of the system. The higher ones are more responsive to changes in the environment, and more autonomous in the selection of their objectives. As Simon (1974) has emphasized, most natural or artificial systems are 'quasi-decomposable', meaning that, over short periods, the behaviour of one part is independent of that of other parts. Even if they act independently, the various co-regulators must collectively perform in a coherent manner; for example in a jazz band (Watslawick *et al.,* 1967), there is no conductor, but each musician cooperates with the others. For this, the co-regulators communicate directly, or indirectly in feedback loops; however, their heterogeneity makes this communication very partial and belated; for instance a modification of a lower level will be reflected at the higher levels only with a delay, so that those levels cannot quickly answer in an adapted way and are likely to make a choice incompatible with some other objectives, to maintain their own homeostasis. When sufficient coherence between all the co-regulators cannot be achieved, dysfunctions of a particular sort (called fractures, discussed in Section 4 below, and in Chapter 7) may occur, to the extent even of being beyond the possibility of repair; thus, risking instability, aging, or even destruction of the system.

1.3. Meaning of Information

In the preceding discussion, we have used the terms information and communication without defining them. Since their specific meanings matter to what follows, we shall do so now (*cf.* Ehresmann and Vanbremeersch, 1997).

The usual meaning of information is something like communicated knowledge or news, and so it supposes some intention in the process. In mathematical theories of information, of which the theory of Shannon (Shannon and Weaver, 1949) is the best known, we have an emitter which produces an encoded message, a channel through which the message is transmitted, and finally a receiver which decodes it. The transmission of the message can correspond to the transfer of some material object (such as a letter via the post office), or to the propagation of a disturbance in *e.g.* an electromagnetic or acoustic medium, thereby in turn modifying the state of the receiver. Its information content is quantified by a statistical evaluation of the probability of each component symbol of the incoming message. The aim of Shannon's theory is to study the number of bits of information that will be faithfully transmitted in the process, without any regard for the meaning of this information. This pre-supposes some convention between emitter and receiver over the class of possible messages. Indeed, without such a convention, any incoming signal would appear as purely a matter of chance, since there are a large number of possible codes and a message may have different meanings depending on the code. It has been proven (Benzecri, 1995) that, as a practical matter, the process of agreeing upon an efficient code, between two actors communicating only through action, without the aid of an already agreed-upon language, is complex.

The Shannon theory can be interpreted in a subjective perspective: what is the information gain for the receiver, as measured by the difference between its uncertainty with respect to the message before reception, and its uncertainty after? More objectively, *i.e.* for an external observer knowing the code, the information content of the message is its specificity: that is, the difference of size between the variety of all possible messages, and the sub-variety of those messages which can lead to the received signal.

However, the notion of information which we consider here is somewhat different, in that it neither presumes intentionality nor resorts to an action language, in which the mirror symmetry of receiver and emitter through a common convention would be destroyed. While the usual theories are concerned with the correct transmission of message-symbols, here the concept of message is enlarged to also comprise endogenous processes (commands, constraints, spontaneous oscillatory processes) or even processes caused by external perturbations; and the characteristics of the messages (or signals) come from the interactions they generate, or in which they participate, and the responses they trigger. The same signal can encode information of different sorts, which can be decoded or interpreted only by certain co-regulators; for example, a colour centre and a shape centre in the brain decode different information from the same visual target. Conversely, several signals can be decoded as the same message, *e.g.* different blue objects are not distinguished by a colour centre. We should also be cautious that the 'transfer of information' can be an artefact which exists solely in the mind of an external observer, who detects a purely coincidental correlation between a particular signal, some configuration of the emitter, and some configuration of the receiver.

In the evolutive system, which models a complex natural system, information transfers are affected along the links of the configuration categories. The transmission characteristics of a signal or a message depend on the structure of the link, on its weight (which can describe, for example, the amplitude or the frequency of the signals), and on its propagation delay. However, a link from a component B to an agent of a co-regulator represents an information transfer only in the following cases:

(i) When the co-regulator can decode the signal (a letter can be written in an unknown language!), and take it into account in its next action. The response can be immediate, delayed, or opposed to the message, or even a non-event (differentiation of a cell results from omitting to express some genes). The signal may be unintentional, such as the traces left by a prey animal, which reveal its presence to a predator, or the effects of some cellular perturbation which gets transmitted to contiguous cells. For a cell, the signal can also result from the diffusion of a product (say, a hormone) secreted in a far-off cell and which diffuses through the circulation or the conjunctive tissue, with more or less delay. Natural selection has led to the development of organisms able to automatically send signals, which are decoded as messages by other organisms, possibly after modification during their transfer (*e.g.* emission of pheromones by a female insect to attract the male).

(ii) When the message is sent to modify the action of the co-regulator, which later receives some feedback from this action. This is the case of a command sent to an effector co-regulator, with the sending and receiving of the signal being intentional or not.

(iii) When the transfer of information is constitutive of the system, meaning that it plays a part in the basic operations of the system, such as the links between the agents of a co-regulator, which permit communication between them.

2. Memory and Learning

We have said that hierarchical natural systems such as biological systems must possess a memory to perform efficiently. However, what do we mean by that?

2.1. *Several Types of Memory*

The word memory can be used with different meanings. From Webster's dictionary, memory is the 'power of reminding or recalling to the mind things that are past'; but 'a memory' is 'something remembered'. Thus, memory can denote either a system that stores and retrieves information, or some particular item of information stored in such a system. This is particularly the case when referring to the mind, when what is retrieved is a memory-image accompanied with a feeling of familiarity (in the terms of Russell, 1949). And of course, memory itself also refers to the power to remember.

Let us give some examples to illustrate the versatility of the term memory.

(i) Simple physical systems already have a kind of memory. A stone tablet records marks chiselled into it. A spring remembers its original shape: if it is stretched within its elastic limit, it elongates, but once released, it springs back. A thermostat retains a set temperature at which it must stop the heating. A computer has several built-in memory devices (RAM, hard drives, and so on), and when they are appropriately filled, such as when the computer has been programmed to serve as an expert system, the items stored in memory can be automatically recalled in the appropriate contexts.

(ii) Similarly, a robot is able to recognize some features of its environment via its sensors, and it remembers several procedures (built-in or learnt) to react in an appropriate manner.

(iii) Living systems have greater capacity for memory. For example, bacteria engage in metabolic activity, reproduce, and repair damaged DNA. All these activities are autonomously controlled by their genetic 'program', which serves as a memory of the organism's ancestry, and thus, of the species. (The term program used for the genetic material is contested by Mahner and Kary, 1997.)

(iv) An animal with a rudimentary nervous system, such as a fish or a lizard, receives information about its environment via its sensory organs. The central nervous system also receives information concerning the animal's internal states (*e.g.* hunger or pain), and may remember them for later recognition. Such an animal has some innate behaviours (instincts and reflexes), but is also able to learn new skills and behaviours, and to evaluate them.

(v) Higher animals, such as mammals, are capable of developing a semantics, which may modulate their actions according to the circumstances, and they exchange information through communication, such as when emitting alarm sounds, or in the education of the youngsters. The storage and the recollection of an item can be unconscious or conscious.

(vi) Societies develop a collective memory that explains, for example, cultural differences among communities of chimpanzees (Whiten *et al.*, 1999). Human language allows for particularly efficient internal representations and communication among individuals, enabling the emergence of conceptual knowledge and the development of rich cultural forms.

2.2. *Different Properties of Memory*

The examples we provide above point to diverse and contrasting aspects of both memory and of things remembered. They correspond to a long-term memory; there can also exist a short-term memory or working memory, but it will never be what we intend when we just speak of memory. The item remembered can be inherent in the system and more or less fixed during construction; this is the case for the proverbial inscription 'set in stone', as described above. Alternatively, for a living organism (example (iii)), memories can be inherited (as a consequence of natural selection). However, additional memories can be acquired as in robots, living systems, societies (examples (ii)–(vi)) by learning, and by combining items already remembered. In animals and societies (examples (iv)–(vi)) the memory is flexible enough to on the one hand provide the stable raw material for approximations and generalizations (*e.g.* by induction), yet on the other to allow for modifications in response to new circumstances.

Memory (as a system) is often distributed, and a given co-regulator may only have access (in either the sense of simple interconnection, or else of literal permission, depending on the co-regulator) to some part of it: a visual module has no access to the shape of an object, and a journalist may only be able to obtain certain documents from an archive. In general, memories are distributed, *i.e.* they cannot be entirely grasped by any single component, just as, for example, no mathematician can grasp the whole of mathematics.

Different types of (long-term) memory can be distinguished. The empirical (perceptual and episodic) memory relates to relatively stable information concerning features of the environment, internal needs and more or less complex situations, as well as significant experiences (for animals). Procedural memory relates to procedures and behaviours for responding in an adapted way to some external or internal events, by anticipating the possible results. In higher animals and societies (examples (v) and (vi), there is also a semantic or declarative memory, in which invariants are formed by taking invariance classes of items. Then conceptual knowledge can be developed by logical processes (disjunction, conjunction, negation), by deduction (formation of chains of links, as in mathematical proofs), or by inference of any sort, in particular abduction in the sense of Pierce (1903). Communication of knowledge within a social group is at the root of culture. Language increases this communication, whether via speech, or via enduring cultural artefacts (books, films, CDs, and so on).

2.3. Formation and Development of the Memory

These different aspects of memory will be reflected in our model for autonomous systems. Unless otherwise specified, the term memory always refers to long-term memory. The memory, *i.e.* the sub-system of the system that does the remembering, is represented by an evolutive sub-system of the overall evolutive system modelling the system, and its components are called *records*. A record has a stability span long enough to maintain its complex identity over a useful period of time. However, as for any complex component of an evolutive system, its internal organization may vary, giving it some flexibility, so that a given record might be recalled in various more or less approximate situations, and later generalized and adapted to modifications of the context. Records can be innate, or they can be formed in the wake of an event, to remember features of the environment, internal configurations, or situations the system does not recognize, or procedures it develops to react in an adapted manner. Items are remembered only if they are persistent enough (*e.g.* repeated several times) or significant: an animal continuously gets new sensory information about the surrounding objects, but discards most of it.

As we will explain in Chapter 8, the formation of the record (or storage) of an item depends on the integration of an internal assembly of components more or less briefly activated (represented by a pattern in a configuration category), into a stable internal component (the record M as the colimit of this pattern). However, this record is flexible: M takes on its own identity as a multifold object by consolidation, meaning that it may acquire different internal organizations (decompositions of M) when confronted to later presentations of the same or approximating items; and it may be recalled later on by the initial pattern or by any near enough pattern figuring another of its decompositions.

The process can be automatic or, in more complex systems (*e.g.* higher animals), an active learning process more or less controlled by some higher co-regulators. An important way to construct more complex records is to combine pre-existing records, for instance to learn new procedures (*e.g.* behaviours) uniting several more elementary ones. Moreover, in these more complex systems, the records are classified in a semantic memory. In any case, the memory plays a central role in the dynamics of the system. Each co-regulator has an access to the records, which have some connection with its own function, and it participates in their development.

To sum up, the model we are going to describe has at least the following properties:

(i) A net of internal regulatory organs (the co-regulators), each with a specific function and its own timing, which collect information, select and implement responses, and evaluate the result of these procedures, at least locally.

(ii) A memory, which develops over time. It stores information and procedures with enough stability so that they may be recalled when needed, yet enough flexibility that they may be modified if the anticipated result is not then obtained.

(iii) In more complex systems, a semantic memory, which emerges by classification of records.

3. Structure of Memory Evolutive Systems

The preceding characteristics of autonomous systems are taken into account in memory evolutive systems, which are particular evolutive systems introduced to model complex natural systems, in particular biological, neural, cognitive or social systems. We begin with a general description.

3.1. Definition of a Memory Evolutive System

Its architecture is a compromise between a parallel processing system with a modular organization (of the multi-agent distributed system type; Minsky, 1986), and a hierarchical associative network (Auger, 1989; Goguen, 1970; Salthe, 1985). Its dynamics is modulated by the cooperative and/or competitive interactions within a net of internal organs of local regulation, the co-regulators introduced above. Each co-regulator has a specific function and operates at its own level of complexity, with its own discrete time scale extracted from the continuous time scale of the system (as indicated in Chapter 5).

Moreover, a memory evolutive system has a central internal memory, which develops over time. It allows a perceived stimulus to be compared with the stored records, and if recognition occurs, it selects an appropriate response by anticipating the probable result of this response from past experience. In this way, the system acts as an 'anticipatory system', in the terminology of Rosen (1985b). Each co-regulator has access to only certain parts of this memory, and participates in its development through a trial-and-error learning process. More formally:

Definition. A *memory evolutive system* (or MES) is a hierarchical evolutive system **K** over a continuous time scale T, and in which the following evolutive sub-systems are distinguished (Fig. 6.1):

(i) The memory: it is a hierarchical evolutive sub-system with the same time scale, which develops over time, possibly by the formation of higher and higher levels (through successive complexifications).

(ii) A net of evolutive sub-systems, called *co-regulators*. The components of a co-regulator are called its *agents*, and their complexity level depends on the co-regulator. Each co-regulator has its own discrete time scale, formed by a sequence of instants of the reference time scale T; and it is assigned particular components of the memory, called its *admissible procedures*, which represent the local operations it may engage in to fulfil its objectives, by sending the corresponding commands to components of the system, called *effectors*.

The notion of a procedure will be defined more precisely in Chapter 8. Here, it is sufficient to know that a procedure controls some action, process or behaviour of the system, via commands sent to effectors. It is stored as a component of the memory (also called a procedure), and its commands, which activate the effectors to implement the procedure, are modelled by links from the procedure to these effectors. Depending on the case, these effectors can operate in a purely internal way or have effects on the environment.

In this section, we give only a rough definition of a co-regulator and of its admissible procedures; it will be made more explicit later on (*cf.* Section 3.2 below and Chapter 8). The net of co-regulators does not form a strict hierarchy, for there can exist several of them of the same level (meaning that their agents are on this level), possibly with different time scales. For example, in a nation, the co-regulator representing a major city has a different

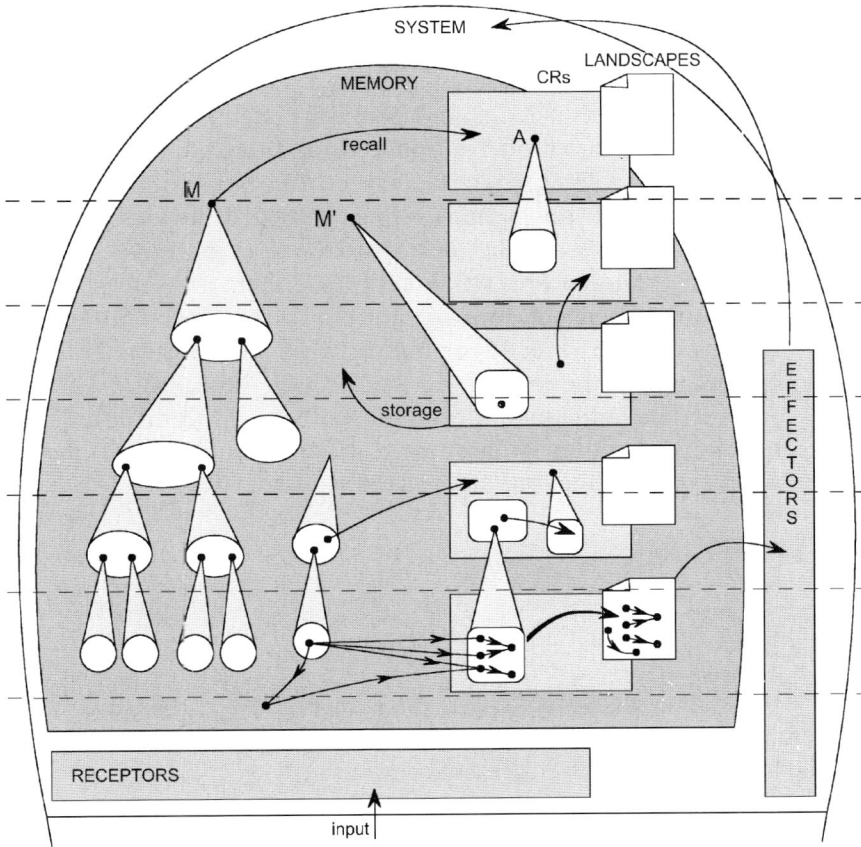

Fig. 6.1 A schematic view of a memory evolutive system.
A memory evolutive system (MES) is a hierarchical evolutive system, and the figure represents one of its configuration categories. Different sub-systems are distinguished. The receptors model components of the system which receive information from the environment or from internal states. The memory stores records of objects met by the system, of behaviours, and of experiences of any kind; these records can later be recalled in similar situations; the figure shows one ramifications of the record M. The co-regulators, or CRs (depicted in front of the memory, the difference of colour meaning they are not included in it) participate in the regulation of the dynamics, and in the storage of records (such as M′) and their recall (depicted for M). Each co-regulator operates at its own complexity level, and at its own pace, on the basis of its landscape, which collects the partial information received from the system. The landscape is not a part of the system, but an internal model of it for the co-regulator. The landscapes are depicted on the right of each corresponding co-regulator; in the bottom-most one, some of its elements are represented. Each co-regulator operates via its admissible procedures whose commands are carried out by effectors. A higher level co-regulator can control a lower one, for instance via a decomposition of one of its agents A. (Some arrows and cones correspond to various constructions explained later.)

timing from that of one representing a small village. There are generally no co-regulators at the lowest level, or they are degenerate; for instance the nucleus of an atom can be considered as such a co-regulator at the particle level, its agents being the nucleons and protons constituting the nucleus.

Lower level co-regulators model specialized modules, sometimes acting as receptors with respect to certain external aspects (receptors on the membrane of a cell, sense organs in an animal). Some co-regulators act as effectors, which carry out the orders of other co-regulators (*e.g.* excretive cells, motor organs, deliverymen, and so on). At higher levels, there are slower operating associative co-regulators, which control and coordinate the activity of a group of lower level ones; either directly, or indirectly by the constraints that they impose on them. A component of the system can be an agent of several co-regulators (*e.g.* an individual can be a member of several associations). A co-regulator may exist only temporarily, for example an association formed only to effect a specific function, after which it dissociates.

3.2. *Function of a Co-Regulator*

Each co-regulator operates stepwise according to its own time scale (which would define its subjective time in the terminology of Bergson, 1889). A step extends between two successive instants of this scale, and its length (or duration) of the step is measured by the difference between these two instants in the reference scale T of the memory evolutive system. At each step, the co-regulator performs three successive actions:

(i) Internal observation: its agents form its landscape, which collects the partial information on the configuration of the system that they receive at this date.
(ii) Regulation: it selects objectives and an admissible procedure to achieve them; the commands of the procedure are sent to effectors.
(iii) Control function: at the end of the step, it evaluates its final results, and takes part to their storage in the memory at the beginning of the next step.

A co-regulator is *cyclic* if its steps are all of the same duration, and if in each step the same process is completed (*e.g.* biological or physiological rhythms, such as the cardiac rhythms), or if in each step the same procedure or plan of action (*e.g.* as specified in the calendar for a university year) is undertaken or implemented. Otherwise, the steps are of variable length depending on the situation.

To operate, a co-regulator must have multiple links, both afferent and efferent, with the other parts of the system. In particular it may receive links from components receiving signals from the environment (called *receptors*), from the components of the system that it controls, and from the part of the memory which it can access, in particular in order to recognize features of its environment, and to search for admissible procedures. It must also have efferent links towards the memory and to effectors, to implement the chosen procedures, activate their commands, and participate, at the following step, in the process of storing the situation and the result, and the procedure too, if it is a new one.

3.3. Propagation Delays and Time Lags

For the various operations of a co-regulator to be effective, it is necessary that information or commands be transmitted fast enough, so that they remain current. Likewise, it is also necessary that the objects intervening in the landscape and in the selected procedures do not vary too quickly. These temporal constraints lead us to incorporate the propagation delays associated with each link into the model as well. Let us be more specific.

We have defined a link g of an evolutive system from the component A to the component B as being a family $g = (g_t)$ of links g_t between configurations A_t of A and B_t of B, formed by an initial link and all its later configurations (Chapter 5); we say that g is *present at t* if it has a configuration g_t. Such a link models an interaction between A and B. The interaction can be permanent or intermittent (*e.g.* a synapse transmits a nerve impulse only episodically).

An *episode of presence* of g is defined as a maximal interval $[t^o, t^\&]$ such that g has a configuration for every t' between t^o and $t^\&$. In other words, t^o is either the initial date of the link, or else the most recent date from which g is continuously present up to $t^\&$, following some interruption. Note that a physical interaction is not instantaneous. If it begins in A at t^o, it only reaches B at some later instant t^*. The period needed for the completion of the interaction, namely the difference $t^* - t^o$, is the propagation delay of the corresponding link.

Definition. In a memory evolutive system we associate to each link of the configuration categories a non-negative real number called its *propagation delay*, which satisfies the following conditions:

(i) The propagation delay of the composite of two links is the sum of their delays.
(ii) If g is a link from a component A to a component B, the function associating the propagation delay of g_t to each instant t of an episode of presence of g is constant during this episode.

The first condition means that the configuration categories are labelled in the additive monoid of the non-negative real numbers (Chapter 1). The second condition implies that the propagation delay is a characteristic of an episode of presence of a link.

The propagation delays impose temporal constraints on the agents of a co-regulator. To effect some joint action, say to collect information or select a procedure from the memory, they need sufficient time to communicate with each other. For example, some time elapses between the perception of a predator by an animal, its cry of warning to its fellows, and the following replies; the alarm will be effective only if these times are not too long. The latency period or time lag of the co-regulator at an instant t of its time scale is the duration necessary for the successive operations it performs during the step beginning at t, taking account of the propagation delays of the links; it must be much shorter than the duration of one step because otherwise the agents would not be able to analyse the information they receive and communicate between themselves to select a procedure during the step. Thus, the formal definition:

Definition. The *time lag* of a co-regulator E at an instant t is the maximum of the propagation delays at t of the links between its agents, and of the links coming from the memory accessible to E. It is supposed to be of a smaller order of magnitude than the length of the current step of E at t.

The time lag is very variable depending on the co-regulator and, for the same co-regulator, it can vary over time. It is generally longer for higher level co-regulators; thus, a co-regulator of the molecular level will have a much shorter time lag than one at the cellular level.

In the applications, not only the propagation delays, but also the strengths of the interactions represented by the links, *e.g.* their weights or the forces involved, are typically also quantified in appropriate units. This is also done by labelling the configuration categories in a suitable monoid (which, depending on the case, could be the monoid of real numbers with multi-plication, or a group of vectors with addition).

4. Local Dynamics of a Memory Evolutive System

We are going to describe one step for a particular co-regulator of a memory evolutive system. The step is divided into several more or less concomi-tant phases combining observation, analysis and action (thus forming an 'epistemo-praxeologic loop' in the sense of Vallée, 1995, p. 100):

(i) Formation of the referent of the co-regulator, called its *landscape*, which collects the information that its agents receive during the first phase of the step, and on which the results of the preceding step are evaluated.
(ii) Selection of an objective on the landscape, and of an admissible pro-cedure to achieve it, taking into account the results of the preceding step, the

various constraints, and the admissible procedures used earlier in similar situations.

(iii) Sending of commands, specified by the procedure, to effectors to realize the objectives. Finally, at the beginning of the next step.

(iv) Evaluation of the final result to determine if the objectives are realized, and the storage of this in the memory.

4.1. Phase 1: Construction of the Landscape (Decoding)

Let E be a co-regulator of the memory evolutive system **K**, and t an instant of its time scale. We denote by d the duration of the step starting at t and by π the time lag of E at t. The first phase will last from t to $t+2\pi$ and it consists in the formation of the landscape of E. The landscape is not a sub-category of the system; it is a temporary model of the system from the point of view of the co-regulator, and contains only the information the agents of E may decode during their present step.

The co-regulator (via its agents) does not have direct access to the system. A component B of the system may be observed by E only through the links its agents receive from B, called the *aspects* of B for E, and it may have several of them. Two aspects of B pass on the same information to E if this information is interchangeable via communication among the agents; that is, if they are correlated by a zigzag of links between the agents of E, so that they are in the same perspective of B for E (in the sense of Chapter 3). Thus, a *perspective* of B corresponds to the formal description, by the agents operating together, of the component seen under various coherent aspects. The landscape of E (to be compared to the perspective space of Russell, 1971) is a category, which has for objects the perspectives that correspond to information which arrives at the agents 'fast enough' (let us say between t and $t+2\pi$), and which remains current until the end of the step, so that it might be analysed and used.

The landscape constitutes an internal frame of reference for the co-regulator, giving it an assessment of the present situation (its epistemological image of the system, in the terminology of Vallée, 1986). It is a selective representation of the system (masking all that is not observable at the level of E), whence the difference between the landscape and the system, measured by what we call the *difference functor* from the landscape to the system. This difference is not directly observable by the agents of E, but its consequences might be revealed at their level at the following step, or later on. For example, molecular changes will be seen only by their overall effect in the landscape of a cell.

Let us define the landscape of E more precisely: the configuration category E_t of E at an instant t of the time scale of E is a sub-category of the configuration category K_t of the memory evolutive system **K** at t, and can be looked at as a pattern in this category. Thus, we can speak of the perspective of an object B_t for (this pattern) E_t and say that two aspects of B_t pass on the same information to E_t if they are in the same perspective. However, this does not take into account the time taken for the information to arrive at the agents, that is the propagation delays of the aspects, nor the possible changes in B during these delays. To model the information really decoded by the co-regulator during a step, we refine the notion of a perspective, to make it less dependent on the instant.

Definition. A *perspective* for the co-regulator E at t is a perspective (in $K_{t+2\pi}$) of a component B for $E_{t+2\pi}$ satisfying the following conditions:

(i) B has the same level or a level adjacent to the level of E, and its stability span at t is of an order of magnitude greater than that of the time lag π of E at t;
(ii) the perspective contains at least (the configuration at $t+2\pi$ of) one link b arriving at one of the agents of E between t and $t+2\pi$.

   These conditions assert that B is 'near enough' to E so that at least one of its aspects b arrives at some agent of E during the present phase, and remains stable enough to last the duration of the step. An aspect b entirely defines the perspective to which it belongs, and this perspective will be denoted by pb. In particular each agent A of E has a perspective for E, defined by the identity id_A of (the configuration at $t+2\pi$ of) A. Any component B may have several perspectives for E. If f is a link from B to C and c an aspect of C for E, then fc is an aspect of B for E.

Definition. The *actual landscape* of E at t is the category L_t whose objects are the perspectives for E at t, while the links $f: pb \rightarrow pc$ from a perspective pb of B to a perspective pc of C are defined by the links f from B to C correlating these perspectives (meaning that fc is in the same perspective as b), and lasting up to the end of the step (Fig. 6.2).

Proposition. *The landscape L_t of the co-regulator E at t is equipped with a functor to $K_{t+2\pi}$, called the difference functor, which maps a perspective pb of B onto the configuration $B_{t+2\pi}$ of B, and a link $f: pb \rightarrow pc$ onto the configuration of f at $t+2\pi$. It contains a sub-category isomorphic to $E_{t+2\pi}$.*

   This sub-category has for objects the perspectives of the various agents of E. Thus, the co-regulator has only partial information about the system, but complete information about itself.
   The difference functor measures the extent to which the internal referent of the co-regulator, namely its landscape, differs from the system (whence its

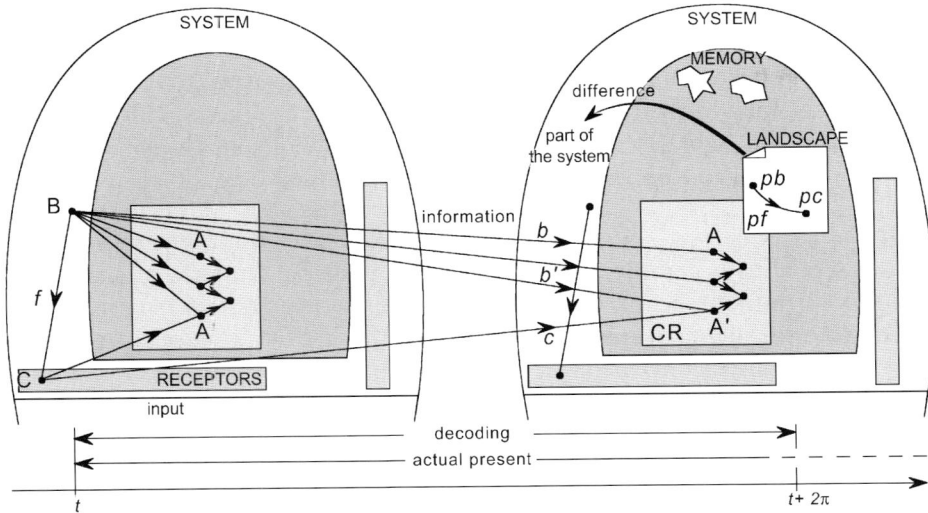

Fig. 6.2 Landscape of a co-regulator.

The landscape of a co-regulator at a given time t is a formal model of the information about the system that the co-regulator can collect and decode during the first phase of its step beginning at t, *i.e.* during the period from t to $t+2\pi$, where π is its time lag at t. A component B of the system presents different aspects b, b',... to its agents, and two aspects give the same information to the co-regulator if they are correlated by a zigzag of links between its agents. The perspective pb of B is formed by aspects so correlated and arriving during the first phase. The landscape at t has for objects these perspectives. A link from pb to pc is defined by a link f from B to C correlating the two perspectives; on the figure, c is an aspect for A$'$ coming from a receptor C. (The aspects b and b' of B for the agents A and A$'$ are represented both in the system at t, and as arrows from the configuration of B at t to the configurations of A and A$'$ at $t+2\pi$ to suggest the propagation delay before their perspective pb is formed; idem for the aspect c of C for A$'$.) The landscape (represented on the right of the figure) gives a deformed model of the system, and the deformation is measured by the difference functor, which maps pb and pf to (the new configurations of) B and f. The white spots in the memory on the right figure depict some admissible procedures for the co-regulator.

name). It would be an isomorphism (*i.e.* no difference) only if E had complete information about the entire system; *e.g.* if E has a unique agent which is a final object (Chapter 1) of the system. In any case, the difference is not accessible internally to the co-regulator, but only to external observers with a complete view of the system and of the landscape. Among the discrepancies that E cannot internally recognize, the difference functor indicates:

• the components of the system which are not observable at all by the agents, *i.e.* which have no perspective for E (they are not in the image of this functor);

• when two perspectives are in fact perspectives of the same component (both have the same image by the functor), though they are treated as completely different objects in the landscape. For instance, over the course of centuries, men in their landscape did not realize that the morning star and the evening star, which they named, respectively Venus and Lucifer, were in fact the same astronomical object, the planet Venus.

Once the landscape is formed (after $t + 2\pi$), it does not vary until the end of the step, thus it cannot take into account possible changes of the system or its environment occurring between these two dates, and may become more and more inaccurate.

Remark. Applying the definition of the field of a pattern, for each t' we have the field $D_{t'}$ of the pattern $E_{t'}$ in $K_{t'}$: it is the category for which the objects are the perspectives of an object B of $K_{t'}$ for $E_{t'}$, and the links are defined by the links of $K_{t'}$ correlating two perspectives. The transition functors map a perspective onto a perspective, and therefore determine transition functors between these fields. In this way we obtain the evolutive system of the fields, denoted $\mathbf{D}$, on the time scale reduced to the interval $[t, t + d]$. The landscape of E at t is a sub-category of the field $D_{t + 2\pi}$.

4.2.   Phase 2: Selection of Objectives

By storing the information collected during the present step of the co-regulator, the landscape of E plays the part of a short-term working memory as defined in Bunge *et al.* (2000):

> Working memory (WM) refers to the temporary storage and processing of goal-relevant information. WM is thought to include domain-specific short-term memory stores and executive processes, such as coordination, that operate on the contents of WM. (p. 3573)

It is through the landscape that, in a second phase, objectives are selected to respond to the situation, as well as a procedure Pr to achieve these objectives via the commands of Pr to effectors. An admissible procedure for the co-regulator E is a procedure that E can partially control, by virtue of it being observable in the landscape of E, and of the agents being able to activate some of its commands. Its selection by E (through its perspective) activates these commands (the procedure can have more commands). The standard consequences anticipated in the landscape are modelled by the objectives of an (possibly mixed) option on this landscape, called the *option associated to the procedure* (Fig. 6.3).

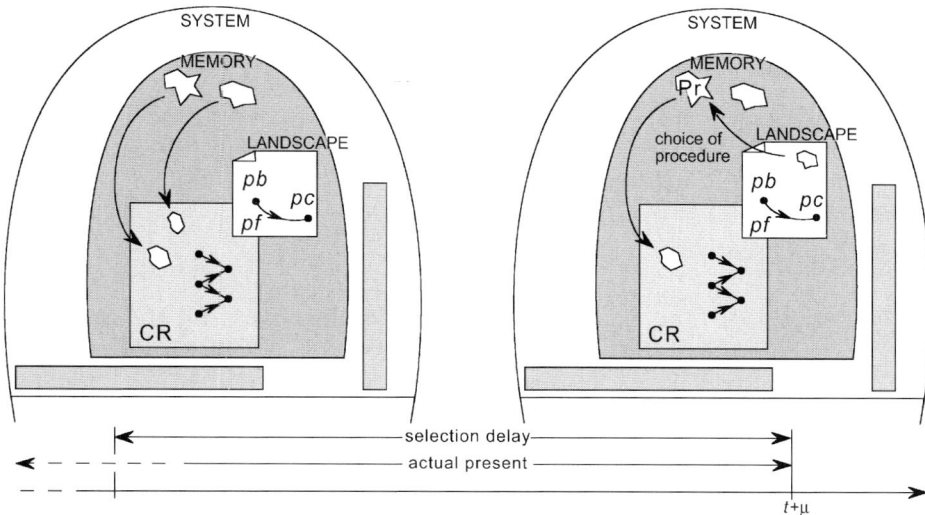

Fig. 6.3 Selection of a procedure.
During the second phase of a step which lasts up to $t + \mu$, the co-regulator will select the objectives to be carried out by choosing an admissible procedure through its landscape. The admissible procedures are particular components of the procedural memory which determine the function of the co-regulator. To this end, they have perspectives in its landscape through which they can be recalled, and some of their commands are controllable by efferent links from the agents to effectors. The admissible procedure most adapted to the context, say Pr, is selected. The selection involves a delay, which must be shorter than the actual present of the co-regulator.

As we discuss in greater detail in Chapter 8, procedures are selected in different ways depending on the co-regulator:

(i) A procedure can be externally forced on a co-regulator, either by observers external to the system or by other parts of the system, generally other co-regulators. In particular it is the case for the commands of procedures transmitted to co-regulators acting as effectors. It can also be the result of a random process.

(ii) A procedure may come from the internal structure of the system, by which the procedure is connected to a specific context. This is the case for cyclic co-regulators, which always adopt the same admissible procedure in response to specific signals (biological clocks), or for specialized co-regulators of lower levels where the choice of procedures is hard wired. If there are admissible procedures which have already been used to respond to a similar situation, one of them is selected by either a local or a central optimization process, and the objectives are those which had been achieved by the commands of this procedure observable in the landscape.

(iii) The selection can be directed by a probabilistic process, or even a random process (for instance a consequence of noise in the system).

(iv) If the situation has not yet been met, or is not connected to an admissible procedure, a higher co-regulator may try to form a new procedure by combining procedures already used in response to parts of the situation. If it succeeds, this procedure will be stored in the memory as a new admissible procedure at the following step. Such a process can be considered as one form of creativity.

The amount of time necessary to select the objectives and the procedure (operation lag in the terminology of Rosen, 1958b) is of the order of several π. The interval, say $[t, t+\mu]$, during which the co-regulator forms its landscape and selects a procedure is called the *actual present* of the co-regulator at t. It is more or less equivalent to what Bohm (1983, p. 207) describes as follows: 'a moment... covers a somewhat vaguely defined region which is extended in space and has duration in time'. It must be smaller than the length d of the step beginning at t. The actual present is generally longer for higher co-regulators than for lower ones.

4.3.   Phase 3: Commands (Encoding) of the Procedure, and Evaluation

The last phase of the step corresponds to transmission of the commands of the procedure selected by the co-regulator E, from the agents to the effectors. Its duration (transport lag) depends on the propagation delays of these links. It is not defined in a precise way; it will be admitted that it also covers several time lags π. Another constraint on the successful implementation of a procedure is that the components intervening in it have their stability span long enough to be still existing at the necessary moment.

The commands of the procedure selected by E are sent to effectors. The objectives of the co-regulator correspond only to the commands which can be activated by efferent links from its agents to the effectors (we recall that the procedure can have more commands). These objectives form the (possibly mixed) option associated to the procedure on the landscape. If they are realized and there are no other changes, the next landscape at the end of the step would be the complexification of the landscape L_t with respect to this option.

Definition. The *anticipated landscape* for the end $t+d$ of the step is the category AL_{t+d} obtained as the (possibly mixed) complexification of the landscape L_t at t with respect to the option formed by the objectives selected by E to be achieved by the commands of the procedure Pr that E can activate (Fig. 6.4).

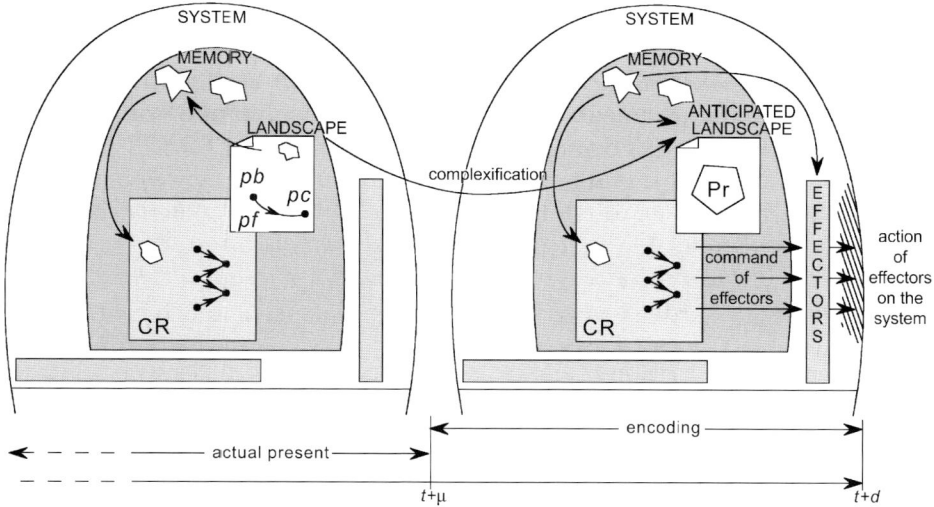

Fig. 6.4 Anticipated landscape.
The commands of the admissible procedure Pr selected by the co-regulator on its landscape at t are sent to effectors. The standard changes expected from the commands controlled by the co-regulator (via efferent links from its agents to effectors) form the objectives of an option on this landscape. If the objectives are achieved, the anticipated landscape at the beginning $t+d$ of the next step would be the complexification of the landscape at t with respect to this option.

The anticipated landscape models what would be the landscape at the beginning of the following step if the objectives of the option, *i.e.* of the commands of the procedure Pr controlled by E, were achieved. However, failure to do so may result from any of several problematic situations:

• The objectives have been selected on the landscape, which is only a more or less deformed representation of the system, while the procedure may be carried out on the system through all its commands (of which some might not be activated by agents of E), and it operates not on the landscape but on the system.
• Moreover, the other co-regulators also interact with the system by implementing their own procedures, which all enter in competition with the procedure Pr (as explained in Section 5).
• The various temporal constraints could not be respected.

All this explains why the objectives may not be achieved, and how the step in progress may even be abruptly interrupted by what we call a *fracture* for the co-regulator.

During the first phases of the following step (which starts at $t+d$), the co-regulator, acting as an organ of evaluation, will control on its newly formed landscape if the selected objectives have indeed been attained. In order to determine that, the anticipated landscape will be compared with the new landscape L_{t+d} at $t+d$. Formally:

Proposition. *There exists a (possibly partial) 'comparison functor' from the anticipated landscape* AL_{t+d} *to the new landscape* L_{t+d} *at* $t+d$. *It is an isomorphism if the objectives of the procedure are realized; otherwise it indicates the errors* (Fig. 6.5).

Indeed, the existence of this functor is a consequence of the universal property of the complexification (Complexification Theorem, Chapter 4).

The comparison functor is explicitly constructed from only the data accessible to the agents of E. It is an isomorphism if the objectives are achieved and if the situation has not changed. Otherwise it gives a 'measure' of the unexpected changes (*e.g.* emergence of unanticipated elements, loss or fusion of components), and so points to possible adjustments to

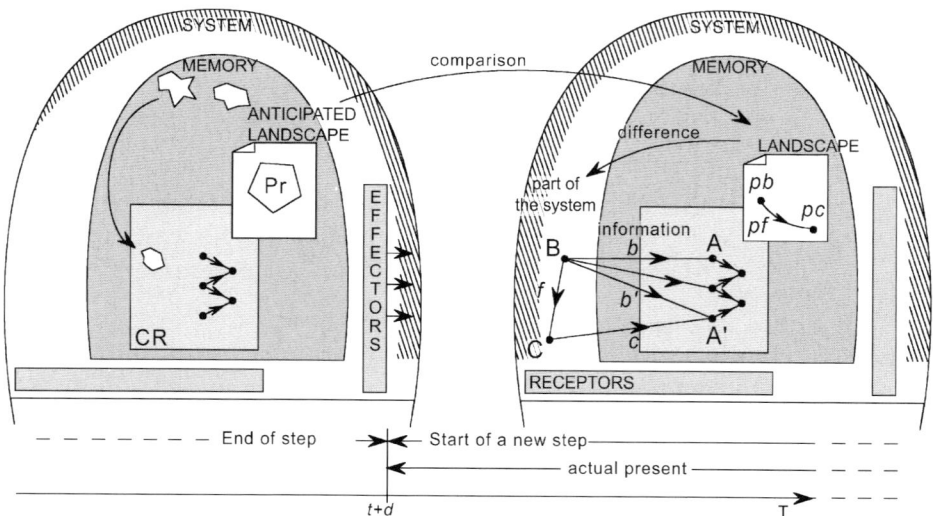

Fig. 6.5   Evaluation.

At the beginning of the next step the anticipated landscape is compared with the actual landscape at $t+d$. The partial functor 'comparison' from the anticipated landscape to this new landscape evaluates the extent to which the objectives of the selected procedure Pr have been realized. It is an isomorphism if the procedure has succeeded, otherwise it indicates what remains to be changed in order to realize the objectives. However, the comparison cannot be formed if the procedure has failed completely.

correct the errors. In this latter case, the comparison functor may be only partial, or perhaps cannot be formed at all. This will be the case if, for example, the binding of a pattern has not been achieved because of the destruction of some components of the pattern, or because it would require more time than allowed during the step. Let us note that the comparison is done at the beginning of the next step, for it requires that the next landscape be formed, so that it may be compared with the anticipated one (Fig. 6.6).

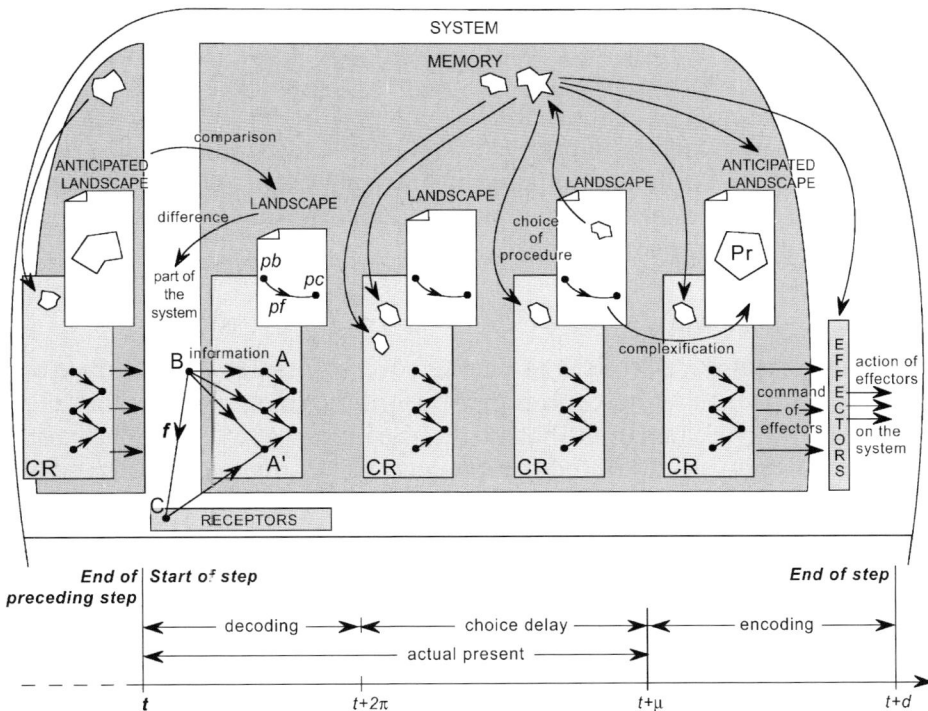

Fig. 6.6 Complete step of a co-regulator.
The successive phases of the step of the co-regulator from t to $t+d$ (Figs. 6.2–6.5) are depicted together. The anticipated landscape shown on the left is that formed during the preceding step, and it is compared with the landscape (at t) of the present step, during the first phase of this step. The anticipated landscape constructed at the end of the represented step will be compared with the landscape (at $t+d$) at the next step, not shown here. The figure indicates the delays required for: the decoding of the information (formation of the landscape at t with evaluation of the preceding step) and the choice of the procedure Pr, which together form the actual present of the co-regulator; and finally the encoding of its commands with formation of the next anticipated landscape.

The co-regulator has a *fracture* if the comparison functor cannot be formed for any of the following reasons:

- because the step has been interrupted for lack of an admissible procedure;
- as a result of the effect of an event external to the co-regulator (*e.g.* external perturbations or new commands sent by a higher co-regulator);
- or because of temporal constraints which we specify below.

The storage in the memory of the situation seen in the landscape, of the procedure Pr which had been selected, and of its result as evaluated by the comparison process, will be one of the objectives of E at the new step. For example, in a neural system with retrograde propagation (as defined by Rumelhart *et al.*, 1986), this functor would measure the difference between the desired outputs and the outputs that are obtained, and is used to improve the training of the system. More generally, the stepwise operation of a co-regulator can be interpreted as a *trial-and-error training process:* the trials correspond to the (imposed or internally controlled) selections of successive procedures and their implementation, and the error is evaluated in the comparison process.

4.4.   *Structural and Temporal Constraints of a Co-Regulator*

From the preceding description of one step of a co-regulator, we deduce that the completion of the step requires that the order of magnitude of the length d of the step be greater than that of the time lag π at t, and less than the minimum stability span of the components which intervene in the actual landscape and in the selected procedure (Fig. 6.7). If these temporal constraints fail to be satisfied, there is a fracture that might be repaired during the next step. If they fail to be satisfied during several successive steps, we speak of a *dyschrony* (see Chapter 7). However, there is no long-term problem as long as these constraints are satisfied except for isolated steps.

Formally, the structural and temporal constraints of a co-regulator relate its period to the propagation delays and stability spans, the period of E being defined as follows:

Definition. The *period* of the co-regulator E at an instant t is defined as the average duration $d(t)$ of the steps of E preceding t, the average being calculated over all the steps which occur during the preceding step of a higher level co-regulator controlling E.

In a cyclic phenomenon, the period is constant and has its usual meaning. In the general case, the period as defined here may vary, since it corresponds to the mean duration of the nearby preceding steps, and relates to the operation of the co-regulator: there is a selection of procedures at more or

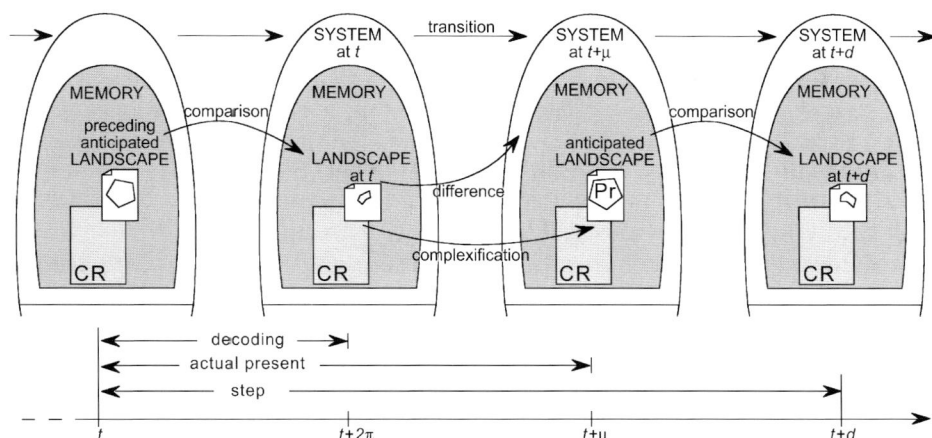

Fig. 6.7 Structural temporal constraints.
For the successive phases of a co-regulator's step to be achieved within its step duration d, their cumulative durations must be strictly less than d. Each phase requires several successive operations, whose durations all depend on the time lag π, so that its total duration must be much greater than π. This imposes the constraint $\pi << d$. On the other hand the components on which the operations are carried out must remain relatively stable during the step, whence a second constraint $d << z$, where z is the shortest stability span of these components.

less regular intervals, dependent on its time scale but possibly also on external disturbances. This 'periodicity' is thus a kind of additional time scale superimposed on the internal time scale of E, as for example the average actual duration of a daily wake/sleep cycle is superimposed on the underlying diurnal rhythms. This concept is not a simple artefact: models of neural systems show that, during a training process, certain units will specialize into internal clocks (Patarnello and Carnevali, 1989). Various authors go further and think that any autonomous natural system has a subjacent cyclic structure: for example in the theory of arithmetic relators (Moulin, 1986), it is the first fundamental principle of an approach to modelling natural systems (the five other principles would also be satisfied in our model); and it is one of 6 characteristics of complexity singled out by Chandler (1991, 1992).

Finally, from the above discussion, we deduce the following temporal constraints:

Theorem 1 (Structural Temporal Constraints of a Co-Regulator). *In order to achieve its current operations at a given time t, a co-regulator must respect the following structural temporal constraints (Fig. 6.7): the order of magnitude of the period $d(t)$ must be greater than that of its time lag $\pi(t)$*

and less than the minimum stability span z(t) of the components intervening in the actual landscape and the selected procedure:

$$\pi(t) << d(t) << z(t).$$

If these constraints fail to be satisfied during several successive steps, there is a dyschrony, which may lead to a later change of the period to bring it back into compliance with these temporal constraints.

In the applications to natural systems, there are also energy constraints, which are related to the strengths of the links. In some cases, the activation of a perspective in the landscape is accompanied with an increase in its strength (calculated as the mean of the strengths of the aspects which make it up), and brings energy to the agents. This energy is then expended in the choice of the procedure and the activation of its commands. For lower level co-regulators (*e.g.* at the molecular level), this mechanism is described by Schneider (1991), under the name of a molecular machine (he gives as an example the action of the protein EcoR1 on DNA which leads to a specific cut in the DNA).

5.   Global Dynamics of a Memory Evolutive System

Each co-regulator of the memory evolutive system operates independently to obtain the best result of the internal measurement process (in the terminology of Matsuno, 1989) that it makes on its landscape. However, the commands of the procedures selected on the landscapes of the different co-regulators are not carried out on the landscapes, but on the system, where they all enter into competition with one another, as well as with possible other independent commands, while not always being compatible. Indeed, all the co-regulators depend on the same total resources and interact directly or indirectly, although their landscapes give only more or less deformed representations of the other parts of the system (as measured by the difference functor).

5.1.   Conflict between Procedures

Let s be an instant of the time scale of the memory evolutive system **K**. As the time scales of the various co-regulators are different, a step of a particular co-regulator, say E, in progress at s can be in any of its phases. Let us suppose that it is in the third phase (the other phases having no effect external to E), command of the selected procedure, this being say Pr_E. Then the commands of Pr_E which can be controlled by E are transmitted to the system to activate the effectors.

The commands so transmitted by the different co-regulators are supposed to all be carried out. This would amount to the formation of a procedure Pr′ which integrates these different commands, and which may also integrate commands of the various Pr_E independent from the co-regulators which have selected them, and other commands coming directly from the system, or imposed by an external observer or by environmental constraints, possibly with the aim of forming new agents. Such a 'global' procedure would be well defined only if the commands coming from the different co-regulators were compatible, as in the case of co-regulators that are independent and operate strictly in parallel, using resources independently from each other. However, this is not the case in a complex natural system, where different co-regulators interact, while being more or less heterogeneous and competing for common resources (energy or information). Thus, conflicts may emerge between their objectives. Among the causes of incompatibility, let us mention:

• Conflict between two procedures: using the same component for two different functions which it cannot simultaneously fulfil (*e.g.* when several workshops need the same repairman at the same time).
• Poor coordination between two co-regulators: when the commands of a procedure chosen by one of them depend on the selection by the other of some specific procedure different from that co-regulator's own choice. For example, if there is forcing of a procedure on one of them by the other (directly or indirectly).
• Problems due to differing time scales, for example, if an accumulation of successive changes at a lower level makes it impossible to implement a procedure chosen by a higher co-regulator (see Chapter 7).

5.2. *Interplay Among the Procedures*

The conclusion is that a procedure Pr′ integrating all the commands sent by the various co-regulators might not exist, and in this case the procedure Pr° finally implemented and stored in the memory at the next step (called the *operative procedure*) will be the result of an equilibration process which we call the *interplay among the procedures*. It results from competition, interference, and mutual compensation between the different objectives of the co-regulators, and possibly objectives specific to the system and/or external influences as well. It is extremely sensitive to temporal constraints. The resulting configuration of the memory evolutive system **K** will be the complexification of the configuration category K_s at s with respect to the option formed by the objectives carried out by Pr°, and as a result higher order components and new properties may emerge.

The interplay is generally not centrally directed, but results from a dynamic modulation between the various procedures and the co-regulators, as they comply with the external and internal constraints (physical laws, energy requirements, temporal structural constraints, and so on), and can be compared to Darwinian selection. It takes into account the respective strengths of the co-regulators and of their procedures, in particular the perspectives activated in their landscapes and coming from the memory. Its flexibility is increased by the fact that complex commands of a procedure can be carried out through any one of their ramifications (if the system satisfies the multiplicity principle, that is generally the case when modelling natural systems), so that there are more possibilities for finding a compromise between conflicting commands.

However, in consequence, the objectives and procedures selected by one co-regulator, say E, will be discarded if they cannot fit in a coherent way with the others, or if the structural temporal constraints cannot be satisfied; in these cases there will be a fracture in the landscape of E. Even if there is no fracture, the result of the operative procedure Pr^o for E can be different from the anticipated one. This difference is evaluated at the next step of E, by comparing the anticipated landscape with the new landscape. Discrepancies can arise from three causes:

(i) Passage from the objectives selected on the landscape of E (from which the anticipated landscape is determined), to the objectives achieved on the system, since the procedure Pr_E selected by E operates on the system itself and not on the landscape of E, and it may have commands not observable by E.

(ii) Interplay among the procedures which may modify or even suppress some commands of Pr_E in the operative procedure Pr^o, the objectives of which are attained in the next configuration of the system.

(iii) Evaluation of the result in the next landscape, and not on the system where Pr^o is carried out.

Which commands sent by the co-regulators will be preserved in the operative procedure Pr^o, and which will be abandoned? In other terms, in the competition between the procedures, how may we determine which co-regulators will achieve their objectives, and which will undergo a fracture in their landscape, obliging them to modify their procedure? These questions do not have a general answer, for the result depends on the structure of the system and of its net of co-regulators, and on external factors. However, let us give some general ideas.

• If the co-regulators form a command hierarchy (as in an army), the longer term commands coming from the higher levels will have priority. There

remains though some flexibility, because complex commands to the effectors can be realized through their different ramifications, and there are delays between the sending of orders and their implementation.
• In parallel distributed systems, there is often a central 'executive' co-regulator, which imposes a priority order on the commands of the other co-regulators operating as parallel modules.
• In autonomous natural systems, such as biological or social systems, the procedures of the lower co-regulators will take over in the short run, but these co-regulators remain under the control of higher co-regulators with longer periods, which may impose procedures for the more or less long term, either to resolve a situation, or to avoid a fracture at their own level; nevertheless there will always remain a certain plasticity, therefore some unpredictability, coming from the plurality of the ramifications via which a complex object can be unfolded.

When the procedures and their results are weighted (as it is often the case in complex natural systems), the interplay among the procedures will take account of their respective weights, in a deterministic or a probabilistic way. The aim might be to optimize a global 'cost' function (energy, negentropy, a satisfaction function in biology or sociology, or utility in economics), to increase the number of beneficial opportunities (prey encounters in ecology), or to stabilize an equilibrium. Thus, this interplay could be modelled using methods related to various domains, in particular: game theory (*cf.* Rapoport, 1947, 1985, for the applications of game theory in biology) or meta-game theory (an operative procedure corresponds to a meta-procedure in the sense of Howard, 1971); dynamic programming (Bellman, 1961, and, in the categorical frame, Bastiani(-Ehresmann), 1967); near-optimization methods (parallel network approaches of Hopfield and Tank, 1985, or of Durbin and Willshaw, 1987); or optimization by simulated annealing (Kirkpatrick *et al.*, 1983). The latter originated in the study of the physics of glasses, and is now applied to a wide range of problems (*e.g.* neo-connectionist models of neural systems; see Hopfield, 1982; Changeux *et al.*, 1986; Toulouse *et al.*, 1986; and others). However, these models cannot take into account interactions between various levels.

6.   Some Biological Examples

The examples given in this section and in the following one illustrate some characteristics of the complex natural systems modelled by memory evolutive systems. These will be analysed more thoroughly in the next chapter, in particular with regard to how co-regulators with very heterogeneous time scales interact, via the interplay among their procedures.

6.1.   Regulation of a Cell

The biological organization of a cell depends on relatively autonomous regulatory modules, made up of several species of interacting molecules and having discrete functions. Examples are: ribosomes, which are the site of protein synthesis; the DNA replication system; a signal transduction system, such as found in bacterial chemotaxis or yeast reproduction; and so on. The operation of these modules depends on the specificity with which the chemical signal binds to receptors and starts a cascade of reactions between proteins. According to Hartwell *et al.* (1999), the integrative properties of the cell come from the interactions between its various regulatory modules, information circulating in two directions: macroscopic signals influence gene activation, which in turn changes the protein products.

The way in which this organization functions is well described by the dynamics of a memory evolutive system modelling the cell: the modules correspond to co-regulators having time scales varying from milliseconds to years, and their interactions to the interplay among their procedures. Let us give two examples.

6.2.   Gene Transcription in Prokaryotes

Using the operon model (Jacob *et al.*, 1963), consider a structure gene whose transcription is under the control of the promoter of the operon, and is initiated (positive regulation) by the binding of an activator (coded by the regulatory gene of the operon) to the operator site. Negative regulation, or the impeding of transcription, is accomplished by the binding of a repressor. The process includes several steps, such as the recruitment and the operation of a transcription module formed by several regulatory proteins, the most important being RNA-polymerase (RNA-p). There are two main segments of RNA-p, namely the sigma subunit that recognizes the promoter site, and the core enzyme that carries out the synthesis of the RNA.

The signal to initiate transcription is the binding of the sigma subunit of RNA-p to the promoter site at the start of the gene, which thus forms a 'closed complex' RNA–p/DNA. There is then a change of conformation of this complex into an 'open complex', by the unwinding of the double helix (via helicases) and the separation of the two strands, forming a 'bubble' of about 17 base pairs in length. The core enzyme begins the base-by-base transcription of one of the strands, and the sigma subunit dissociates the two strands. At each step a new RNA nucleotide is hydrogen-bound to the complementary DNA base and joined to the preceding nucleotide, and the transcription bubble moves down the DNA to synthesize the next base. The elongation of the chain is rapid. It ceases when the RNA-p arrives at the terminator sequence of the gene (this step may also involve a Rho enzyme).

To model this process, we consider two co-regulators: the promoter-CR, modelling the promoter of the gene, and a transient transcription-CR formed at this end. One step of the promoter-CR covers the number of steps needed by the transcription-CR to transcribe the successive bases. The information that the promoter-CR decodes in its landscape during one of its steps is the absence or presence of an activator protein on the operator, and in both cases there is a unique admissible procedure:

(i) in the first case, 'do nothing';
(ii) in the second case, formation of a transcription-CR, to which a transcription procedure is imposed by binding RNA-p and DNA. This procedure will extend until all the bases of the gene are transcribed (end of step).

The transcription is then carried out in several steps by the transcription-CR, which keeps its identity while gliding along the strand. In each of these steps the procedure will be: to unwind the double helix and separate its strands by cutting the bonds between them (decomposition of a colimit); to bind a complementary ribonucleotide to one base of the gene, and to the preceding one on the new RNA (formation of colimits); and to proceed to the following base. When the terminator sequence is reached, the transcription procedure is stopped, and one step of the promoter-CR is completed.

6.3.   Innate Immune System

The innate immune system of an animal is a part of the immune system, which has developed effective methods to respond quickly to micro-organisms. One of the main immediate responses is the detection of a pathogen by a certain pattern of receptors on the surface of the immune cells, the toll-like receptors (TLRs). For instance viral genomes with double-stranded RNA are detected by a pattern of TLR3 receptors, while the lipo-polysaccharides of a bacterium stimulate a pattern of TLR4. The activation of such a pattern is signalled to 'adaptor' proteins, the start of a signalling pathway (*cf.* Hoebe *et al.*, 2003).

In the memory evolutive system modelling an immune cell, the adaptor proteins will be represented as co-regulators. Let us study the case where a TLR4 pattern is activated. It is recognized by a co-regulator modelling the MyD88 protein, by the anchoring of the tail of TLR4 to the TIR domain of the protein. The procedure of the protein is to activate the transcription factor NF-kB, which starts the production of cytokines. It is also recognized by the co-regulator Trif (modelling the Trif protein), whose procedure is to activate the transcription factor IRF3 controlling the production of interferon beta. The latter in turn activates its receptor, which is recognized by the co-regulator (modelling) STAT1, which selects a procedure to activate various genes. Thus, the interplay among the procedures leads to a rapid response. If one of the co-regulators, say Trif, is missing (as in Trif-deficient mice), the response is impaired, causing a fracture for (viral infection of) the cell.

6.4.   *Behaviour of a Tissue*

Now, let us take a cellular tissue; for example, a section of skin. Its different cells, with all their internal organization and their links, form a memory evolutive system with various levels: molecular, sub-cellular, cellular. There is a whole network of heterogeneous co-regulators: beside the co-regulators modelling the cells which have a long period (corresponding to the control of genes), there are co-regulators with shorter periods corresponding to the organelles or the membrane, and so on down to the molecular level. The memory corresponds to programmed processes; *e.g.* in the event of a surface attack (for example, scraping), keratinization by synthesis of certain fibrous proteins, followed by associations between molecules.

Although there is no central co-regulator, the behaviour is coherent in the following way. Each co-regulator forms its actual landscape, in which the information coming from other parts of the system appears. Thus, adjacent cells, maintained together by adhesion molecules, communicate through desmosomes and/or squasmosomes (depending on the level), through which various molecules circulate and act as messengers. In this way, an external change will be communicated to the cells contained in increasingly deeper layers, more or less quickly. Each one of their co-regulators will react by a suitable choice of procedure, and the total response will be the result of the interplay (cooperation and/or competition) among these procedures.

Let us examine in more detail the case of an irritation of the skin caused by scraping. It will involve modifications of the surface cells of the external layer, of which some will be destroyed, while others respond to the attack by increased keratinization. The modification of these cells involves in its turn that of the lower layers, down to the basal layer, in which the speed of division and keratinization will be increased to compensate for the losses. These procedures of the cells do not result solely from the higher level co-regulators, but are obtained by the interplay among the procedures of the various levels. For example, the modifications of the membrane will quickly activate cytoplasmic messengers (at the molecular level), which will transmit an order to transcribe certain genes, which were before repressed or at least not very active. This transcription depends on co-regulators with long periods, and cannot abruptly stop once engaged. When scraping stops, the surface cells stop sending new transcription orders. Although this stoppage happens quickly, transcription still continues until the end of the cellular cycle, so that the new landscape will be prolonged more than necessary.

Thus, even in the case of a simple tissue formed by cells aggregated under the effect of adhesion molecules, global events occur, though there is no central executive co-regulator. This is thanks to the interconnections between the cells, as seen in their respective landscapes, which promote the interplay among their procedures. Moreover, the process depends heavily on temporal constraints.

7.   Examples at the Level of Societies and Ecosystems

Here, we give examples at higher levels, where individuals and social groups of various levels participate.

7.1.   Changes in an Ecosystem

An ecosystem is modelled by a memory evolutive system with several levels, among them the level of individuals, several levels of social groups, and the overall ecosystem level, each with its own time scale. A higher animal forms a co-regulator at the individual level. Its landscape is restricted to its close environment (or Umwelt; von Uexküll, 1956), and its procedures are selected to respond in a well-adapted way. It keeps its identity in spite of the deep transformations of the ecosystem (*e.g.* climate change, extinction of species), which occur too gradually to be observed on its time scale. A change of procedure of such an individual can be signalled to other individuals, and affect the higher co-regulator modelling their group, without any deep modification of the ecosystem.

For example (*cf.* Cyrulnik, 1983), amidst a group of Japanese macaques, an investigator deposited sweet potatoes in a sand heap. At first, the macaques ate them covered with sand, until one (a female) had the idea to wash them in the sea before eating them. At this stage, the change was only in her own landscape, with the introduction of this procedure (to form the colimit potatoes + water). Later an exchange of information took place, and so more and more macaques discovered that after washing, the potatoes had a better taste, and thereby recognized the value of this procedure, and adopted it. In this way, the innovation extended gradually to the whole group (cultural evolution). However, the dominant males refused to adopt it: they chose to maintain their prestige rather than to accept a change proposed by a dominated member, even though it was an improvement.

On the other hand, for a co-regulator at the group level, the time scale has much longer periods, and during its time lag, evolution at the individual time scale can be neglected as a first approximation. The evolution of the ecosystem results from the interplay among the procedures of the various co-regulators. For instance, the equilibration thus ensuing between different species of the ecosystem (modelled as co-regulators) brings us closer to co-evolutionary theories of the 'Red Queen' type (Van Valen, 1983; Kampis, 1991) than the neo-Darwinian point of view, for which evolution is entirely directed by natural selection acting as an external agent.

However, an unforeseeable, and for a long time unperceivable, modification of the ecological relations in an area can result from an evolutionary phenomenon secondary to the interplay among the procedures of the various co-regulators, and possibly contrary to their interests, thus causing fractures at their level later on. For example, a progressive extension of the Sahel has been the result of choices made long ago by local populations, for economic and social reasons. The selection of a procedure of intensive cultivation of groundnut, and of settling of the nomads, has caused unwanted, major, and ever greater modifications of the system, at a different scale and in domains very far removed from the initial motivations. These changes were recognized only much later by the political and agricultural decision centres (higher co-regulators) of the local populations in their landscape, and have obliged them to modify their procedures to try to fight the perverse effects of the initial choices.

7.2. *Organization of a Business*

We have defined an evolutive system associated to a business enterprise of a certain type (Chapter 5). In fact, it is a memory evolutive system: its co-regulators represent the control units of the various units, from simple workshops to design offices, or to commercial, technical, production and management units; and the memory models the knowledge necessary for correct functioning (different procedures, production strategies, supplies on hand, and so on), as well as the written reports of any nature, and the archives.

The period of a co-regulator corresponds to the average duration of its cycle: daily for the offices and workshops, weekly for higher level units, monthly for the departments, up to several years for design or management units. A step is divided into a phase of analysis and preparation, leading to the formation of the current landscape; a phase of design and decision, where a procedure is chosen; and a command phase, for the execution of the procedure followed by its evaluation. The relative importance of the various phases varies according to the nature of the services provided. The phases of analysis, design and decision are longer in the higher level executive co-regulators, where the procedures correspond to development strategies, while the production phases are more important in the workshops (see Fig. 6.8).

For example, in a production department, a step corresponds to a cycle of production; the preparation depends on a review of the orders and of the

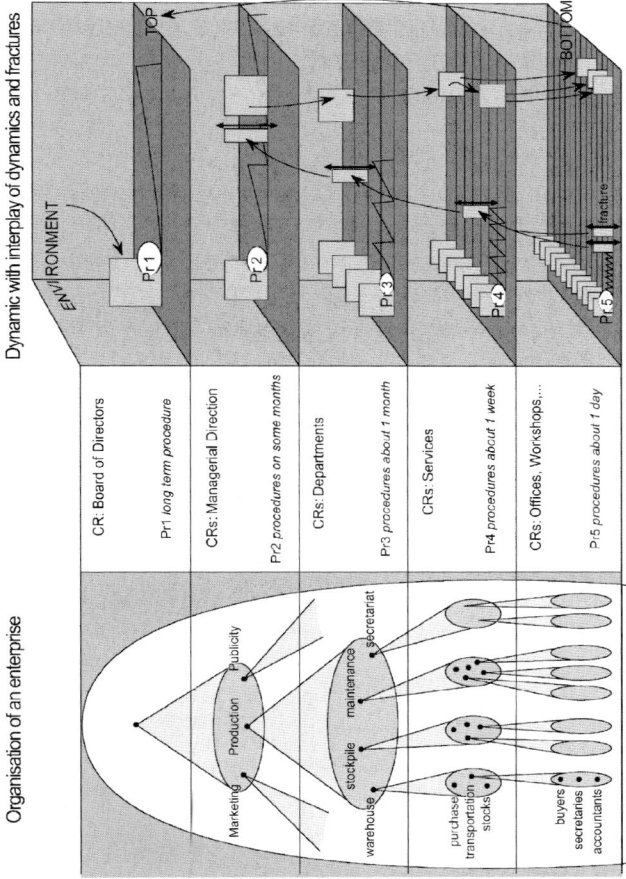

Fig. 6.8 A business enterprise.

On the left, a memory evolutive system modelling a business enterprise is depicted, with the hierarchy of its co-regulators of different levels. On the right, the dynamics in their various landscapes are depicted, and the differences in the durations of their steps are indicated by the relative sizes of the relevant rectangles. While the directorial procedure Pr_1 is a strategy which encompasses the long term and has long-term repercussions for all the levels, the duration of the various procedures decreases with level. The arrows indicate the inter-actions, which can go in any direction, possibly causing fractures for one or more co-regulators. A fracture (indicated by the symbol $\updownarrow$) can induce a change of procedure at a lower level, to repair it, but also a fracture at a higher level. Here the fracture caused at the workshop level induces a fracture at the higher levels much later, and then the managerial co-regulators may impose a complete change in the organization, from top to bottom.

stock; the design phase programs the work, by taking account of the experience gained, to make up for possible delays and to select new objectives. In the personnel department, the procedure will aim at the recruiting or the dismissal of employees; the purchasing department will order the purchase of supplies or of raw materials, and so on. At the higher executive level, the procedure may become a long-term strategy, aiming to reorganize the activities, by suppression or creation of certain units.

In any case, the procedure of a co-regulator is confronted with the procedures of other units, because the resources, human as well as material, are limited and must be distributed in the most effective way for the enterprise as a whole. Moreover, the implementation must take into account the temporal constraints, related to the propagation delays of information flow, and the time lags necessary to prepare the work, to acquire, manufacture or transport the products. A certain stability or consistency is required: the maintenance of sufficient supplies on hand, and of an adequate number of employees (*e.g.* if several units have recourse to the same maintenance or repair department). The consequence is that a procedure may be impeded for various reasons, and a unit may even be forced to halt the procedure because of a fracture in its landscape. It will then try to repair this fracture at the following step; if it does not succeed quickly enough, a re-organization will have to be considered at a higher level, such as the creation of a new unit, or a different distribution of the tasks between units.

The situation can be illustrated by the case of a workshop which produces nuts necessary for the manufacture of machines on an assembly line. A short delay in the workshop will be easily made up in the following days, before the users have exhausted their reserves of nuts. However, if the situation cannot be remedied soon enough, let us say as a result of a staff shortage, the fracture of the workshop will be reflected at the level of other units, with the progressive exhaustion of their stocks of nuts. These units will have to slow down or even temporarily stop their production. If the situation recurs, it can be resolved by the intervention of other co-regulators (personnel department, or management), which will hire more employees in the workshop, or decide to reorganize the factory by automation, with concurrent reduction of the workforce.

7.3. *Publication of a Journal*

The memory evolutive system associated with a journal has several co-regulators, modelling in particular the board of editors, the editor-in-chief, the referees, the publishers, the secretaries in charge of the relations with the external world (authors, readers, printer), and the subscription department.

Let us describe one step of the co-regulator modelling the board of editors, consisting in one of their meetings. In its first phase the editors will register the various papers submitted since their last meeting, the referees' reports on papers previously received, the letters sent by readers, and verify that the decisions (procedures) taken at their preceding meeting have been correctly carried out. In a second phase the board will select the papers to be published in the next issue of the journal, the referees for newly received papers, and the intended schedules. For this, they will recall preceding reactions of readers, the referees who have previously done a good job, and previous delays of the printer. The procedure can be literally chosen by the editors (intentional action), or imposed on them by other co-regulators (*e.g.* if the journal has a higher level management body), by external constraints (excessive cost for lengthy papers), or else be a known automatic answer to the given situation (always the same referee for papers of a certain type). Finally, the board sends the commands of the procedure to the secretaries, who will send the papers for the next issue to the printer, and the new ones to the chosen referees. The results are evaluated at the next meeting, when the editors will check whether the issue has been well printed and the expected reports of referees received; they will note if the printing delays have been respected and no complaints have arrived.

However, in some cases the meeting can be interrupted (fracture) if no procedure can be found (the editors cannot agree on which paper to publish), or if the selected procedure cannot be implemented (the printer refuses to continue printing the journal). Even if the meeting succeeds, there can be problems. The selected procedure has been relayed to the effectors of the journal, where conflicts may occur between procedures of various co-regulators. For instance the secretaries may not have enough time, the editorial policy may clash with the economic constraints of the publishers, whence an interplay among the procedures of the various co-regulators: the editors can argue with the publishers and ask that the journal be better financed.

The various structural temporal constraints play a major role: a paper or report not received by the date of the meeting cannot be examined, the printing of the issue can be delayed if the secretaries are overworked, or likewise if the printer does not keep to schedule. If these constraints cannot be satisfied, a fracture occurs in the landscape of the corresponding co-regulator, and if it is not quickly repaired, it leads to a long-term dyschrony, the regular publication schedule cannot be resumed for several issues. However, fractures can have a creative role, by requiring a complete review of the situation. If the printer cannot respect the delays, a new printer can be chosen, one who will perhaps do a better job.

Because the time scale of the editors is more limited than the longer term policies of the publisher, fractures can become manifest only with delays at this higher level. For example, if the editors progressively modify the contents of the journal, the publishers might perceive the change only later, but then it can displease them, and they may react by dismissing some editors. This may have other consequences for the readers, whence an example of the 'dialectics' between heterogeneous co-regulators, which we describe in the next chapter.

Chapter 7

Robustness, Plasticity and Aging

In the preceding chapter we defined the notion of a memory evolutive system, which provides a model for a natural autonomous system embedded in an environment to which it must adapt in order to survive. We showed how its dynamics are modulated by cooperative and/or conflicting interactions amongst a net of overlapping local regulatory modules, its co-regulators: each co-regulator operates with its own rhythm, but an equilibration process (the interplay among the procedures) intervenes to harmonize their procedures. Despite this, there remains a risk of local dysfunctions; that is, of fractures or synchronization errors (called dyschrony) for particular co-regulators. This equilibration process is characteristic of complex systems, and here we will study it in more detail. In particular, we will show that it leads to a 'dialectics' between heterogeneous co-regulators unable to communicate in real time. What are the consequences for the evolution of the system? How can the system preserve its homeostasis despite a more or less changing environment, and possibly develop new aptitudes, leading to better adaptation?

We examine these various problems and develop two applications. One analyses the replication and repair process in a cell; the other proposes a theory of aging for an organism, as a 'cascade of re-synchronizations' at increasing levels.

1. Fractures and Dyschrony

First we analyse the problems which may arise if the dynamics of the whole system, or of particular co-regulators, are disturbed (we speak of a *system dysfunction*)—for example, when the operative procedure, obtained after the interplay among the procedures, discards the procedure of some particular co-regulator.

1.1. Different Causes of Dysfunction

System dysfunction can result from a loss of components, of links or of colimits, or conversely from the presence of unrecognized elements. It can also result from a slowing down or an acceleration of certain processes. There are many possible causes:

(i) a perturbing external event (*e.g.* in an automobile traffic system, an obstacle in the roadway may create traffic congestion); the return to a normal situation will strongly depend on the specific time scales of the different co-regulators (such as the reaction times of the drivers at the resumption of the traffic);

(ii) conflicts between co-regulators that are not solved by the interplay among their procedures (competition between two co-regulators for a non-shareable resource in any system);

(iii) inadequate selection of procedures (out-of-stock condition in consequence of bad management in a business), or the impossibility for one or more co-regulators to find an adapted procedure, or to carry one out.

A moderate dysfunction may have no perceptible consequence, at least over the short term. Beyond a certain threshold of tolerance, or if there are several troubles which add up, a number of co-regulators will be affected, directly or indirectly.

Let us consider in more detail how a particular co-regulator may respond. If the trouble does not halt the step in progress, it will be perceived by the co-regulator only at the beginning of its next step, when it evaluates the situation, by comparing the anticipated landscape with the new actual landscape. A simple discrepancy can be fixed during this new step by the choice of an appropriate procedure (*e.g.* headache relieved by aspirin). However, there will be a fracture for the co-regulator if the disturbance is such that it halts the step in progress because the landscape cannot be formed, or if no procedure is found, or if the selected procedure cannot be carried out (a worker too ill to go to work).

1.2.   *Temporal Causes of Fractures*

The cause of a fracture for a co-regulator is often a conflict between the operations it has to accomplish, and its structural temporal constraints (Chapter 6). The latter relate its period d, its time lag (maximum propagation delay) π and the (minimum) stability spans z of the intervening components, these three factors often acting in synergy. Let us examine again the various phases of a step (see Fig. 7.1).

Phase 1: Formation of the Landscape.   If the propagation delays of the links arriving at the agents increase, the information they transmit to the landscape risks being outdated by the time it arrives, especially if the components from which they originate are unstable (thus if their stability span is small). The landscape will then be a less faithful representation of the system, and possibly will overlook some important data (letter arriving too late). On the other hand, if the propagation delays of the links between

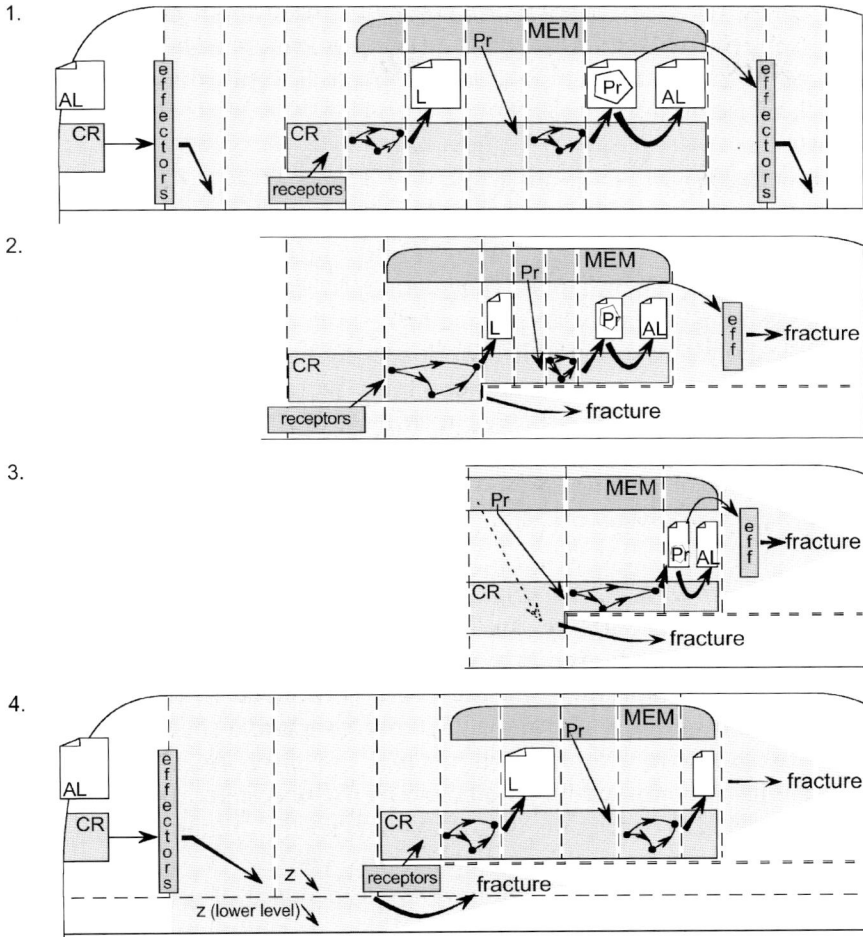

Fig. 7.1 Different causes of fractures.
Time is horizontal, from left to right.

1. Regular step for the co-regulator, showing the lengths of the different phases of a step: formation of the landscape L, selection of a procedure Pr on it, sending of its commands to effectors and formation of the anticipated landscape AL.

2. The propagation delays increase, the landscape is less extensive, and either no admissible procedure is found, so that there is a fracture before the end of the step, or the procedure is poorly adapted, causing a fracture at the end of the step.

3. Either the time lag is too long for a procedure to be found, whence an immediate fracture, or the procedure when found is no longer adapted, causing a fracture at the end of the step.

4. The fracture comes either immediately, or at the end of the step, as a result of the decrease in the stability spans *z*, which makes the information, both in the landscape and about the procedure, outdated at the time of its use.

agents increase, the perspectives are poorly formed, and if the length d of the step is fixed, the co-regulator might not even have enough time to form the landscape (*e.g.* the cellular cycle is stopped if replication has not been achieved in due time).

Phase 2: Selection of the Procedure.   If the propagation delays of the links between the memory and the agents increase, recourse to the memory to find an admissible procedure becomes difficult, perhaps impossible (*e.g.* poorly filed documents). The same is true if the communication between agents is too slow, in particular if some agents also belong to other co-regulators having more or less opposed interests (*e.g.* a group unable to reach a decision). Thus, when the time lag π increases but the duration d of the step remains fixed, a fracture may occur at this step because of an inability to find a procedure, or possibly later on, if the procedure is poorly adapted (*e.g.* a decision taken too hastily, without taking account of important data received too late).

Phase 3: Commands to the Effectors.   The commands of the procedure are carried out on the system and not on the landscape, and the interplay among the procedures can lead to their modification, or to the cancelling of some of them (as when an opposing order is received from a higher level). Even if the procedure is carried out, it may not have the expected effects, because of excessive propagation delays, or of changes in the internal organizations of components, and consequent shortening of their minimum stability spans z. Then there is a risk of fracture, either during the step in progress, or at the following step. This will be the case particularly if z decreases (*e.g.* delivery made impossible because of a disruption in the supply chain). Let us note that the reduction in z can come from a shortening of the propagation delays at a lower level.

1.3.   Dyschrony

A fracture can often be repaired at the following step, or in a small number of steps (*e.g.* a cold cured in a few days). However, an important fracture, the result of external disturbances, a poor apprehension of the total situation, or important conflicts between co-regulators, might not be repaired rapidly.

Definition. We speak of a *dyschrony* for a co-regulator if several of its successive steps are interrupted by a fracture.

A dyschrony risks being propagated to other parts of the system, and to be corrected it may require external intervention. This intervention can come from other co-regulators (*e.g.* in the event of an infection, the doctor who administers antibiotics). The correction can be externally imposed on the agents of the co-regulator, in particular in the case of a blockage that may or may not be known at the level of these agents, and it can appeal to known procedures (*e.g.* call the plumber to repair an outflow).

If the situation is unknown, however, more effective procedures will have to be found, leading to structural and/or temporal changes in the system (*e.g.* adoption of a new way of life for a diabetic). Nevertheless, the structural temporal constraints $\pi < < d < < z$ (Chapter 6) of the co-regulator must still be respected, these constraints imposing that its time lag be much lower than its period, in turn much lower than the minimum stability span of the corresponding components. If these constraints are at risk of being violated, it is the period d which is most likely to be modified in response: we speak of a *de-synchronization* for the co-regulator. For example, if z decreases, the period d can be decreased; or if π increases, d could be increased, on the condition that it remains lower than z (see Fig. 7.2).

A de-synchronization can be temporary, or enduring. In the latter case we will speak of a *re-synchronization*. As we will see, a re-synchronization of one co-regulator is likely to cause fractures for co-regulators of other levels, and, if they are not repaired quickly enough, to also cause a dyschrony at those levels, possibly necessitating re-synchronizations.

2. Dialectics between Heterogeneous Co-Regulators

One of the characteristics of a complex natural system that is taken into account in the memory evolutive systems is that its local regulatory modules (its co-regulators) operate with different rhythms and at different complexity levels, yet nevertheless must act coherently together. Thus, if a dysfunction occurs in one co-regulator, because of the global interdependence, it is likely to be propagated sooner or later to others. Here, we will study the functional loops of interactions (with fractures and repairs) between two co-regulators on widely different levels, and having widely different time scales. We will also study the effects on their landscapes, which result in very divergent descriptions of the system (*cf.* Vanbremeersch *et al.*, 1996).

2.1. Heterogeneous Co-Regulators

Consider an ecosystem. The pace of change at the population level is much slower than at the individual level, and small modifications to individual organisms do not affect the ecosystem immediately at its level. They may though cause a fracture later on, if their effects on the ecosystem are

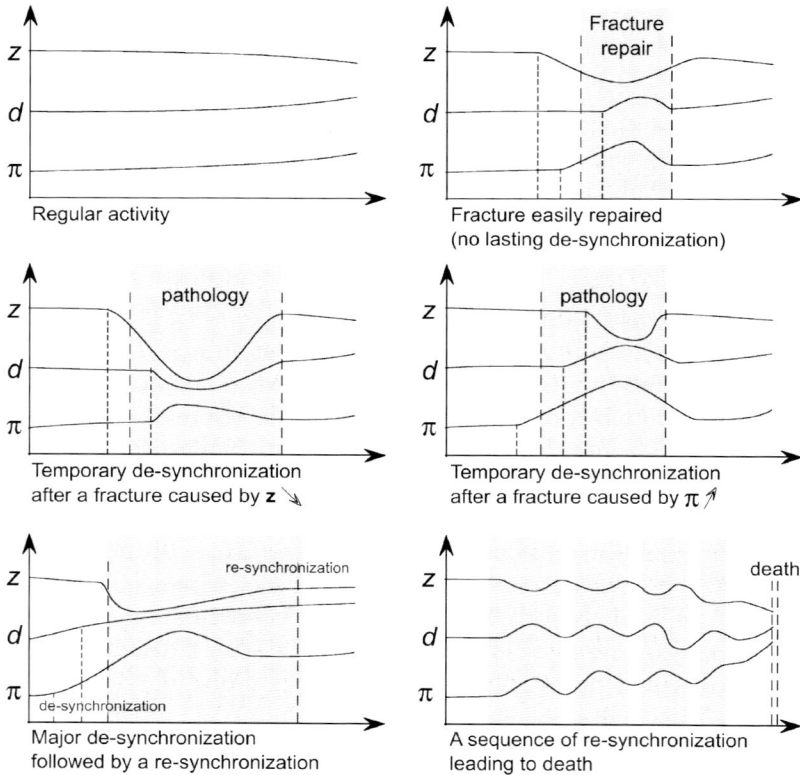

Fig. 7.2 Different types of dyschrony.

Comparison, for a given co-regulator, of the evolution over time of the time lag π, the period d, and the minimum stability span z of the components intervening in its landscape. Successively:

1. During a regular activity, the three curves remain well separated, with that of d in the middle.

2. If there is a fracture easily repaired, the period can be slightly and temporarily increased or decreased, while remaining in the middle; we speak of a de-synchronization. Once the fracture is repaired the regular regime is resumed.

3. If z shows a large decrease, d will be decreased for a longer time, but it must remain greater than π.

4. Alternatively, if π increases greatly, d will be increased, as long as it remains less than z.

5. In a major de-synchronization, there will be a lasting change in d (re-synchronization).

6. Finally, an un-interrupted sequence of re-synchronizations with π progressively increasing and/or z decreasing will lead to death when no conclusive re-synchronization is possible.

cumulative. This situation is general for heterogeneous co-regulators, which we defined as follows.

Definition. Two co-regulators are said to be *heterogeneous* if their complexity levels are different, and the period of the lower level one is of an order of magnitude less than the period of the higher level one.

Given two heterogeneous co-regulators, we associate to the co-regulator of the higher level the prefix 'macro', calling it the macro-CR, and we associate to the other one the prefix 'micro'. For instance the macro-CR could be a cell, and the micro-CR could be at the molecular level. During one step of the macro-CR, let us say from T_1 to T_2, there will be a large number of micro-steps (from t_1 to t_2, from t_2 to t_3, and so on). The changes that they produce are not observable in real time in the macro-landscape, because of the propagation delays. However, their accumulated effects may finally modify the macro-landscape, and so, for example, change the internal organization of some of its components (Fig. 7.3).

Indeed, let B be a component which has a perspective in the macro-landscape L_1 at T_1. If B has a part of its internal organization at the micro-level, the micro-changes may progressively modify it by destroying some of its components, or binding others in new ways. As a result, B may lose its complex identity, and so disappear from the macro-landscape. This loss can take place before the end T_2 of the macro-step, in which case it will become known by the macro-agents when they perform their evaluation at the beginning of their next macro-step. On the other hand, the loss can occur only during this following step. In this case, B will still appear in the new landscape L_2, and the next procedure Pr_2 of the macro-agents may still include B. However, the disappearance of B will prevent Pr_2 from being implemented, and this step will be stopped by a fracture. To repair it, the macro-CR will choose a procedure Pr_3 which can retroact (more or less quickly) on the micro-CR, possibly imposing a fracture at its level.

Thus, the micro-level, though not directly observable by the macro-CR, can, after a small delay, influence the latter's behaviour, by its successive actions on the internal organizations of the different components of the macro-landscape. And conversely, the need for stability by the macro-CR may also cause fractures for the micro-CR. For example, the macro-CR's procedure may require the disappearance of an intermediate-level component, one used by the procedure of the micro-agents. Or, the macro-CR may even impose a direct modification at the micro-level.

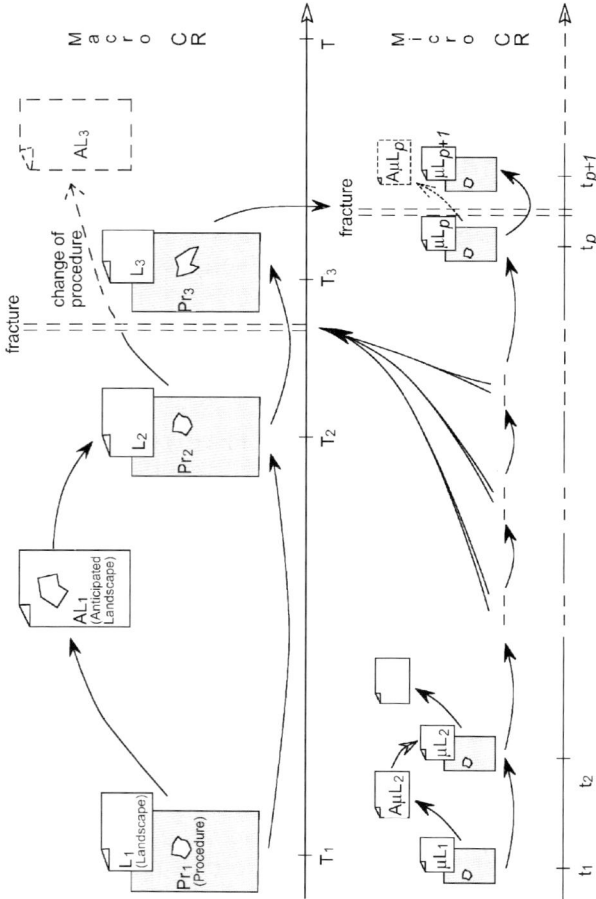

Fig. 7.3 Interactions between two heterogeneous co-regulators.

We have two co-regulators with different complexity levels; the higher level one, or macro-CR, has much longer steps than the lower level one, or micro-CR. During one macro-step from T_1 to T_2, there are many successive micro-steps, from t_1 to t_2 to t_3, and so on, during which the micro-CR forms its successive landscapes μL_1, μL_2, … Each micro-step may modify elements intervening in the macro-landscape L_1, without the macro-CR being able to observe them in real time, because of the propagation delays between the two levels. But the accumulation of changes will eventually be observable in the macro-landscape and may cause a fracture, either at the end of the macro-step (when the anticipated macro-landscape AL_1 is compared with the new landscape), or possibly only in the macro-landscape L_2 formed at the next step (from T_2 to T_3), forcing the macro-CR to begin a new step with formation of a macro-landscape L_3 and selection of a procedure Pr_3 on it. This new procedure may sooner or later react on the micro-CR, by imposing a new procedure on its agents which causes a fracture in its micro-landscape μL_p.

2.2. Several Cases

The result is that various situations may occur during a macro-step:

• *Synergy between the two co-regulators*: The successive micro-changes are coherent with the procedure selected by the macro-CR, and even help its realization (*e.g.* the procedure of DNA replication, selected at the cell level and carried out at the molecular level).

• *Fracture at the micro-level without immediate consequences for the macro-CR:* The procedure of the macro-CR, or events independent from the two co-regulators, may impose on the micro-agents a procedure that they cannot carry out. This generates a fracture at the micro-level; to repair it, an intervention external to the micro-level might be necessary. This intervention may come from another part of the system, or from the environment, which will play a stabilizing role. Though not necessarily related to the macro-CR, the intervention may resolve the situation, but at the price of a fracture at the micro-level, which may in turn cause a macro-level fracture later on.

• *Fracture to the micro-level that reverberates to the macro-level* (*e.g.* important damage to the DNA changes the metabolic activities of the cell): As before, there is a fracture in the micro-landscape, but it is observable by the macro-CR before the end of its step, and impedes the carrying out of its procedure. Then there is a macro-fracture, which may necessitate an external intervention, if the macro-agents have no other admissible procedure.

• *Fracture at the macro-level caused by a sequence of micro-procedures:* Successive micro-steps are completed, but they accumulate micro-changes that conflict with the objectives of the macro-procedure, and so a fracture for the macro-CR results, as explained in Section 2.1 above. The next procedure adopted to repair this fracture may react on the micro-CR by imposing on it a different procedure, in agreement with the new macro-objectives.

• *Fracture at the macro-level independent from the micro-level:* Changes external to the two co-regulators (whether in the system, or in its environment, which is then destabilizing) may cause a fracture for the macro-CR, with more or less important consequences. The next procedure adopted to repair the situation may later impose a change of procedure on the micro-CR.

In each of these cases, the fractures can result from a lack of information by some co-regulators, and can be easily overcome. They may also point to a real failing in the system; then the repair may require either a structural modification, such as an alteration of the time lags, or of the stability spans; or possibly a temporal adjustment by re-synchronization (change of the period of one or of both co-regulators). And the same situation repeats between co-regulators of different levels (Fig. 7.4).

2.3.   *Dialectics between Co-Regulators*

The preceding analysis shows that the dynamics of a memory evolutive system set up what we call a *dialectics between heterogeneous co-regulators,* which modulates the evolution of the system. If the levels and periods of two co-regulators are very different, successive micro-steps will not be distinguished and only their all-inclusive result will have effects at the higher level. Consequently, fast micro-changes will be reflected on the macro-landscape only later on, once they have accumulated. There they may cause a fracture, and to repair it the macro-CR may impose new procedures on the micro-CR, which may trigger a new cycle of fractures and corrective procedures. This can be represented in the form of functional loops of interactions and feedback, each possibly accompanied by a fracture:

$$\text{micro-level} \rightarrow \text{macro-level} \rightarrow \text{micro-level}$$

In the example of a journal (Chapter 6), if the editors progressively modify the contents of the journal, the publishers will perceive the change only after some delay. At that point they may be displeased, perhaps enough to dismiss some of the editors.

Since the effects of the fractures and corrective procedures cross levels, such cycles of fracture and repair may propagate, more or less quickly, to arbitrary levels, and thus force structural or temporal changes. In particular a re-synchronization for a co-regulator is likely to trigger fractures to co-regulators of other levels, and possibly even a dyschrony, which will also have to be repaired by structural or temporal modifications. In the event of major disturbances, this can involve a *cascade of re-synchronizations* affecting higher and higher levels.

In a developmental process, such changes are creative. The re-synchronizations will reduce the periods of certain co-regulators, thanks to their direct access to increasingly complex components, and to more elaborate procedures in the memory, which allow faster and more effective reactions to known situations (*e.g.* once learnt, a movement is executed more quickly). Conversely, during degeneracy or aging, an accumulation of damaging external events not repaired in due time can increase the time lags at the lower levels; then, the periods of higher level co-regulators will be lengthened, in particular to allow for the completion of the lengthier lower levels operations intervening in their procedures. This process stops when too many errors have accumulated. We will elaborate on this in Section 6, when we develop a theory of aging of an organism.

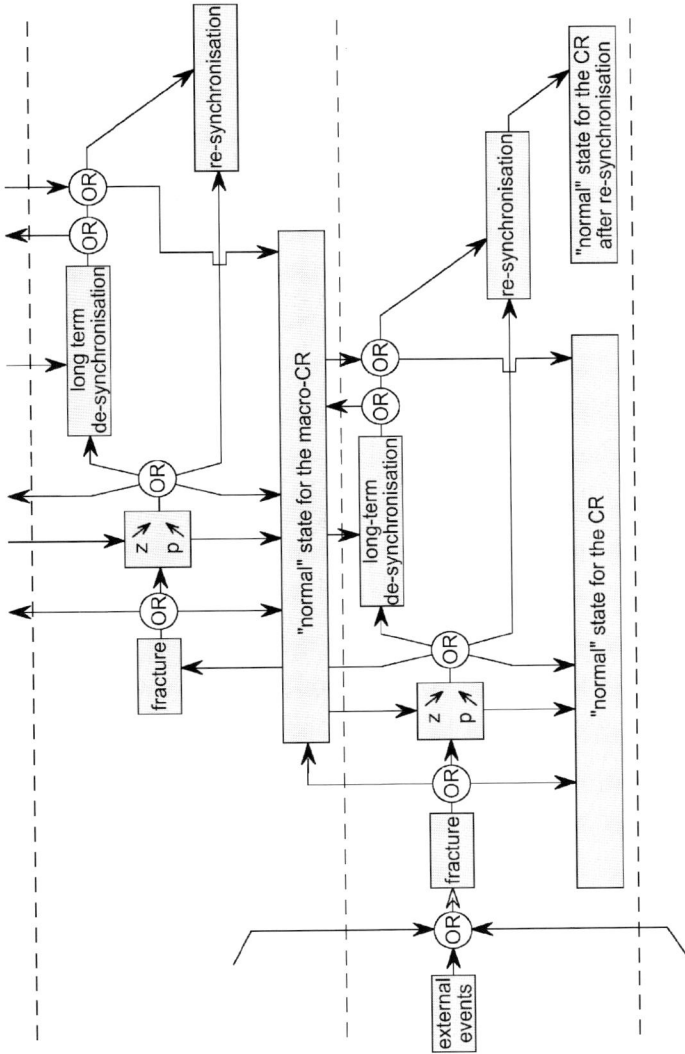

Fig. 7.4 Fractures and their repairs.
Flowchart of the many possible outcomes from fractures in co-regulators interacting between and across levels. Pattern repeats to higher and lower levels and along levels. The possible consequences within a single co-regulator on one level (the lower one on the figure) are: immediate repair allowing return to the normal configuration; structural changes of the time lags π and of the stability spans z without affecting the period; long-term de-synchronization (change of the period), followed by a return to the normal period or by a re-synchronization with another period. These events can be reflected to a higher level co-regulator, the macro-CR, initiating the same alternatives at its level; they can also be reflected to lower levels.

3. Comparison with Simple Systems

The dialectics between heterogeneous co-regulators has consequences for the entire system. It is as if the needs of the system, imposed on its components of different levels, had a boomerang effect on its various co-regulators, with more or less unforeseeable effects in the long run. This explains why long-term prediction is impossible for a complex system (for instance, in meteorology). In practice, although the changes of higher levels occur more slowly, they are more critical for the system, because they have greater consequences for the other levels. Fractures can have a destructive role when they are propagated to higher and higher levels; but they can also have a creative role, if they force a complete re-examination of the situation, and a search for original procedures, in turn possibly leading to the emergence of more complex objects and adaptive emergent complex phenomena. For example, in an animal, extensive development of the memory allows for the emergence of higher order cognitive processes.

3.1. *Classical Analytic Models*

This behaviour is very different from that of the systems described by the classical analytic models. As indicated in Chapter 5, these models rely on observables defined on a fixed state space (position, momentum, concentration of certain products, statistics on populations, and so on), whose values are the solutions of systems of differential or partial differential equations, and occur as trajectories of a *dynamical system* (among many authors, *cf.* Poincaré, 1951; Smale, 1967; Prigogine et Glansdorff, 1971; Thom, 1974; Arnol'd, 1984; Winfree, 2001). Examples are most varied, from the Schrödinger equation in quantum physics, to the Maxwell's equations, to the Lorenz equations (Lorenz, 1963) for meteorological systems (which are the origin of the present interest in the field). In biological systems, such models are mostly used to describe specific process, at a particular level, and even in this case they are not so easy to handle. For example, Novak and Tyson (1993) have given a mathematical model of M-phase control in eukaryotic cells; this control, which relies on two positive feedback loops and one negative feedback loop, necessitates ten differential equations.

   In multi-level systems, the observables of interest and the laws change according to the level (Farre, 1994, speaks of a 'cartesian cut' at the interface between levels). Thus, the classical models are not applied to the whole system, but only at a particular level (*e.g.* molecular level in chemical kinetics models used in biology; individual level or population level in ecology; sub-symbolic level in neural systems, following Hopfield, 1982). There are some dynamical systems models in population genetics (*e.g.* Akin, 1979), which describe an interplay of processes occurring at both the genetic and organismal levels, but in general it is difficult to study the interactions between heterogeneous levels. This is the reason why there are not many mathematical multi-level models, though the importance of hierarchical systems has been stressed, and such systems described on a more informal basis in various domains, from philosophy to science and engineering. Among the authors who have tried to adapt the traditional models to study hierarchies, let us cite Auger (1989), where differences in description according to level, and couplings between levels, are taken into account by the use of observables related to the number of individuals and populations of the various levels. Although this approach has quantitative applications in ecology, biology (connections between the molecular and cellular levels) and economics (connections between the regional and national levels), even in these models, the framework is too restricted to support the dialectics between heterogeneous levels that we have seen in memory evolutive systems, and to develop its consequences, such as the interferences between the structural temporal constraints at different levels.

3.2. Comparison of Time Scales

As said above, one of the differences between the complex systems we study and those considered by classical analytic models is the existence of the dialectics between two heterogeneous co-regulators. Over one of its steps, a co-regulator acts as a simple system. If the time lag of the macro-CR is much greater than the period of the micro-CR, then in the macro-landscape, the micro-level time scale might be confused with a continuous one. Thus, the observables for the macro-CR could be taken as continuous functions describing some global behaviour on the micro-level. For example, if the micro-level is that of individuals (molecules, cells, animals, and so on) and the macro-level that of their populations, the observables could be population statistics related to the individuals (some examples include chemical reaction rates, protein concentrations, population kinetics). In this way, the macro-representation evens out the discontinuous evolution of the micro-level.

However, the micro-level discontinuity may later feed back into the system by causing a fracture at the macro-level, once too many micro-changes have accumulated for the continuous approximation to remain valid. This fracture, which forces a change of procedure on the macro-CR, is, in terms of the macro-observables, a change of parameters (and not just a change in the initial conditions, as stated by Atlan, 1975). Everything proceeds as if the micro-changes, relayed with a delay to the macro-level, are inputs to a continuous orbitally stable state (in the sense of Teodorova, 1985), with a fracture representing the transition between two orbitally stable states. For example, the growth of a child (taken as a macro-step) breaks up into many micro-steps, separated by fractures, which are evident as distinct stages on the plot of body mass versus age (see Thouvenot, 1985, where the role of the various stages is discussed).

The difference between levels is well illustrated in evolutionary sociology, which underlines the difference between the individual level, where the changes are more or less chaotic, and the societal level, where the curve of the socio-economic history (in the terms of Ribeill, 1974) is an average curve, possibly broken off by discontinuities. The fractures of the macro-evolution represent major changes due to a disturbance in an insufficiently adaptable society (as is often true of traditional societies). On the other hand, a less rigid society manages to maintain an overall balance in spite of crises, thanks to an adapted choice of procedures.

Various conceptions of history depend on the level the historian chooses to describe:

• At the individual level, individual facts will be emphasized, as in biographies.
• A scientific history (such as that professed by Marxism; or that elaborated by the Annales school with M. Bloch, Braudel (1969), Le Goff (1988), and others; or cliometrics) studies the societal level and tries to find 'laws' by using a quantitative approach, the observables being for instance statistics on populations.
• On the other hand, the dialectics between levels is illustrated by the 'existential history' recommended by Ariès (1986), who shows that the historian, in his own landscape, may consider as identical situations or procedures which for the protagonists are distinct—whence the difficulty of finding universal laws in sociology. For example, a given reform can be accepted by one society and rejected by another, depending on their state.

3.3. *Mechanisms vs. Organisms*

As discussed in Chapter 4, Rosen has proposed distinguishing simple systems, which he also calls 'mechanisms', and which are suitable for classical models, from autonomous hierarchical anticipatory systems, such as biological or social systems, which he refers to as complex systems, or 'organisms'. He argues that the distinction relies in particular on their causal behaviour: in mechanisms, Aristotelian material causation can be split off from efficient or formal causation, and final causation is rejected; in complex systems, the causal categories are mingled, and some anticipation is possible. And he suggests that a complex system could be approached 'locally and temporarily' by simple systems, the approximation being changed 'when the discrepancy becomes intolerable' (Rosen, 1985a, p. 193).

Using the causality attributions with respect to complexifications given in Chapter 4 (Section 7), we are going to show that the mathematical model of complex systems given by a memory evolutive system, fulfils this program. As causal attributions depend on the point of view from which they are considered (Mackie, 1974), we'll have to examine the situation both on the level of a co-regulator and on that of the whole system, and for various lengths of time.

First, at the level of a particular co-regulator, its landscape must be analysed, since the agents operate through this referent, not only as agents, but as observers as well. During one of its regular steps (*i.e.* without fracture), and once the objectives are selected, the behaviour on the landscape has for its material cause the initial state, for formal cause the chosen procedure, and for efficient cause, the action of the commands of this procedure on the effectors. Thus, during a single step, the co-regulator operates as a simple system, and its dynamics can be described by the analytic models. In particular the formation of a colimit can be interpreted as the convergence to an attractor of the dynamics. In fact, the categorical description of the step and its analytic model through observables are complementary, the first one emphasizing how the dynamics are regulated (selection of a procedure) and their results (evaluation at the next step); while the second describes how the observables vary during the intermediate period, where these selected dynamics unfold.

If we consider the situation during several successive steps of the co-regulator, the dynamics of the landscape evolve by a sequence of complexifications. If we suppose that the multiplicity principle holds, we have shown that such a sequence leads to the emergence of increasing complexity, and cannot be replaced by a unique complexification of the first landscape (Chapter 4, Theorem 3); and further that the material, formal and efficient causes have to be updated at each step, so that they are completely intermingled in the global transition from the first to the last landscape (Chapter 4, Section 7). Thus, the long-term behaviour of the landscape, even restricted to its own level, cannot be identified with that of a simple system. In any case, the difference functor from the landscape to the system shows that the landscape is only an 'approximation' of the complete system, valid 'locally' (at the level of the co-regulator) and 'temporarily' (during one of its steps), and it becomes more and more unreliable over time. In particular, if a fracture occurs, it introduces a singularity (either a *bifurcation* or *chaos*, or a *catastrophe* in the sense of Thom, 1974), and even the observables may have to be modified.

For the whole system, the point of view will be that of an external observer taking in the global evolution, without acting on it. At a given moment, the current operative procedure is the result of the interplay among the procedures of the various co-regulators (which might be at different stages of execution). The result of this equilibration process cannot be ascribed to specific causes, but depends on the structure of the network of co-regulators, and on the various constraints in effect at the time. It can be structurally imposed (necessity), controlled by higher level co-regulators or by external agents, dependent on specific rules or optimization processes, or purely random (hazard for Monod, 1970, or noise for Atlan, 1979).

The causes of the long-term evolution of the system as a whole are even more difficult to disentangle than for one co-regulator. Indeed, the accumulated changes, in particular the emergence of complex processes, come from a succession of complexifications, in which the selection of the options at each step depends on a more or less indecipherable interplay among the various procedures. As explained above, there is emergence of increasing complexity, and the material, formal and efficient causes are totally intermingled.

On the other hand, the choice of procedures in response to external constraints, and the interplay among them, are internal processes, partially directed by the memory, which allows for some degree of anticipation by comparison with former experiences. In this sense, a memory evolutive system is an anticipatory system in which the causal interactions between all the levels are continuously merged into the dynamic flow.

In summary, a memory evolutive system can properly be qualified as an organism in the terminology of Rosen, with local and temporal anticipatory behaviour relying on the memory, with emergence, robustness and plasticity coming from the multiplicity principle.

4. Some Philosophical Remarks

Above, we have shown how the behaviour of a memory evolutive system depends on the combined effect of the successive choices of procedures by the different co-regulators, and their interplay. This may lead to the emergence of increasingly complex components and links. We have also shown that the material, formal and efficient causes are intermingled, but we have not spoken of final cause. Since the memory allows for some anticipation, can we speak of 'final' causes in the choices?

4.1. On the Problem of Final Cause

Let us recall that in philosophy, holistic explanation, starting from the whole, is opposed to atomistic explanation starting from the parts. The first, the top-down explanations, are often framed in terms of goals and intentions, while the second, bottom-up explanations, are usually stated in terms of matter, causes, and effects.

Aristotle granted final causation, or teleology, to organisms and to nature, which he considered as an organism; this was a view disputed by the Epicureans. For Kant, it is necessary to distinguish external causality: 'By external purposiveness, I mean that by which one thing of nature serves another as means to a purpose' (Kant, 1790, Section 8.2), from inner causality, valid only for organized beings 'one in which every part is reciprocally purpose, [end] and means' (Kant, 1790, Section 6). Thus, he thinks that 'teleological judgement' has an objective existence only if it is related to an organism or a work of art, for they are the only cases where the parts are understood only in view of the whole. Science sought to remove teleological language from its discourse. For Renan (1876), the problem of the final cause escapes science and is only solved in poetry.

As physical systems are subject only to material and efficient causes, possibly formal ones as well, but not to final causes, Rosen (1985a) deduces that they are less complex than organisms, for which it is also necessary to take account of final causes, in particular through their anticipatory behaviour. For self-organized systems, cybernetics introduced the concept of *teleonomy,* 'which retains the idea of finality while replacing the concept of "final cause" by a causality with understandable loops' (translated from Piaget, 1967, p. 189). Monod (1970) speaks about teleonomy in terms of a project that objects at the same time represent in their structures and achieve by their performances.

4.2.   *More on Causality in Memory Evolutive Systems*

What can we say for a memory evolutive system (*cf.* Ehresmann and Vanbremeersch, 1994)? Though we have adopted a vocabulary (action, selection of procedure, of objectives and so on), which might appear as teleological, it does not mean we accept final causation at all the levels. Indeed, we have emphasized several times that these 'actions' were not supposed to be intentional.

For a co-regulator, a final cause could only come up in the selection of an admissible procedure. If this choice is imposed on the agents by another co-regulator, or by structural or temporal constraints of a general nature not directly observable by them (this is the case at least for lower level co-regulators), there is no question of a final cause at their level. On the other hand if the selection is made by the original co-regulator, by recourse to the memory to anticipate the results of an admissible procedure (as will be explained in Chapter 8), this anticipation can play the role of a final cause, even if it remains implicit for the co-regulator. There might be a final cause

in the literal sense only for higher level co-regulators which can act intentionally, and which in some way can impose their procedure on the system
(as is the case for the intentional co-regulators of a cognitive system, to be
considered in Chapter 10).

For the system as a whole, a final cause could intervene in the selection of
the operative procedure, through the interplay among the procedures, but
we have seen that there is no general rule covering that situation. However,
the fact that a compromise is reached, allowing the system to function
(possibly with fractures for some co-regulators), allows us to speak about
inner causality, in the sense of Kant. Over the long term, we have seen that
the multiplicity principle is at the root of emergence of complex objects,
non-reducible to lower levels, constructed through the unfolding of a ramification from the lowest level up (Chapters 3 and 4), and which later take
their own identity; we have also discussed the role of the memory allowing
for anticipation in memory evolutive systems. The presence of such qualities
in a self-organized system could be sufficient to qualify it as a teleonomic
system in the sense of Monod (1970). To speak of teleology, the system
should develop consciousness (*cf.* Chapter 10), to select objectives on the
long-term.

4.3.   Role of Time

The preceding discussion shows that memory evolutive systems offer a
relational model, incorporating time, allowing for 'function' (at the level of
particular co-regulators), and in which a complex causation and even a kind
of final causation (via the memory) is present. Now Rosen (1986) has said
that there cannot be such a model: is there a contradiction? No, because for
us time is not just a physical parameter (as it mostly is for Rosen), but
contributes to a complex multifold dynamical process: each co-regulator
operates on its own time scale; the interplay among the procedures depends
on the differences and constraints introduced by these time scales; and
finally the memory, which in some sense subsumes the past and the present,
allows for some—more or less exact—anticipation of the future, so that it
may influence the present state. Thus, consideration of the memory evolutive
systems suggests that there is no need for a new science to study life, but only
for a thorough reflection on the nature of time and organization.

In a memory evolutive system we take simultaneously into account
the diversity of the co-regulators, with possibly conflicting objectives, and the
unity of the system. It is not a synthesis between different approaches, since
it allows for their co-existence and their temporally modulated interactions.
It could be compared to the discrepancy which Hegel sees between the
diversity of the 'limited opposites' of the beings in conflict with one another,

and their unification in the 'Being', the conflict supposing itself a preliminary unification (Hegel, 1807). More specifically, the dialectics between co-regulators represents a kind of 'conflicting complement' between descriptions relating to very different levels, of the same type as the duality between waves and particles in quantum physics: each one is valid on its own level, but together the two are more or less antagonistic (this is not to be confused with the ago-antagonist couples of Bernard-Weil, 1988, where the members of a couple are on the same level).

This dialectics could allow for a new approach to many previously insoluble paradoxes and philosophical problems that arise in the search for a unified description of such heterogeneous landscapes; in particular the problem of the irreversibility of time: is the time-arrow an artefact? Even if we could reverse time at a micro-level, an accumulation of micro-changes generated toward the past would be transmitted to a macro-CR only with some latency, and we cannot see how there could be a simultaneous reversal for two heterogeneous co-regulators. This is even more the case at the system level, since a complex system has a whole net of co-regulators, depending on very different time scales, which interact through the interplay among their procedures, imposing still more constraints (in particular through the switches between ramifications of complex objects), and making a global reversal unlikely.

5. Replication with Repair of DNA

A cell has developed a multitude of adaptive mechanisms for homeostasis, and to maintain its genetic material, despite a variety of disturbances. This section is devoted to a model of the *replication with repair* process in the memory evolutive system modelling a cell (*cf.* Chapter 6).

5.1. Biological Background

Let us first recall the specific mechanisms for DNA replication and repair available to a bacterium, such as *Escherichia coli* (*e.g.* Pritchard, 1978; Friedberg, 1985). The two strands of DNA are replicated separately and simultaneously, base after base. Initiation of DNA synthesis is dependent upon a variety of cellular and environmental factors, such as the rate of RNA synthesis and the energy resources. It seems to involve control via regulatory proteins, be they 'initiator' proteins associated with the membrane (replicon model, Jacob *et al.*, 1963), or effector/repressor proteins signalling a critical cell mass (Pritchard, 1978). The replication is directed by DNA polymerase (DNA-p), which binds to the base to be replicated, while helicases unwind the helix, and single-stranded-DNA binding (SSB) protein binds to the following bases, to separate the two strands.

Replication requires a strict pairing of the bases, but mismatches can be made, and DNA bases are subject to damage from several sources (thermal fluctuations, reactive metabolites, ultraviolet light, and so on). Various efficient repair mechanisms eliminate immediately most of the errors, which otherwise could have disastrous consequences for either the cell, or the organism as a whole. Let us describe some of these mechanisms.

To begin with, DNA-p itself functions as a self-correcting enzyme, correcting its own polymerization errors as it roves along the DNA helix. Additionally, patrol enzymes 'walking' along the DNA helix recognize a fragment of a single strand that has been damaged, say by alkylation agents, and act in one of two ways: direct elimination of a lesion affecting one base (*e.g.* elimination of an additional methyl group by a dealkylase); or multiple-step excision of a larger region of the strand (*e.g.* DNA glycosylases remove altered bases, while nucleases cleave the phosphodiester backbone next to the damaged site), the missing fragment being then restored by the action of specific enzymes during replication, or by post-replicative recombination.

However, if the damage is too great, these repair systems are incapable of correcting it, and DNA replication is blocked, with possible lethal consequences for the cell. Still, the cell can be rescued, but only at the cost of an alteration in its genetic material. A complex regulating system, called the SOS system (Radman, 1975), can make the replication complex tolerant to an incorrect base-pairing. This system consists of several genes which control the transcription of sequences involved in DNA repair functions, cell division inhibition, and a variety of DNA reorganization activities such as recombination. It is usually repressed by the LexA protein (*cf.* Little and Mount, 1982). When replication is blocked at a lesion, signal molecules, probably dependent on the amount of SSB protein (see Moreau, 1985), activate the RecA protein, which releases the SOS system by inhibiting its repressor LexA. Replication can thus be resumed, possibly with an error in the base-pairing, leading to a mutation in one of the daughter cells (Little and Mount, 1982; Maenhaut-Michel, 1985).

5.2. The Memory Evolutive System Model

The different co-regulators.   In the memory evolutive system modelling a bacterium, the replication process (to be compared with the transcription process modelled in Chapter 6) with possible DNA repair will be dependent on three heterogeneous co-regulators:

(i) A replication-CR, whose agents model the different proteins that operate DNA replication. These are DNA-p, helicases, SSB protein, and the patrol

enzymes. When replication is not underway, its procedure is 'do nothing'.
(ii) A SOS-CR modelling the SOS system, having two admissible procedures: 'do nothing' if LexA is observable in its landscape (and thus is repressing its activity), 'repair' otherwise.
(iii) A co-regulator playing the role of a macro-CR, formed by the sub-cellular systems implicated in the cellular cycle. These include in particular operons, functional regulatory modules such as the SOS system, and various protein populations (SSB, RecA, LexA and others). This macro-CR controls the two preceding micro-CRs.

Simple Replication.   Depending on the information decoded from its landscape, the macro-CR may impose the replication procedure on the (previously inactive) replication-CR, and its macro-step will cover the whole replication process. Each (micro-)step of the replication-CR corresponds to a procedure of replication and verification of bases, or else to one of simple repair. To replicate the n-th base, the procedure consists in forming colimits, representing the bindings of DNA–p with the base n, of helicases and SSB protein with the bases following n, and of corrector enzymes with the appropriate DNA sub-units. If the procedure succeeds (*i.e.* if the base n is correctly replicated), the next micro-step will bind the new base to the bases already replicated; and then the same for base $n+1$. If there is an error, the next procedure will try to correct the erroneous bases, before the replication-CR goes on to base $n+1$. Hence, its successive steps are determined by the loop:

$$\text{replication} \leftrightarrow \text{preliminary control of base pairing}$$
$$\rightarrow (\text{if necessary, repair} \rightarrow) \text{ replication}$$

SOS Intervention.   In the case of very extensive lesions, the simple repair system of the replication-CR is overwhelmed, and the replication procedure imposed on it by the macro-CR is blocked. Thus, there is a fracture in its landscape, which in turn imposes a fracture on the macro-CR, which interrupts its macro-step. The macro-CR may then adopt a new procedure to recruit the SOS-CR. This procedure combines the DNA fragments linked to SSB with RecA (formation of a colimit binding the triple complex), and this complex cleaves the LexA protein that repressed the SOS system. The change in the landscape of the SOS-CR (disappearance of LexA) triggers a new procedure to repair, possibly with errors, the damaged part. After the repair, the macro-CR halts the SOS procedure (via repression by free LexA), causing a fracture to the SOS-CR; and its new procedure is to reactivate the replication-CR, which resumes the replication process after the site of the lesion (thus binding DNA-p with a base $n+k$).

Cell Level.   At the still higher level of the cell, this process allows it to maintain its genetic identity and metabolic function, at least over some steps of its life cycle, the possible mutations so generated causing no immediate fracture to the cell. However, over the longer term, further processes may come into play, resulting in deep changes. For example, the SOS intervention might (through the activation of RecA) allow the transfer of mobile genetic elements which promote the dissemination of antibiotic resistance genes (Beaber *et al.*, 2004).

6.   A Theory of Aging

We have proposed (Ehresmann and Vanbremeersch, 1993) a theory of aging for an organism, based on the characteristic dynamics of a memory evolutive system. We hypothesize the main cause of aging to be the progressive loss of synchronization between the different co-regulators as a result of external events, or of greater and greater internal dysfunction, which impose a cascade of re-synchronizations at co-regulators of increasingly higher levels. We proceed to briefly describe this theory, first in a general memory evolutive system, and then in that modelling the organism of an animal.

6.1.   Characteristics of Aging

We recall (Chapter 6) that, to achieve its current step at a time t, a co-regulator of a memory evolutive system must respect the structural temporal constraints

$$\pi(t) << d(t) << z(t)$$

where $\pi(t)$, $d(t)$ and $z(t)$ denote, respectively, its time lag (related to propagation delays) at t, its period at t, and its stability span (related to the minimum stability span of the components intervening in its actual landscape and its selected procedure). These constraints can also be expressed by the inequalities

$$z/\pi >> z/d >> 1$$

where π, z and d are all functions of time, to be satisfied except possibly during isolated steps of the co-regulator. If one of the above constraints is not respected at a certain instant for a certain co-regulator, for example following a fracture, normally the deviation will be rapidly compensated (during the following step) via repair mechanisms controlled by the same co-regulator or by other ones.

The ratio z/π may decrease as a result of a decrease of z (the components decay more rapidly), or from increases in π (communications slow down). As the ratio z/d must remain between z/π and 1, a large decrease of z/π may necessitate that of z/d, which may be achieved by an increase of d (the activity of slows down), hence by a de-synchronization of the co-regulator. Changes in z/d and z/π must respect the structural temporal constraints for each co-regulator; as the higher levels are more or less dependent of the lower ones, a decrease in the ratios at a certain level will be propagated to the ratios corresponding to higher levels, with progressive de-synchronization of different levels.

With these facts in mind, and relying on the various experimental facts to be given in following sections for organisms, we propose the following:

Proposition. *Aging for a memory evolutive system comes from a progressive decrease of the ratios z/π, related to more and more of its co-regulators of increasing levels, to be compensated by an increase of their periods.*

In normal aging, the minimum threshold of 1 for z/π and z/d is gradually approached. Problems occur when the inter-levels compensations cannot be achieved rapidly enough, so that it is impossible for the organism to maintain homeostasis. Ultimately, this may lead to death.

Graphically, these considerations lead to a representation of z/d and z/π as functions of time of the form depicted in Fig. 7.5, where the descending part corresponds to aging. The rate of decrease is related to the rate at which the organism ages. The rising part of the curves would correspond to the growth

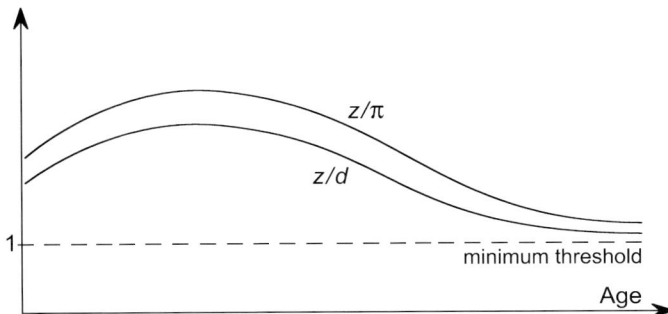

Fig. 7.5 Aging curves.
The curve of z/d as a function of time slowly approaches 1 during senescence, while the curve of z/π also approaches 1, but remains above the first curve. The increasing regions of the curves correspond to growth of the organism.

phase, having a rapid rhythm of structural transformations. z is initially low, but gradually increases, the time lags π remaining relatively stable. An important dysfunction, such as a pathology for an organism, would then correspond to the case where, for some particular co-regulator, the curve z/π changes direction so as to pass under z/d, even though the latter also decreases (Fig. 7.6). Depending on the rapidity of this change in direction, it may be possible, after repair, to recover the initial configuration. Failing this the slope of z/d may also decrease suddenly, so as to go beyond the minimum threshold; this would correspond to the case where it is impossible for the system to recover its equilibrium, thus resulting in death.

Analytically, d may be interpreted (Ehresmann and Vanbremeersch, 1996) as the period of an oscillatory process for a limit cycle in a dissipative system with multiple regulation (for description of such systems, see Prigogine and Stengers, 1982). The successive procedures of the co-regulator lead to changes in z and π, which correspond to a modification of the initial

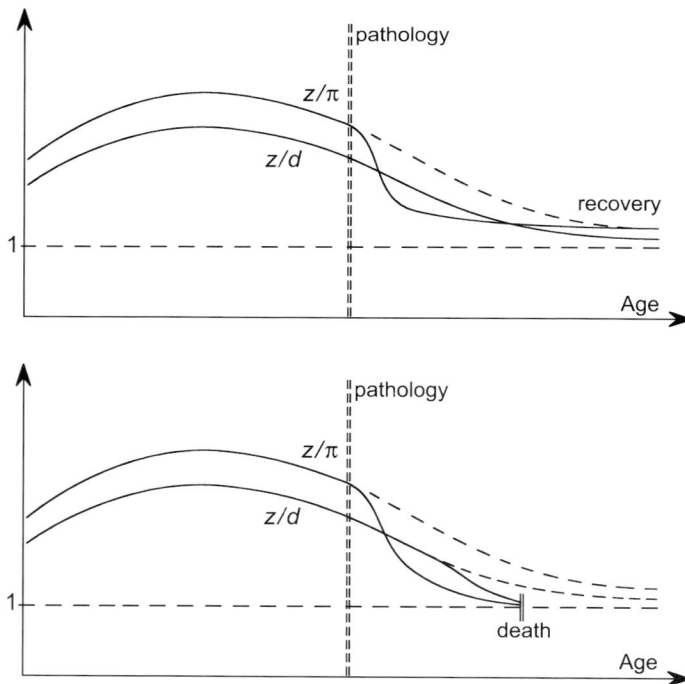

Fig. 7.6 Pathology and recovery during aging.
During pathology, z/π may cross under z/d for a while. For recovery, z/π must climb back above z/d, though both eventually decrease with age. Death will ensue if this is not possible.

conditions. A de-synchronization leads to a brief chaotic state, before reversion to an oscillatory process having another limit cycle with a longer period, which is that of the co-regulator after re-synchronization (the equations are similar to those of Goldbeter, 1990).

To summarize, we propose the following:

Theory of aging. *Aging is due to a cascade of re-synchronizations of co-regulators at increasingly higher levels, to compensate for progressive decreases in the ratios z/π at lower levels.*

These re-synchronizations act as mechanisms to return the system to within the bounds of the temporal constraints of its different co-regulators, in the wake of external disturbance or internal dysfunction.

This leads to the realization that aging is quasi-absent in a certain number of cases. If a system has a restricted number of hierarchical levels, and so is relatively simple, and has in addition few external functional relations, then it is possible for z/d to remain almost invariant. Thus, such a system will undergo little aging (except perhaps if external disturbances become too strong). This could be the case with cancer cells, and with cells of certain specific types, such as spermatogonia.

Several physiological theories of aging for an animal have been proposed, which attribute it to phenomena occurring at various levels. We are going to show that these theories may be interpreted by an accelerated decrease of the stability spans z or an increase of the time lags (or propagation delays) π between levels, forcing re-synchronization of different cycles and leading to fractures at higher and higher levels, and thus are special cases of the theory of aging presented above. As such, they help to justify the theory proposed here, both conceptually, and via any empirical support which they enjoy.

6.2.   *Theories of Aging at the Macromolecular Level*

Most authors base the phenomena of aging at the macromolecular level. Aging is seen as the result of many factors, including decreasing stability of components, instability of the genome, increased delays in replication leading to fractures. Each of these is discussed in detail below.

Decrease in the Stability of Components.   It is recognized that with age, the complex identities of proteins and of DNA are less well maintained (z decreases at the macromolecular level). This change occurs as a result of an accelerated DNA mutation rate (Sinex, 1977), and an accumulation of defective enzymes (having lost 30–70% of their catalytic activity; Gershon, 1979). Orgel's (1963, 1970) theory attributes cell aging to poor translation of mRNA, causing amino acid substitutions, and thus defective proteins. This would result in a 'catastrophe of errors', related to poor enzyme functioning and timing errors. Using our theory, this would be modelled by poor performance of procedures selected by the RNA, which would be reflected back to the other levels, thus beginning the cascade of re-synchronizations. Since then conflicting results have been obtained. Thermolabile enzymes were detected in fibroblasts (Holliday *et al.*, 1981, Goldstein and Moerman, 1976), in particular in the course of diseases imitating accelerated aging (Progeria and Werner's syndrome; Goldstein, 1978). However, other experiments did not corroborate the hypothesis of a routine increase in errors of protein synthesis with age (Harley and Goldstein, 1980; Wojtyk and Goldstein, 1980).

Instability of the Genome.   More frequent DNA mutations have been implicated by various authors as the basis of aging. Errors could be stochastic (Szilard, 1959; Burnet, 1982; Holliday *et al.*, 1981), or the result of an 'aging program' (Sinex, 1977). They could also be due to defects in the mechanisms of DNA repair, although there is no evidence that these are modified with age (Goldstein, 1978; Gensler, 1981). Chronogenetics (Gedda and Brenci, 1975) attributes senescence to a progressive and differential decrease of the stability span, and of the activity of genes (measured by their 'chronon' and 'ergon', which can be interpreted as being proportional to the stability span of the gene and its products).

Other causes of genomic instability (decrease in its stability span z) have been implicated in aging. Thus, a loss of differentiation of certain genes that would result from a decrease in DNA methylation at specific regions close to the gene (theory of Cutler, 1985; Holliday, 1987a). A similar cause would be at the basis of the reactivation of the inactivated X-chromosome (Holliday, 1987b). Repetitive DNA sequences may facilitate the formation of extra-chromosomal circular DNA molecules (Shmookler *et al.*, 1986), which increase the genome's instability (decrease its z value). Thus, (in *Podospora*), DNA functioning is disturbed by one of the introns of the mitochondrial gene encoding for cytochrome oxidase, a plasmid that would be reinserted into the nuclear DNA and block gene expression (*cf.* Esser, 1985).

Increased Delays.   Insufficient maintenance of the complex identities of DNA, RNA and proteins (decrease of z, see above) results in a lowering of their activity and, consequently, an increase in the time lags π, and in the period d of the higher level. This results in a re-synchronization of the molecular, macromolecular and cellular levels, leading to fractures that are more and more difficult to overcome.

The rate of protein synthesis decreases with age, and abnormal proteins accumulate, as shown for example in the mouse liver (Lavie *et al.*, 1982). It appears that reduced protein production results from inadequate elongation rather than a defect in the initiation of the synthesis (*cf.* Danner *et al.*, 1990; Webster, 1985). A reduction of 20–25% in the catalytic activity of elongation factor EF10, possibly as a result of post-translational modifications, has been observed (Rattan *et al.*, 1986). RNA synthesis also decreases as a result of aging (*cf.* Danner and Holbrook, 1990), especially that of mRNA (80–90% decline; Richardson and Semsei, 1987).

Kirkwood and Austad (2000, p. 235) cite many studies which support the idea that 'it is the evolved capacity of somatic cells to carry out effective maintenance and repair that governs the time taken for damage to accumulate to levels where it interferes with the organism's viability and hence regulates longevity'.

6.3.   *Level of Infra-Cellular Structures*

Here, the changes are essentially a slowing down of intra-level communications and communications with lower levels, that is, a lengthening of the time lags π of the corresponding co-regulators. Combined with the decrease of z at the macromolecular level (*cf.* above), a re-synchronization between these levels results, and is amplified over time, causing dysfunctions due to reciprocal fractures of different components (decrease in their z).

Degradation of Mitochondria.   In the mitochondria, aging is manifested at the micro-level by poor functioning of the ion transport system, and a loss of its capacity to carry out the ATP-generating Krebs cycle (Robert, 1990). This results in an increase of the propagation delays (increase of π). It leads to membrane 'fatigue' (due to poor control of Ca^{2+} and Mg^{2+} ions) and an increase in the proportion of incompletely reduced oxygen (2% in normal circumstances), which appears in the form of free radicals (see LeBel and Bondy, 1992; Pacifici and Davies, 1991). Accumulations of point mutations, and deletions of mitochondrial DNA, have both been shown to reduce the lifespan of mice (Trifunovic *et al.*, 2004).

Theories Implicating Free Radicals.   Numerous authors have implicated free radicals as responsible for aging (see, *e.g.* Harman, 1986; Laborit, 1987). They are one of the factors intervening in most of the cell deterioration processes mentioned above, and the activity of the defence system capable of neutralizing them (incorporating molecules such as superoxide dismutase, glutathion peroxidase, and vitamin E) may decrease with age (Pacifici and Davies, 1991). The accumulation of free radicals results in damage to cellular components, in particular membrane degradation (thus, a decrease in the corresponding z value). For example, in erythrocytes and the brain, free radicals attack the protein Band3, which is a key component of mem-branes. One of the by-products of this protein alteration is termed 'aging cell antigen', as it binds an auto-antibody IgG, which marks the cell for death (Bartosz, 1991; Kay, 1991). The degradation of membrane lipids by free radicals also results in the accumulation of lipoprotein particles, or lipofuscin (up to 6–7% of the intracellular volume at 90 years of age), in particular in the myocardium and the central nervous system, with a con-sequent slowing down of the renewal of organelles, such as mitochondria (Robert, 1983). The importance of oxidative stress in aging is emphasized by the fact that longevity-influencing genes recently identified have been shown to reduce this stress; and by the fact that caloric restriction, which decreases oxidant generation, substantially increases longevity (see the reviews by Finkel and Holbrook, 2000; and Nemoto and Finkel, 2004).

Changes in the Cytoskeleton.   Macieira-Coelho *et al.* (1986) implicate another harmful loop, this time between the macromolecules and the cyto-skeleton: a reorganization of the genes for cytoskeleton proteins results in modifications to the links between membranes and the nucleus, which func-tion more slowly (thus a lengthening of the time lag π for the organelles), and thus a poor anchoring of the DNA and a slowing down of intracellular communications, which in turn increases the duration of the cell cycle (considered below), by delaying the initiation of its first phase. Progressive malformation of nuclei occurs with age: Haithcock *et al.* (2005) have shown that, in a nematode worm, the morphology of the nuclear scaffold known as the lamina changes; as the lamina is involved in the DNA repair pathway (Kamiuchi *et al.*, 2002), its alteration compromises the signalling and repair of damaged DNA, and thus decreases genome stability (decrease of its z).

6.4.   Cellular Level

All the lower level changes mentioned above have repercussions at the cel-lular level, via an increase in the cell's time lag π and period d. The length-ening of the cell cycle (thus of d) can result from a longer latency phase G_0 (Bazerga, 1977; Hayflick, 1977), or a slower DNA replication. Dysfunctions of the cycle timing produce messages that are poorly transmitted to the chromatin, and the cells will divide before DNA replication is completed, resulting in a loss of volume (for the situation in ciliates, see Smith-Sonneborn, 1990). For numerous types of cells, a clear decrease in the number of divisions was observed with age (loss of 0.2 divisions per year of age for normal human fibroblasts; Martin *et al.*, 1970). This led to the idea that cellular aging is encoded in the genome, by an appropriate clock, and that the number of divisions of cells in continuous mitosis would be limited (model of Hayflick, 1977; *cf.* Robert, 1983, 1990).

In the immune systems of aged subjects, a greater proportion of T-lymphocytes do not enter into mitosis in the presence of activators, perhaps subsequent to defects in Ca^{2+} signals (Miller, 1990), or as a consequence of a lesser density of interleukin IL-2 receptors (*cf.* Burns and Goodwin, 1991). In contrast, the changes are less marked for B-lymphocytes, antigen-presenting cells and neutrophils, at least in the healthy aged subject (Kennes *et al.*, 1986; Corberand *et al.*, 1986). At present, it is not known if cell death is a programmed phenomenon, internal to the cell (so-called 'death genes'), or if it is the result of external causes, or if it intervenes in a stochastic manner (see Raff, 1992).

6.5. *Higher Levels (Tissues, Organs, Large Systems)*

Here, the phenomena are the consequence of modifications of lower levels. The decrease of their stability spans z and the lengthening of the time lags π and period d of their co-regulators combine to degrade inter-level communications, by increasing the propagation delays (π at the tissue level), and by causing subsequent fractures at different levels, resulting in more or less considerable changes, whether aging is associated with disease states, or not.

Degradation of the Extra-Cellular Matrix. Several authors (*cf.* Robert, 1983, 1990) emphasize the faster changes in the components of the extra-cellular matrix (thus, z decreases) during aging. The *bypass theory* of Verzar (1957) is based on the increase of inter-catenary bonds (bypasses) of collagen, in part due to attack by free radicals (Rigby *et al.*, 1977). The elastin/glycoprotein ratio of the aorta routinely decreases with age (Moschetto *et al.*, 1974). The thickness of the basement membranes increases (Robert *et al.*, 1977). Generally, degradation of the matrix results from an unfavourable balance between the deterministic mechanism producing the matrix, and stochastic events (UV radiation, nutritional factors), which degrade it (Robert, 1990).

Obviously, tissues which are particularly rich in extra-cellular matrix are most affected by this process. Thus, the vitreous humour, the gelatinous substance that fills the eyeball, composed especially of collagen fibrils and hyaluronic acid, progressively liquefies with age (Robert, 1983). In the normal lens, the SOD activity decreases with age (Ohrloff and Hockwin, 1986), as does the membrane potential, while the total concentration of sodium increases (Duncan and Hightower, 1986). These changes are more pronounced in the case of cataracts. Similarly, bone matter undergoes a decrease in its density due to an increase in catabolism of the calcified matrix; this results in osteoporosis, in particular in women after menopause, or over the age of 70 (Riggs and Melton, 1986). The water content decreases while the collagen content increases, with an increase in the number of bypasses (Meynadier, 1980). The speed of cicatrization decreases exponentially with age (Lecomte de Noüy, 1936).

Slowing Down of Communications.   Less effective hormonal or neurally mediated exchanges between cells and tissues (lengthening of the corresponding π) results in less effective metabolic regulation. In response to external or internal stimuli, there will be an increase in the latency period (π for the tissues co-regulators), and a decrease in the amplitude of the responses of effectors (Adelman, 1979). The result is that homeostasis is more difficult to re-establish at the level of the large systems (longer period d). For example, between the ages of 30 and 90, nerve conductivity decreases 15% and the cardiac index at rest decreases 30%; renal circulation decreases 50%, and maximal respiratory capacity, 60–70%, by around 50 years of age; while muscle strength decreases markedly only around the age of 65–70 (Robert, 1983). These changes are greater in the case of disease, such as arteriosclerosis, which occurs frequently. The adrenocortical axis becomes more vulnerable with stress, having a tendency towards excessive secretion of glucocorticoids, which conversely can accelerate aging and give rise to diseases (Sapolsky, 1990). A similar slowdown of communications has already been discussed in the context of the aging of the immune system.

Aging of the Nervous System.   The weight of the brain decreases 2% per decade past the age of 50, with glial atrophy and dilation of the ventricles (Horvath and Davis, 1990), the loss of white matter becoming preponderant over the age of 50 (Miller *et al.*, 1980). Multi-synaptic transmissions slow down, following an insufficient renewal of neuromediators, especially the catecholaminergic ones; but the GABA-ergic system does not appear to vary (Robert, 1983). Conflicting results have been obtained concerning the role of free radicals in cerebral aging (LeBel and Bondy, 1992; Aston-Jones *et al.*, 1985). Senile plaques with amyloid substance between the neurons (especially in the cortico–cortical connections of layers II and III, Duyckaerts *et al.*, 1987), and cerebral vascular amylosis (infiltration of cerebral blood vessel walls by amyloid substance) can be found in the normal aging brain (Hauw *et al.*, 1987). They are, however, more developed in the case of dementias, in particular Alzheimer's disease, where they are associated with neurofibrillar degeneration, and the presence of abnormal filaments that can fill the neuron (Lamour, 1991).

From the cognitive viewpoint, in normal aging, it is essentially the central processes, such as attention, that are affected (Signoret, 1987). There can be progressive deterioration of the frontal lobes, in particular in the right hemisphere; performance in tasks implicating the orbitofrontal cortex is impaired more than in those implicating the dorsolateral prefrontal cortex (Lamar and Resnik, 2004). Perceptuo-motor integration worsens, and language and memory disorders arise, although these are not as severe as usually believed (Albert, 1987). The α-rhythm slows (Robert, 1983).

Slowing Down of Biological Rhythms. Changes in biological rhythms with aging have been detected. The circadian cycle lengthens, and the amplitude of the circadian variations of certain endocrine variables decreases (resulting in increased propagation delays and time lags at different levels), but the amplitude of these changes is difficult to evaluate (Richardson, 1990). For Robert (1989), there is a hierarchy of biological clocks (perhaps even one in each cell), up to a 'master clock' at the level of the hypothalamus (do we return to the pineal gland of Descartes?), synchronizing the others through hormonal action. The decrease in the number of cell divisions (see above) decreases the synthesis of the relevant hormones, and modify their genetic expression, resulting in progressively greater disorder of the ensemble.

Global Events. More theoretical studies, to which the cascade of re-synchronizations we described above can be related, have suggested that aging would not be due to local errors (such as accelerated degradation of proteins), but would be inherent in the global organization of a complex system. Thus, Rosen (1978) showed that any system that, like an organism, has feed-forward loops, can show global modes of failure: the different sub-systems function correctly as far as can be ascertained according to local criteria, but each loop has only a limited lifetime, so that the system as a whole becomes less and less adapted over the course of time: 'It is possible for a complex system to exhibit global modes of failure which are not associated with local sub-system failures. It is a priori conceivable that senescence is associated with such non-localized or global modes of failure' (Rosen, 1978, p. 580). An example has been found in rats whose lifespan has been extended by caloric restriction; post-mortem, one-quarter evidence no precise cause of death (Shimokawa *et al.*, 1993).

Other authors (Baas, 1976; Sutton *et al.*, 1988a,b) have implicated the hierarchical organization of the organism in aging. Laborit (1987) also considers the role of the socioeconomic level: he claims that the inhibition of gratifying actions is one of the factors accelerating the aging process (which he believes is genetically programmed). A great number of experiments prove that caloric restriction substantially increases lifespan in several animal species (*e.g.* Shanley and Kirkwood, 2000). There is as yet no single explanation for this, although one hypothesis already mentioned is that it results in fewer free radicals.

Lastly, several *evolutionary theories of aging* have been proposed. For example, Williams (1957) proposes that genes having good effects early in life are selected, even if they are deleterious later; while the 'disposable soma theory' explains senescence by a trade-off between reproduction and somatic maintenance (Kirkwood, 1996).

Chapter 8

Memory and Learning

In a memory evolutive system, its 'memory' plays an essential role. Let us recall that it is a hierarchical evolutive sub-system which models a *long-term memory* of the system; by recording features of the environment and internal states, it allows to recognize them when they occur again later. This memory is not rigid, like a computer memory, but flexible. In this chapter, we study how a flexible memory of this sort can develop in a memory evolutive system satisfying the multiplicity principle, and adapt to changing circumstances during the life of the system. Then its components, called records, represent classes of patterns of internal representations which, depending on the context, can be recalled via any one of them, or their various ramifications.

A sub-system of this memory—the procedural memory—allows the system to record procedures for responding to particular situations. The different co-regulators can later recall their admissible procedures to regulate the dynamics more quickly and efficiently. We describe the relations between a procedure, the commands of its effectors, and the contexts in, and objectives for, which it can be activated. Finally, we show how more complex systems are able to in addition classify the records in memory, to extract invariants (invariance classes of records) which form a semantic memory.

1. Formation of Records

We have seen that the components of a memory evolutive system are not fixed once and for all. The composition, like the structure of the system, changes over time, with some components disappearing, and others emerging. In particular, patterns of components which act in synergy, repeatedly or intensely, can give rise to new more complex components. The development of the memory rests on the emergence of such components.

1.1. Storage and Recall

As we wrote in Chapter 6, the memory evolutive systems are able to learn to identify various features of their surroundings and internal states, and to develop a repertoire of adapted responses. These abilities are dependent on an internal memory, which is a hierarchical evolutive sub-system of the whole evolutive system. This memory is at once robust (meaning that it maintains its contents in spite of disturbances), and plastic enough to adapt to the context. It plays an essential role in the dynamics of the system, by allowing it to recognize objects and events met previously, and to select procedures already used, taking into account the previous results. In this way the system can act like an anticipatory system (Rosen, 1985b; Dubois, 1998), but here the anticipation comes only by projection of past experiences into the future, assuming a certain invariance of the environment. In the presence of changes, this anticipation is likely to lead to errors.

Consider some particular memory evolutive system **K** which satisfies the multiplicity principle (so that it has multifold components, *cf.* Chapter 3). The components of the memory, called *records,* are multifold components with long stability spans, and can represent items of a varied nature, such as signals, stimuli or inputs of any kind, features of the environment, events, internal processes, more or less complex situations, behaviours or procedures. Let S be one such item. For S to be distinguished by the system at an instant t, it must more or less briefly activate (the components and links of) a pattern R of the configuration category K_t at t. If S comes from the environment (*e.g.* if it is a molecular signal for a cell, food or some unfamiliar object for an animal), R will be a pattern of receptors having connections with the external milieu, *i.e.* the components of the pattern will be components of the system in direct contact with the environment (such as membrane receptors of a cell, sense organs of an organism). For an internal signal (*e.g.* a communication between cells within an organism, or between the members of a social group), R will be a pattern of internal receptors (such as promoters of a gene, secretaries of an office). R may also be a pattern of components in relation with internal receptors, such as already formed records recognizing certain parts of S, that make it possible to record increasingly complex events or behaviours.

A record of S is formed only if S occurs several times, or is meaningful for the system (*e.g.* in connection with 'pregnancies', in the terminology of Thom, 1988); in these cases, the record M of S (and also of R) will be a component of the memory which integrates R and becomes its colimit (constructed by a complexification process). The record internally represents S; this word 'represents' does not signify that M gives an internal image of S, depicting it accurately as a photograph would do, but only that it will be possible to recognize S later by recalling M dynamically, when the pattern R which gave birth to M is reactivated.

The construction of the record is done in the configuration category of the system at a given time, but the record becomes a component of the system's memory. It maintains its identity over time, with a long stability span, but allowing for gradual revisions due to changes in the context. In particular, S may activate more or less different patterns at different times, and M will become a multifold object which is the record of each of them, so that M could be recalled later through the reactivation of anyone of them.

It is assumed that the memory develops by a succession of complexification processes with respect to mixed options (Chapter 4), starting from an *innate memory*, *i.e.* a sub-system of the memory given initially. The storage as well as the later recall of a record result from the synergistic action of the different co-regulators through their respective landscapes, and from the interplay among their procedures. Each co-regulator takes part in a trial and error training process, by trying at each of its steps to evaluate the outcome of the preceding step, and to remember the results of that outcome. This last objective is relayed to the system and will be integrated as one of the objectives of the operative procedure, obtained after the interplay among the procedures of the different co-regulators.

Conversely, the procedures of the co-regulators are selected by taking account of the part of the memory which is accessible to them, in particular the admissible procedures, which characterize their function (Chapter 6). According to the nature of the items, their record will be formed as either a colimit or a limit of the pattern activated by the item (the limits corresponding to records in the procedural or the semantic memory; see Sections 3 and 7 below). In the two cases, the processes are similar, and we will describe the case for a colimit.

1.2. Formation of a Partial Record

Let S be an item to be remembered (*i.e.* to be stored in the memory under the form of a record). At a given time, S activates a pattern R of linked components R_i of the configuration category. This pattern is not necessarily a sub-graph of the category, for there may exist two indices i and j with $R_i = R_j$. This occurs if for example the item S is a complex object containing two identical objects (say, a drawing with two identical triangles), which thus activate the same component. S may have a record which pre-exists in the innate memory or has already been formed; otherwise the record may emerge through a complexification process. In this last case, let us explain how the co-regulators participate in its emergence.

A particular co-regulator, let us say E, distinguishes only some attributes of S (*e.g.* its colour or shape, if E is an area of the visual cortex dealing with colour or shape; the epitope of an antigen, if E is a cell of the immune system). These are represented in its actual landscape through a pattern r_E of perspectives coming from a sub-pattern (possibly void) of R: its components are the perspectives for E of the components of R having aspects for at least one agent of E, and its distinguished links come from the distinguished links of R correlating two such perspectives. This pattern r_E is identified by the fact that it appears in the landscape L_E of E and maintains itself synchronically over the actual present of E. One of the objectives of the next procedure of E will be to bind the pattern r_E, to remember it. This will be reflected to the system by the request to bind the pattern, say R_E, which is the image of r_E via the difference functor from the landscape to the system (see Fig. 8.1).

This pattern R_E is related, but not necessarily identical, to a sub-pattern of R. More precisely, let us recall that a component R_i of the pattern R is a pair consisting of an index i (a vertex of the sketch of R), and the component of the system associated to it by R, and thus implementing i; if we forget the indices, the components of the system so implemented by R correspond to some of the components implemented by R_E, but if a component of R has several, say n, different perspectives belonging to r_E, it will figure with n

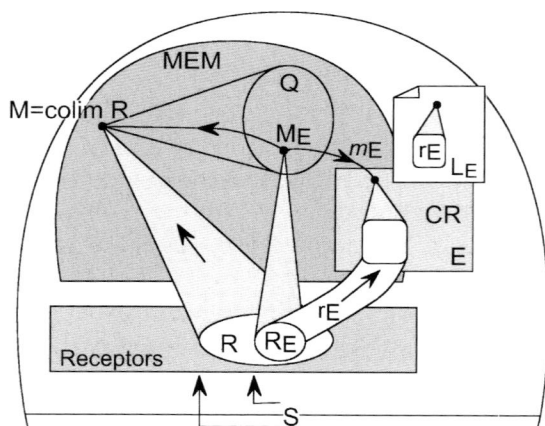

Fig. 8.1 Formation of a record.
An item S activates a pattern R of (internal or external) receptors. Only some of its components are observable in the landscape L_E of a particular co-regulator, say E, by the pattern r_E of their perspectives. (This pattern is depicted twice: in the landscape; but also in the system, by the aspects defining the different perspectives.) One of the next objectives of E will be to bind this pattern, in order to remember it. This objective is relayed to the system, where it becomes the objective of binding the pattern R_E which is the image of r_E by the difference functor from the landscape to the system. Thus, the operative procedure on the system should form the colimit M_E of R_E, called the partial record of S for E. If M_E has a perspective m_E in the landscape of E (it is the case in the figure), it is called an internal E-record; in this case m_E is a colimit of r_E in the landscape.
The various partial records M_E of S for the different co-regulators form a pattern Q. The item S has a record M if this pattern Q admits M as its colimit.

different indices in R_E (in the example of the drawing above, if E is a co-regulator dealing with shape and already recognizing a triangle, this triangle will appear with two indices in R_E). The partial record of S for E is obtained by integrating R_E. More precisely:

Definition. We say that S (or the pattern R) has a *partial record* M_E for the co-regulator E if the pattern R_E has a colimit M_E, where R_E is the pattern image by the difference functor of the pattern r_E of perspectives of R for E. This partial record is called an *internal E-record* if M_E has a perspective m_E which is the colimit of r_E in the landscape of E.

S may have a partial record for E, but no internal E-record; in this case S cannot be recognized by E in its landscape. In general, lower level co-regulators have no internal records. If S has an internal E-record M_E, the perspective m_E may arrive at an agent whose existence predates the formation of the record, or be added by the operative procedure in which M_E emerges. This perspective allows the later recall of S via E itself (in its landscape).

1.3.   *Formation and Recall of a Record*

At the same time, other co-regulators having access to perspectives of R will have for objectives the formation of corresponding partial records. Thus, the operative procedure Pr°on the system resulting from the interplay among the procedures will integrate all these objectives. This leads to the formation of the pattern Q of partial records of S: its components are the partial records of S for the various co-regulators, and the distinguished links between them result by complexification from the distinguished links of R. As this pattern Q is synchronously activated as long as S persists or is repeated, Pr° (or a later operative procedure) will bind it so that it acquires a colimit M, and this colimit emerges as a new record in the memory (see Fig. 8.1).

Definition. S (or the pattern R) has a *record* M if it admits partial records and if the pattern Q of its partial records admits a colimit M.
(This definition makes more precise the definition alluded to in Chapter 6.)

A subsequent presentation of S will reactivate the (new configuration of the) pattern R, and consequently, by the interplay among the procedures, its various partial records and their pattern Q; whence the *recall* of the record M of S, which is the colimit of Q, and the recognition of S. Let us notice that M itself can have no 'global' perspective for any co-regulator, and so its recall requires a genuinely synergistic action of them all, each one recognizing those attributes of S from which it receives some aspects. For instance, the bees of a hive 'know' how to construct the hive, but each one participates in this process in a very fragmentary way, thanks to some instinctive procedures. The 'construction of the hive', which is a consequence of the temporal combination of all these procedures, is a component of the memory of the system, but it remains hidden to the individual bees and can be attributed to their society only by an external observer.

1.4. *Flexibility of Records*

A record assumes a distinct identity as it is adjusted to fit changing conditions. Since **K** satisfies the multiplicity principle, as said above the record may be(come) a multifold component, detached from the particular data from which it was constructed, and allowing it to take account of new data. Indeed, once the record M of an item S has emerged in the memory (through a complexification process), it will take on its own complex identity, representing all the items activating a pattern functionally equivalent to R (that is, homologous to R, in the general sense of Chapter 3). Thus, the record M will recognize not a unique item S, but the whole class of items which can activate patterns homologous to R, and it can be recalled via the activation of any of these patterns. These items (possibly appearing at different times) differ from S by the values of various parameters, and there may be complex switches between them. In this way a record, though having a long stability span, is not a rigid static object, like a photograph, but a component which interacts with the context and may evolve if the context changes.

Let us notice that a record, as a component of the memory, is not part of the 'real authentic furniture of the world' (in the terms of Goldstein, 1999, p. 49) but an object of an epistemological nature. However, it plays a functional role by the intermediary of the class of patterns by which it may be recalled later, which model part of the world. Which of these patterns is reactivated at a given moment depends on the present context. Here we use a characteristic of natural complex systems, namely that a complex component may have several (non-connected) decompositions through which it can be activated (multiplicity principle). For example, the record of a friend's figure will slowly change over time as he ages. The changes are more or less rapid, and their pace will be measured by the stability span of the record. As we have indicated, this span is long during equilibrium, whereas it is shorter in a variable environment, and during development or decline (*e.g.* memory declines during aging).

2.   **Development of the Memory**

The memory of a memory evolutive system **K** is an evolutive sub-system, denoted by MEM. Its development requires not only the formation of its components, the records, but also of links between them modelling their interactions.

2.1. Interactions between Records

The records represent features, signals, behaviours, procedures or information of any nature. The links between them represent their more or less temporary interactions. They are created or strengthened during the complexification process in which the records emerge, so that we can describe them explicitly (as done in Chapter 4). In particular, if M and M' are records binding patterns Q and Q' respectively, there are (Q, Q')-simple links from M to M' which bind a cluster of compatible links between the components of Q and Q'. There also exist complex links obtained by iterating the following processes: formation of the composite of simple links binding non-adjacent clusters (or pro-clusters, when limits are taken instead of colimits); and binding of clusters (or classification of pro-clusters) of such links.

Once the links between records have been defined, the process of formation of a record can be iterated, to obtain records of increasing complexity. The record M representing an item S may thus later participate to the storage of a more complex item that admits S as one of its parts, and which has therefore a record of a higher order.

2.2. Complex Records

A complex record C will be formed in several stages; in the case of colimits, as iterated colimits of patterns whose components are simpler records. At each stage the formation is described as above, by a complexification process with respect to the operative procedure, obtained after the interplay among the procedures of the different co-regulators. Once formed, C will assume a distinct identity, making it possible to apply the record in situations more or less different from the initial context. Thus, (as said in Chapter 3) C plays the part of a frame in the terminology of Minsky (1986), the slots of which can be filled with various parameters (related to the components of its various ramifications), and which applies to various cases by adapting the parameters to the context.

Indeed, the later recall of C will be accomplished by the unfolding of the particular one of its ramifications best adapted to the present context, say $(P, (P^i))$ for a ramification of order 2. For that, different co-regulators simultaneously will activate (through their actual landscapes) components of a pattern P having C for its colimit; then, at a following stage, each one of the components P_i of P will in turn activate one of the patterns P^i of which it is the colimit via lower level co-regulators. The choice of the ramification is determined gradually, from top to bottom: initially choice of the pattern P, then for each one of its components P_i, choice of a pattern P^i, and so on if the ramification goes 'deeper'. Which ramification is finally unfolded will depend on the constraints, of whatever nature, imposed by the context, and the choice will be specified or changed at each level while the action is underway, under the effect of the interplay among the procedures of the various co-regulators, by taking account of their own temporal structural constraints. If these cannot be respected, the action will fail. For example, a general order to take an object on a table will mobilize different muscles according to the distance from the object, and will fail if the object moves before the end of the motion.

2.3.   *Examples*

Let us see how *an orchestra* will learn a new piece of music S, a copy R of which is presented to its members. Each musician E of the orchestra acts like a particular co-regulator in the memory evolutive system modelling the orchestra; in his actual landscape he retains only the lines R_E relating to his instrument, and translates them into a succession of notes and chords which he integrates into a whole. The part, once learned, is stored by a record M_E of the memory of the orchestra; this M_E is an internal E-record of S since E has access to this object in his landscape and can replay his part when he wants. Independently, the other musicians similarly learn their parts. The interplay among the procedures of the various musicians will allow them to harmonize the way in which they play their various parts M_E, and to in- tegrate them into a total performance M, which will be remembered as the record of S. During successive repetitions, this record of S will assume a distinct identity, the piece being recognized even if there are small variations of detail in the play of the musicians or the instruments used. When later the orchestra wants to play this piece of music from its repertoire, it will suffice for each musician E to recall his part, therefore to reactivate his E-record, and, by coordinating their actions, the musicians will all together interpret the piece by reactivation of its record M. Let us note that each musician has only his particular perception of this record, and only someone who can observe the whole (an external observer, such as the conductor) will be able to interpret the result as belonging to the memory of the whole orchestra.

   The immune system of an animal consists of the innate immune system, which reacts quickly, and of the adaptive immune system. It can be modelled by a memory evolutive system which contains as sub-systems the memory evolutive systems of its different cells; other components are populations of sub-cellular or cellular components. The cells form particular co-regulators. Every cell is covered with peptides bound to major histocompatibility complex (MHC) molecules glued to surface receptors. It is through these that the memory will recognize 'self' and distinguish pathogens; while the procedural memory contains different defence procedures against these pathogens.

The response of the *innate immune system* (see Chapter 6) depends on what we call the innate memory (*i.e.* the sub-system of the memory which exists at birth). General purpose defence cells (*e.g.* macrophages and natural-killer cells) have, in the form of pattern-recognition receptors, records of molecular patterns associated with many different micro-organisms, and they have admissible procedures to destroy them.

The *adaptive immune system* response (see, *e.g.* Germain, 2001) expands the memory over time, so that the organism may react to specific antigens. The antigen-presenting cells, such as dendritic cells, macrophages and B-cells, capture and process antigens for presentation to T-cells. They engulf antigens, degrade them into peptides and bind them to MHC molecules. An MHC molecule with a bound peptide epitope is anchored on the cell membrane, where it will be recognized by 'naïve' T-lymphocytes that have complementary shaped surface receptors. This activates transcription factors in the T-cell so that it proliferates, and differentiates into an effector T-cell, having a partial record of the antigen, corresponding to the specific receptor. An effector T8-cell will respond by a procedure to destroy cells with the targeted antigen, through apoptosis. The procedure of an effector T4-cell will be to produce various signals, such as cytokine molecules, that activate other cells. Some activate macrophages that directly destroy the antigen. Others help with the clonal selection and expansion of memory B-cells, producing antibodies or immunoglobulins. These represent a long-term record of the antigen, and allow for antibody-dependent destruction by, *e.g.* natural-killer cells. Finally, the (total) record of an antigen will correspond to the populations of T-cells and memory B-cells which have matured to recognize it through its epitope (see *e.g.* ASSIM, 1991).

3. Procedural Memory

Records of various types can be formed, and correspondingly, we distinguish several evolutive sub-systems of the memory MEM of the memory evolutive system:

• The *empirical memory*, denoted by **Memp**, whose components are records of signals, features of the environment, internal states or particular situations; they are obtained as iterated colimits starting from patterns of (external or internal) receptors. In the case of the memory of an animal, it covers both the perceptual memory and the episodic memory (Chapter 9).

• The *procedural memory*, denoted by **Proc**, has for components the various procedures; we use this term in a very wide sense, to cover actions, behaviours and procedures, in particular those allowing certain objectives to be achieved. These have been called in other contexts *programs, scenarios, maps* (Edelman, 1989), *abstractional schemes* (Josephson, 1998) and so on. In our own previous work, we used the term *strategies*.

• The *semantic memory*, denoted by **Sem**, has for components the concepts as defined in Section 7 (below). It is only present when the system is able to classify its records in invariance classes.

• Finally, in higher animals, the *archetypal core* **AC** integrates their most important experiences (see Chapter 10).

Here we will describe how the procedural memory **Proc** is built up, starting from a sub-system $Proc_0$ of the innate memory, composed of 'basic procedures' (those representing innate procedures or behaviours), by formation of iterated limits (and not colimits) of patterns whose ultimate components are these basic procedures.

3.1. *Effectors Associated to a Procedure*

Just as a signal requires receptors to be perceived, a procedure must be carried out by effectors. Indeed, a procedure represents a formal unit (like a program), whose activation will trigger various commands, to be implemented by various components of the system, the *effectors*. Depending on the case, effectors can operate purely internally, or have effects on the environment. For example, in a cell, a procedure can consist in the activation of a whole network of proteins which act internally (signalling network) or externally (exporter pathway). Likewise, for an organism, a procedure can activate a group of ideational neurons (internal action), or a system of motor neurons (having external consequences).

In a memory evolutive system, a procedure is stored by a component Pr of the procedural memory **Proc**, to which is associated a pattern OPr of its commands toward effectors (modelled by links from Pr to effectors); this pattern of commands is also stored in the empirical memory **Memp** by its colimit effPr. More precisely:

Definition. A *procedure* is a component Pr of the procedural memory, to which are associated (Fig. 8.2):

(i) A *pattern of commands* OPr, which is of the following form: its indices are links f from Pr to an effector OPr_f, called a *command* of the procedure; and the distinguished links from a component $OPr_{f'}$ to $OPr_{f'}$ are the links x between these effectors, which correlate the two commands f and f'.

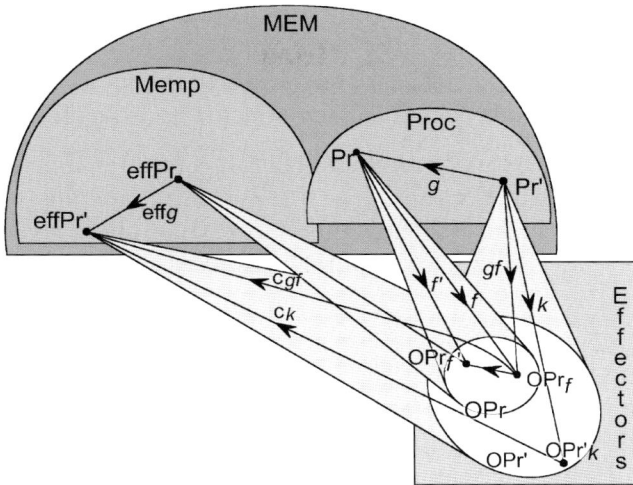

Fig. 8.2 Procedural memory.
The procedural memory **Proc** is an evolutive sub-system of the memory MEM of the memory evolutive system, whose components are called procedures. Each procedure Pr has a pattern of commands OPr, and a component of OPr is an effector OPr_f, where the index f is a command which links Pr to that effector. The pattern OPr itself has a colimit in the empirical memory **Memp**, called the effector record of Pr and denoted by effPr. A link g from some other procedure Pr′ to Pr transforms a command f of Pr into a command gf of Pr′, and it defines a link eff$g$ from effPr to effPr′ (the direction is inverted by the passage from **Proc** to **Memp**). The link eff$g$ binds the collective link (h_f) from OPr to effPr′, where h_f is the composite of gf with the binding link c_{gf} from OPr_f (looked at the component of OPr′ with index gf) to the colimit effPr′ of OPr′.

(ii) A record in **Memp**, called the *effector record* of Pr and denoted by effPr, which is the colimit of OPr.

Thus, the commands of the procedure Pr are the indices of its pattern of commands OPr, while the corresponding components of the system implementing them are effectors. And the effector record is the colimit of this pattern OPr, so that it binds its different effectors. As defined in Chapter 5, a pattern in **K** (such as OPr) is a family of patterns in the configuration categories, formed by an initial pattern and all its later configurations; and a colimit of this pattern (here, effPr) is a component whose successive configurations are colimits of the corresponding configurations of the pattern.

We construct the successive configuration categories of **Proc** by iteration, starting from the basic procedures as follows:

(i) We assume given an evolutive sub-system $\mathbf{Proc}_0$ of the memory MEM, the components of which are called the *innate basic procedures*, and for each basic procedure Pr, the pattern OPr of its commands.

(ii) **Proc** is constructed from **Proc₀** by a sequence of mixed complexifica-tions with adjoining of limits. The construction is such that, if Pr is a pro-cedure that is the limit of a pattern of procedures Pr_i, its effector record effPr will be the colimit of the pattern of the effector records effPr_i.

Here the exchange between limit and colimit is natural: a procedure that is limit of a pattern of procedures has for its commands all the commands of the different procedures. The explicit construction is given in the two following sub-sections, which are purely technical.

3.2. Basic Procedures

The basic procedures represent the procedures which the system possesses from birth, such as automatic responses to some features of the environment (*e.g.* replication procedure for a cell, heliotropism for a plant, grasping reflex for a baby). The basic procedures possess the following properties.

Definition. If **K** is a memory evolutive system, we assume the existence of a sub-system **Proc₀** of its innate memory, satisfying the following conditions:

(i) To each of its components Pr, called a *basic procedure*, there is associated a pattern of commands OPr: a component of OPr is an effector OPr_f indexed by a 'command' f from Pr to this effector; the distinguished links from OPr_f to $\text{OPr}_{f'}$ are the links x between them such that $xf = f'$.
(ii) The pattern OPr has a colimit effPr in **K** (belonging to **Memp**).
(iii) If g is a link from Pr′ to Pr in **Proc₀**, the map which associates to each command f of Pr the command gf of Pr′, defines a homomorphism Og from the sketch of OPr to that of OPr′.

Theorem 1. *There is an evolutive functor* eff₀ *from* **Proc₀** *to the system* **K**ᵒᵖ *opposite to* **K** *which maps a basic procedure* Pr *onto the colimit* effPr *of the pattern of its commands* (Fig. 8.2).

Proof. Let g be a link from Pr′ to Pr in (a configuration category of) **Proc₀**. By hypothesis the patterns of commands OPr and OPr′ admit colimits effPr and effPr′. If k is a command (hence an index) of Pr′, we denote by c_k the corresponding binding link from OPr'_k to the colimit effPr′ of OPr′. In par-ticular, if f is a command of Pr (hence an index of OPr), by hypothesis gf is a command of Pr′ arriving at OPr_f, and the binding link c_{gf} associated to gf goes from OPr_f to effPr′. These links c_{gf}, for the various indices f of OPr, form a collective link from OPr to effPr′, which binds into a link eff$g$ from the colimit effPr of OPr to effPr′. It is easy to verify that the map associating eff$g$ to g defines an evolutive functor eff₀ from **Proc₀** to **K**ᵒᵖ. [More formally, if g is a link from Pr′ to Pr, let Og be the map associating to the command f of Pr the command gf of Pr′. By mapping g to Og, we define a functor O₀ from the configuration category of **Proc**$_0$ at a time t to the opposite of the category of patterns in K_t; this category has for objects the patterns P in K_t, and the arrows from P to P′ are defined by the homomorphisms G from the sketch sP of P to sP′ such that P′G = P. The configuration at t of the functor eff$_0$ is obtained by composing O$_0$ with the partial functor colimit.]

3.3.   *Construction of the Procedural Memory*

The procedural memory **Proc** will be constructed by successive (classifying) complexifications of **Proc**$_0$ with respect to options which have as objectives only to classify some patterns. Thus, the complexification will only add limits (and not colimits). The problem is to describe the pattern of effectors associated to a procedure.

Theorem 2. eff$_0$ *extends into an evolutive functor eff from* **Proc** *to* $\mathbf{K}^{\mathrm{op}}$ *which maps a procedure* Pr *onto its effector record* effPr. *If* Pr *is a procedure that is the limit of a pattern* P *of procedures* P$_i$, *the commands of* Pr *come from those of the various* P$_i$, *and* effPr *is the colimit of the pattern image of* P *by eff* (Fig. 8.3).

Proof. The proof is by iteration of the number of complexifications necessary to construct Pr. We indicate it for the first one, the proof being similar for the following ones.

1. We suppose that Pr is an object of a classifying complexification **Proc**$_1$ of (a configuration category of) **Proc**; that is, Pr is the limit of a pattern P of **Proc**$_0$. Let d_i denote the projection link from Pr to a component P$_i$ of P. The commands of Pr (hence the indices of its pattern of commands OPr) are all the links of the form $d_i f$, where f is any command of one of the procedures P$_i$; and the distinguished links of OPr come from those of the different patterns OP$_i$. Then the colimit effPr of OPr is also the colimit of the pattern eff$_0$P (the image of P by the functor eff$_0$); this pattern has for its components the various effP$_i$ (Fig. 8.3). More precisely, OPr is the pattern colimit, in a category of patterns, of the pattern O$_0$P (image of P by the functor O$_0$, cf. preceding proof). It follows that the colimit effPr of OPr is also the colimit of the pattern eff$_0$P, because of the commutativity of colimits:

$$\mathrm{effPr} = \mathrm{colim\ OPr} = \mathrm{colim(colim\ O_0P)} = \mathrm{colim}_i(\mathrm{colim\ OP}_i)$$
$$= \mathrm{colim}_i(\mathrm{eff\,P}_i) = \mathrm{colim(eff_0P)}$$

2. To define the evolutive functor eff$_1$, we must map a link g from Pr′ to Pr in **Proc**$_1$ onto a link from effPr to effPr′. This link will bind the collective link from OPr to effPr′ defined as follows: it associates to the command $d_i f$

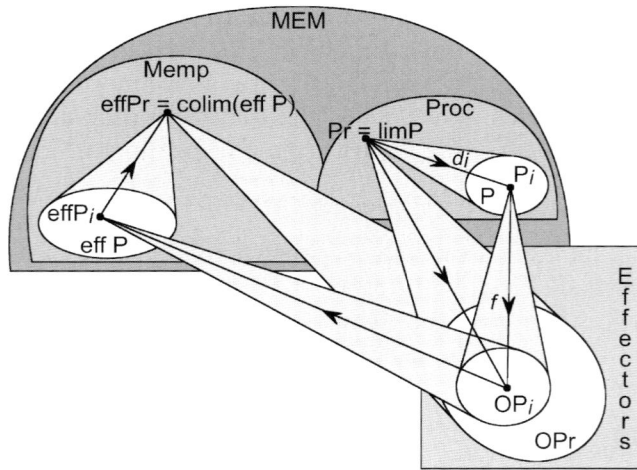

Fig. 8.3 Effector record of a limit of procedures.
The procedure Pr is the limit of a pattern P of procedures, and d_i denotes the projection link from Pr to P_i. The commands of Pr unite those of the various P_i, meaning that they are all the links of the form $d_i f$, where f is a command of one of the P_i; they are the indices of the pattern OPr, whose distinguished links are deduced from the distinguished links of the patterns OP_i (*cf.* the text). The effector record of Pr, which is the colimit of OPr, is the colimit of the pattern eff P of the effector records $effP_i$ of the various procedures P_i. Thus, the functor eff from **Proc** to **Memp** transforms limits into colimits: the effector record of a limit of procedures is the colimit of their effector records.

of Pr, the binding link to eff Pr′ associated to the index $gd_i f$ of Pr′. Here it remains to prove that $gd_i f$ is an index of Pr′, hence a command of Pr′, and the proof depends on the form of g. Let us recall (Chapter 4) that the links of a complexification with respect to a classifying procedure are: pro-simple links classifying a pro-cluster, their composites, and the links obtained from these by the operations of composition and of classification of pro-clusters. We consider only the case of pro-simple links, the other ones being deduced from them. So we suppose that Pr′ is the limit of a pattern P′, and that g is a (P′, P)-pro-simple link. The projection link from Pr′ to a component P_j' is denoted by d_j'. By the definition of a pro-cluster, to i is associated at least one index j such that there exists a link g_{ji} of the pro-cluster from P_j' to P_i (if there are several such links they are correlated by a zigzag of P′). Since g classifies the pro-cluster, we have $gd_i = d_j' g_{ji}$, hence $gd_i f = d_j' g_{ji} f$ for each command f of P_i. As g_{ji} is a link in **Proc**$_0$, we have seen that its composite $g_{ji} f$ with a command of P_i is a command of P_j'. It follows that $d_j' g_{ji} f$ is a command of Pr′ (from Part 1 applied to Pr′).

4. Functioning of the Procedural Memory

The procedural memory allows the system to develop quicker and more efficient responses to situations already met, improving its adaptation to its environment. The system remembers not only the procedures it has used in response to specific items (such as external signals, events or internal processes), but also their result. Thus, a later recognition of the same situation will trigger the procedure which (best) succeeded, and procedures which have failed will no longer be used.

4.1. Activator Links

The recall of a procedure might be accomplished by the intermediary of a link in the memory, called an activator link, from the record of the situation to the procedure.

Definition. An *activator link* h from a record M to a procedure Pr is a link from M to Pr that has the following property (Fig. 8.4):

> The composites hfc_f, where f is a command of Pr and c_f is the corresponding binding link from the effector OPr_f to the colimit of OPr, are all equal to a unique link h' from M to effPr.

(Note that this condition is automatically realized for any h if the pattern OPr is connected).

   More roughly, an activator link is a link h from M to Pr which factorizes a link h' from M to effPr, so that the recall of the procedure can be done either directly through h, or indirectly by activating its effector record through h'.
   Initially, activator links are given in the innate memory, which connect the specific situations for which the system has an innate response to the corresponding admissible basic procedures. For a living organism, their existence will result from natural selection, which retains the best adapted individuals. For example, following the detection of an antigen, it is destroyed via a cascade of events in the innate immune system; a hunger sensation starts an innate procedure to eat or to seek prey. Later, we have seen that the memory develops by a succession of mixed complexifications, which add to the innate memory iterated colimits of records and iterated limits of basic procedures (under the effect of the different co-regulators and the interplay among their procedures). New activator links will be formed during these complexifications, starting from the innate activator links given above. The process of mixed complexification (see Chapter 4) describes how these links are constructed.

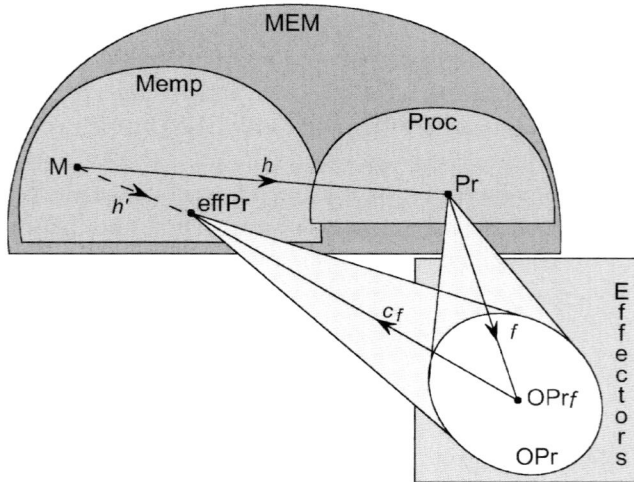

Fig. 8.4 Activator link.
M is the record of a situation, and Pr a procedure which has been successfully used for responding to it; OPr is the pattern of commands of Pr. An activator link h from M to Pr is formed, which allows Pr to be recalled if the same situation is repeated. The (dotted) link h', from M to the effector record effPr of Pr, is the composite of h with a command f of Pr and the corresponding binding link c_f from OPr_f to the colimit effPr of OPr; for h to be an activator link, the link h' must be independent of the command f.

4.2.   Recall of a Procedure

The activator links allow the recall of the procedures, either by a co-regulator or more globally by the system. When the system (via its co-regulators) recognizes a certain situation having a record M that is connected to a procedure Pr by an activator link h, the activation of M activates h, and consequently may recall Pr. This leads to the activation of the pattern OPr of commands of Pr, and thus the recall of its colimit effPr; and that in turn imposes the commands of the procedure on effectors.

What does it mean if the procedure Pr is the limit of a pattern P of procedures? Its recall leads to the activation of each procedure P_i that is a component of P, via the projection link d_i from Pr to P_i. It follows that the pattern of commands OP_i of each procedure is activated, and consequently also its effector record $effP_i$. Now the $effP_i$ are the components of the pattern that is the image of P by eff. Consequently, their simultaneous activation will recall the colimit of this pattern; and we have proved above that this colimit is also the effector record effPr of the limit Pr of P. Thus, the end result will be the activation of the effector record of Pr. In complex systems, the result of the procedure can be stored through weights on the activator links, which increase if the procedure succeeds, and decrease otherwise. In particular, these weights may measure the probability of the procedure's success during previous encounters with the same situation.

The recall of a procedure goes from the conceptual level of the procedural memory, to the operational level of the effectors, possibly iterating the operation for more complex procedures constructed as iterated limits, while going down to lower and lower operational levels. For example, for an animal, a procedure Pr to capture its prey could be chosen by a co-regulator which recognizes the prey in its landscape; the commands of Pr will activate a pattern OPr of complex muscular commands, a pattern memorized during earlier, similar experiences by its colimit effPr. And effPr will activate the corresponding lower motor co-regulators to complete the movement.

4.3. *Generalization of a Procedure*

We have said that the memory of a memory evolutive system as a whole is robust and flexible. This is also true for the procedural memory, and allows for adaptation of the procedures to the context. As for any record, once formed a procedure assumes a distinct identity, in the sense that a procedure Pr constructed as the limit of a pattern P of procedures can be, or later become, the limit of another pattern P' of procedures; and its recall can be accomplished through the activation of one or the other. As the effector record of a limit is the colimit of the effector records of its components, effPr will be at the same time the colimit of the pattern that is the image of P by eff, and of the pattern that is the image of P' by eff, and it can be recalled via one or the other of these decompositions, with the possibility of a complex switch between them. In other words, the procedure may be carried out with various parameters (those relating to P or those relating to P') according to the context, the parameters intervening at the level of the procedures themselves as well as that of their effector records. And this extends to the case of procedures constructed as iterated limits from basic procedures, with the possibility of a choice of new parameters at each stage.

This flexibility of a procedure makes it possible to generalize the conditions wherein it can be applied: if a procedure has been effective in a certain context, and if only a few features change, different co-regulators will try to apply the same procedure, or modify it as little as possible.

• If it succeeds, its domain of application is extended; for instance, a dog learns to bring back a ball to his master in a garden; later on, if another person sends him another object, in a different place, he will apply the same procedure to bring back the object.
• Conversely, if it fails, and creates a fracture for some co-regulators of the system, the repair process may modify some of the corresponding

commands to adapt them to the new context. For example, before knowing how to walk, a child learns how to coordinate various more or less innate simple movements, to stand straight and advance a leg when he is held; that leads to the memorizing of the corresponding procedures in the motor areas. Then these procedures themselves are coordinated in patterns which acquire their own colimits, making it possible to take a step without losing balance. This continues until the formation of a complete procedure to walk, usable under the most varied conditions (Josephson, 1998). However, if the child tries to advance too quickly, his motor coordination will be lost, and he is likely to fall.

In all cases, there will be a process of generalization, so that a procedure can apply later on to various particular cases by adapting the parameters to the context. This process may consist simply of slight variations in the initial procedure, or else of more widespread modifications. For example, a traditional printer, who initially worked on only small projects, can simply use his know-how to embark on more important ones; but he can also make fundamental changes to his workflow, such as by computerizing his operations.

5.   Selection of Admissible Procedures

Here we analyse how a particular co-regulator participates in the selection of procedures, both those intervening in its landscape and those arising through the interplay among the procedures. This analysis leads to a more complete definition of a co-regulator.

5.1.   *Admissible Procedures*

We have said that, in a memory evolutive system, a co-regulator has a set of admissible procedures which characterize its function. Now we have all the tools to specify what this means formally, and thus to give the following definition of a co-regulator.

Definition. A *co-regulator* of a memory evolutive system is an evolutive sub-system satisfying the following conditions:

(i) Its time scale is a discrete subset of the time scale of the whole system.
(ii) Its *time lag* at time t is the maximum of the propagation delays at t of the links between its agents, and of the links arriving to them from the memory. This time lag must be of an order of magnitude smaller than the length of the step at t (which extends between two successive times of the time scale of the co-regulator).
(iii) It is equipped with a set of *admissible procedures* Pr in **Proc**, with afferent links from Pr to its agents and efferent links from agents to some of

the effectors of Pr, so that the composite of an afferent link and of an efferent link (when it exists) is a command of Pr whose effector can thus be controlled by the co-regulator.

This last condition means that the admissible procedures are observable in the landscape by their perspectives, and that the co-regulator can activate some of their effectors through the efferent links, namely those corresponding to a command obtained as the composite of an afferent and an efferent link. It is only these commands which the co-regulator can control. The admissible procedures can vary over time; in particular new ones can be formed to respond to new situations.

As specified in Chapter 6, for each step of a co-regulator, objectives are selected on its actual landscape, as well as an admissible procedure to carry them out. The objectives form an option on the landscape. If the procedure succeeds, the objectives should be realized, and the anticipated landscape for the next step should be the complexification of the landscape with respect to this option. The result is evaluated and stored in the memory at the next step. This raises the following questions, which are considered in the next sub-sections:

(i) How are procedures to carry out particular objectives constructed, and how are they remembered so that they become admissible procedures?
(ii) How may the existence of admissible procedures help in the selection of objectives by a co-regulator?

5.2. *Procedure Associated to an Option*

Let us recall that an *option* (Chapter 4) on a category is a list of objectives of the form: patterns to bind (and also patterns to classify, if the option is mixed), elements to be absorbed, and objects or colimits to be eliminated. We want to associate a procedure to the option so that the commands of the procedure achieve the objectives of the option. To this end, we first give some ideas for how to associate a procedure to each objective of an option. The procedure associated to the option will be the limit in the procedural memory of the procedures so associated to each objective.

Binding. The objective to bind a pattern P is associated to a procedure Pr which, for each index i of P, has a command to the component P_i (taken as an effector), these links being correlated by the distinguished links of the pattern P, so that the pattern OPr of commands of Pr be isomorphic to P. Thus, the effector record of Pr, which is the colimit of OPr, will also be the colimit of P; and the storage or the recall of the procedure Pr will correspond to the binding of P.

Classification.   The objective of classifying a pattern P is associated to the procedure limit of P, in which each component P_i is looked at as a trivial procedure having itself as its sole effector.

Absorption.   The absorption of an external element depends on the operation of a pattern of components through which the element will be introduced inside the system (such as receptors on the membrane of a cell, sensory cells for an animal, or the like). This pattern will be (as above) the pattern of commands of the absorption procedure; its recall will activate this pattern, thus allowing for the absorption of the element.

Elimination.   Similarly, the elimination of an object (*e.g.* decomposition of a colimit) will result from an internal process, to which we associate the commands of the elimination procedure.

• A pattern acting as an inhibitor or destroyer may become the pattern of commands of the procedure. For example, the synthesis of a protein will be suppressed if its transcription is inhibited.
• An (internal) pattern of components may act as an 'exporter'; the commands of the procedure will activate this pattern and its effector record will be the colimit of this pattern. For example, the exocytosis of the content of a vesicle depends on the opening of fusion pores on the membrane of the vesicle.
• The suppression of the colimit of a pattern may also just result from the fact that it had been formed by another procedure that was interrupted, so that the command for this colimit is not maintained. In this case, there is no suppression procedure as such, but the pattern simply no longer figures among the patterns to be bound.

5.3.   *Selection of a Procedure by a Co-Regulator*

At each of its steps, a co-regulator selects objectives and a procedure to achieve them. The objectives are related to standard changes and form an option on the landscape. Once the objectives are selected, the above construction associates a procedure to this option, and if it succeeds this procedure will be later stored as a new admissible procedure. However, if similar circumstances have been met earlier and a successful admissible procedure devised, the co-regulator, instead of first selecting objectives, may directly resort to the memory to recall this admissible procedure (through its perspectives in its actual landscape); and its objectives will be those achieved by the commands of this procedure which it controls via its efferent links to effectors.

Let us study more thoroughly the case where the actual situation S has already been met, and has a record M (the case of a new situation is studied below). For a particular co-regulator, say E, we have several possibilities.

• First, a procedure may be imposed on E; for instance, if E is an effector co-regulator, and if the record M is connected by an activator link to a unique procedure, some of the commands of which are forced on E. The procedure can also come from a random process dependent on the system or its environment.
• Alternatively, E may observe in its landscape an activator link from a partial record M_E of S to an admissible procedure Pr_E. The recognition of M_E activates this link and recalls the procedure via the perspective of this link. The objectives of the co-regulator E will then be the standard changes which should be achieved by the effectors of Pr_E controlled by E. Let us recall that it is possible that only some of the commands of Pr_E are so controlled; in this case the other commands are not taken into account in the anticipated landscape of E, though they will operate on the system, whence they will be a cause for error.
• In the preceding situation, if there exists activator links to several admissible procedures previously used to respond to the situation in different cases, all will be activated; if these links have different weights, the strongest one will be selected, and the corresponding procedure relayed to the system; otherwise, one will be randomly selected.
• Lastly, E may observe in its landscape a partial record of S, but it is not connected to any admissible procedure. Thus, there will be either inaction, or, especially for higher level co-regulators, the eventual formation of a new procedure, to be stored in memory as an admissible procedure.

In all cases, the anticipated landscape is the complexification of the actual landscape with respect to the option formed by the objectives of the co-regulator. At the following step, the co-regulator evaluates whether the anticipated landscape is really the new landscape (by trying to form the comparison functor between them; see Chapter 6). If it is not, the new objective will be to compensate for the difference.

6. Operative Procedure and Evaluation

The admissible procedures which had been selected by the different co-regulators through their perspectives are relayed to the system by the difference functors, and they enter in competition. Moreover, the system or its environment can also transmit commands of procedures independent from the co-regulators, in a deterministic, probabilistic or even random manner (*e.g.* in presence of 'noise'). Then an equilibration process, the interplay among the procedures (Chapter 6), determines the procedure effectively carried out on the system, called the operative procedure Pr°.

6.1.   Interplay among the Procedures

If all the procedures are compatible, the operative procedure should be the limit of a pattern having for components these different procedures. However, the procedures selected by the different co-regulators must be carried out by effectors which operate not on the landscapes but on the system, and, as explained in Chapter 6, there are several potential sources of incompatibility between their procedures, and also with procedures transmitted independently from the co-regulators. In such cases the operative procedure may reject or modify some of the commands, with a possible fracture for the corresponding co-regulators.

There is no general rule to determine the result of this interplay, which depends on a kind of Darwinian selection between the various commands. The process has great flexibility, since the activation of a complex procedure can be accomplished with different parameters (see above), corresponding to the different patterns that it classifies (*i.e.* which admit it as a limit). In particular the compatibility of the procedures selected by two co-regulators may depend on an adequate choice of their parameters. For instance, in a restaurant, one's choice of wine will depend on the food one selects.

If the procedure of a particular co-regulator, say E, is not preserved, and if the partial record M_E of the situation S is linked (even with a small weight) to a procedure by an activator link observable or not by E, the interplay can recall this procedure rather than the selected one, if it is more compatible with the others. Let us note that, if S has a record M which is itself connected to a procedure Pr by an activator link h, such activator links always exist; indeed, each partial record is then connected to Pr by the composite with h of the binding link from the partial record M_E to the record M of S. The commands of the procedures which have the greatest weights (often those selected by the higher levels) are generally preserved. For example, in a business, if two different departments call upon the same repairman at the same time, he will go first to the department whose needs are the most urgent or important.

6.2.   Formation of New Procedures

In the discussion above, we have considered the case of a situation already known. If the situation is unknown, both the various co-regulators and the global system will try to recognize some of its features, to recall procedures previously used for each one of them, and to construct a new procedure by combining them in an appropriate way. For example, an animal learns a complex motion by decomposing it into simpler ones already known. In the discussion below, we describe this process for the system as a whole (the steps are similar for an individual co-regulator).

Let us suppose that the situation S activates a pattern Q of records which has no colimit, but in which each component Q_i has an activator link h_i to a procedure Pr_i. Then the objective will be to form a procedure Pr such that its commands join together those of the various Pr_i; the situation S will be stored in memory by the colimit cQ of Q, and an activator link from cQ to Pr will be formed, binding the different activator links h_i. However, this will be possible only if these activator links are sufficiently compatible between themselves, and likewise with the distinguished links of Q. In particular, such a procedure can be constructed if there is a pattern P having the Pr_i for components, and satisfying the following condition: let h'_i be the link from Q_i to the effector record $effPr_i$ of Pr_i (we have already noted that it exists as the composite of h_i with the commands of Pr_i and the binding links from OPr_i to $effPr_i$); then these links h_i' generate a cluster from Q to the pattern of the effector records of the Pr_i (which is the image of P by the functor eff; see Fig. 8.5). In this case, the situation S will be stored in the memory by the colimit M of Q, and the procedure by the limit Pr of P. The cluster binds into a simple link h' from M to the effector record of Pr (which is the colimit of eff P), and an associated activator link h will be formed from M to Pr (all this will result from a process of mixed complexification).

6.3. *Evaluation Process and Storage in Memory*

The operative procedure Pr^o obtained after the interplay among the procedures can fail, and cause a fracture for some co-regulators. This is especially the case if their structural temporal constraints cannot be satisfied. For example, the regular schedule for the publication of a journal cannot be respected if the printer has not printed it in due time. The failure can be due to a poor apprehension of one of the parameters of the context, either locally (at the level of a co-regulator) or globally (in the interplay), or to some unknown external cause, without calling into question the adequacy of the selected procedures. It may also be attributed to the fact that, over time, the system and its environment change. Procedures which once succeeded can later become ineffective. It will thus be necessary to evaluate the result of the selected procedures and to remember this result. This will still be done via the intermediary of the co-regulators, which evaluate the result in their new landscape at the following step (by seeking to form the comparison

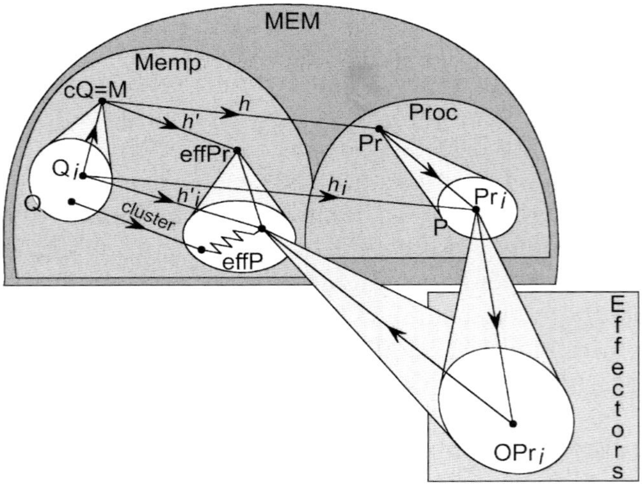

Fig. 8.5 Formation of a new procedure.

A new situation is met, which activates a pattern Q of records Q_i having each an activator link h_i to a procedure Pr_i; let h'_i be the corresponding link from Q_i to $effPr_i$ (constructed as in Fig. 8.4). We suppose that the Pr_i are the components of a pattern P of procedures and that the links h'_i generate a cluster from the pattern Q to the pattern effP (image of P by the functor eff). Then the situation will be remembered by the colimit M of Q, the selected new procedure Pr will be the limit of P, and an activator link h will be formed from M to Pr, with the corresponding h' binding the cluster.

functor from the anticipated landscape to the new landscape). One of the objectives of the procedure of a co-regulator at this following step will be to store the result of the procedure chosen in response to S, possibly introducing an activator link or modifying its weight, in particular if the weight measures the probability of success of the procedure. There are two possibilities:

(i) If the objectives have been realized, there will be an activator link from the record M of the situation S, to the operative procedure Pr^o or, if the link already exists, its weight will be strengthened. In particular, if Pr^o is obtained as a limit of the procedures Pr_E selected by some of the co-regulators, and if for each one there exists an activator link h_E from a partial record M_E of S to Pr_E, these links will be strengthened, and an activator link h from M to Pr^o will be constructed (as in the preceding Section 6.2), or strengthened if it already exists (Fig. 8.6).

(ii) On the other hand, if the procedure fails, no link will be created; or, if an activator link already exists, its weight will decrease, or the link may even be suppressed, if the failure is repeated several times.

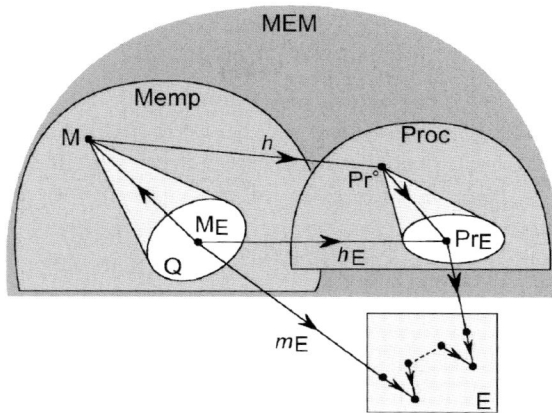

Fig. 8.6 Storing of a procedure in the memory.
The situation S activates its record M, which is the colimit of the pattern Q of its partial records. E is a co-regulator for which M has a partial record (and even, on the figure, an internal E-record) M_E; it selects the procedure Pr_E in its landscape, through an activator link h_E from M_E to Pr_E. The operative procedure Pr^o is the limit of those procedures Pr_E which have succeeded, for the various co-regulators. An activator link h is then constructed from the record M of S to Pr^o, or strengthened if it already exists.

7. Semantic Memory

The development of the memory allows the system to give better adapted responses to specific stimuli. To have more independence from the context, and to act more quickly and in a more efficient way, proficient systems are able to recognize similarities among different stimuli. They can react in a specific way not only to a particular context, but also to a whole class of similar contexts. For example, a frog will jump after a fly, or after any flying object of approximately the same size. This involves a classification of records, so that similar items are grouped into the same class, thus leading to a finite number of invariance classes, or *invariants*, which hold despite changes in circumstance. These invariants are internally represented by records of a specific type, which we call *concepts*. We also determine which of the interactions between concepts confer some meaning with respect to the functioning of the system. The concepts, with the links so defined between them, form a sub-system of the memory, the *semantic memory*. The literature on semantics is very diverse; we adopt the view that a concept is a flexible internal representation of a class of records having a family resemblance, to borrow the terminology of Wittgenstein (1953). A memory evolutive system develops an elaborate semantic memory only if it is complex enough.

7.1. How Are Records Classified?

A semantic memory allows objects to be identified by certain of their attributes, and then classified. For example, similar geometric figures, such as a series of differently sized cubes, will all be identified as cubes, regardless of their size. The development of such a memory in a memory evolutive system relies on solving a hierarchy of classification problems under the control of the co-regulators, and is a three-stage process (Ehresmann and Vanbremeersch, 1992b). In this section, we give a brief overview of this process, which will be developed in the following sections.

Let us recall that a co-regulator has a specific function determined by its admissible procedures. It classifies items (external inputs or internal states) or their records in a pragmatic way, by reacting similarly to items similar with respect to its function. The result is a comparison of items with respect to the attribute corresponding to its specific function. For example, a librarian may sort books according to their subject. This classification by a co-regulator, say E, is purely pragmatic. It takes an internal 'meaning' only if it is internally detected at a higher level, and reflected by the formation, for each of its invariance classes, of a record of a specific type (we call it a *concept* for the attribute associated to E, or E-concept). The items (or the records) of the class of which the concept gives an abstract representation are called the *instances* of the concept. Once the concept is formed, it evolves and takes its own identity, and newly formed records may become new instances. The later recall of a concept will recall one of its instances, if possible the most adapted to the current context (as determined by the interplay among the procedures); but it may also lead to the recall of different instances and to a shift between them.

The above classification depends on the specific attribute associated to a co-regulator. More complex concepts are formed by combining together concepts involving different attributes, thus leading to the whole hierarchy of concepts that form a semantic memory. The existence of this semantic memory gives the system a double degree of freedom to interact with the context, since a concept can be recalled by any of its instances, and then by any of the ramifications of that instance. It does not require the development of language, though language allows the semantics to be extended, through the communication of concepts.

7.2. Pragmatic Classification with Respect to a Particular Attribute

The first stage is the comparison of two objects with respect to a specific attribute (for example: colour in the human visual cortex; some epitope for cellular receptors; size of individuals in some statistical survey of a population; authorship in the catalogue of a library). The comparison depends on the way in which the items behave with respect to the particular co-regulator associated with the treatment of this attribute.

Let E be a co-regulator in a memory evolutive system **K**. Only some aspects of the items the system recognizes can be observed in its landscape; these are the attribute(s) (or features) of the items that are 'treated' by E. For example, in the human visual cortex, a colour-CR treats only the colour of an object; alternatively, searching a library database for a particular author's name will retain from the collection only the books by this author. As said above, E carries out what might be called a 'pragmatic' classification of the items, according to the attribute(s) which it recognizes, by reacting in the same way to items of a same class. This classification will be made via the records of the items. If an item S (say, a signal) has a record M, it is recognized by E at a given time t only through the pattern of agents activated by the aspects of M recognized by E, which we call its E-trace. Another record M' will be treated by E as being in the same class as M if the E-traces of M and M' decode the same kind of information, that is, receive distributed links from the same objects; more precisely, if these traces are *pro-homologous* patterns (Chapter 4). This pragmatic classification can be observed from 'outside' the co-regulator. From this, we derive the following:

Definition. If M is a record, we define the E-*trace of* M at a given time, denoted by $Tr_E M$ or simply TrM, as the following pattern: its sketch is the graph which has for vertices the aspects b of M that have perspectives in the actual landscape of E, and for arrows from b to b' the links a between agents such that $ba = b'$; and TrM associates to b the agent TrM_b which receives the aspect b. Two items or two records are said to be in the same E-*invariance class* if their records M and M' have pro-homologous E-traces (see Fig. 8.7).

In other words, the trace of M has for indices the aspects b of M for some agent of E, the component TrM_b associated to b being this agent; while the distinguished links are defined by the links a between agents correlating their indices. The E-invariance classes form a partition of the set of records having an E-concept.

We note that Cordier and Porter have proposed a slightly stricter definition, within the framework of the 'shape theory' of Borsuk (1975): namely, that the E-traces be isomorphic in the category of patterns, thus isomorphic in an adequate shape-category of Holtsztynski (Cordier and Porter, 1989). However, this definition seems too restrictive in practice; *e.g.* the assemblies of neurons activated by two blue objects in the visual cortex are not exactly the same, but only fulfil the same operation.

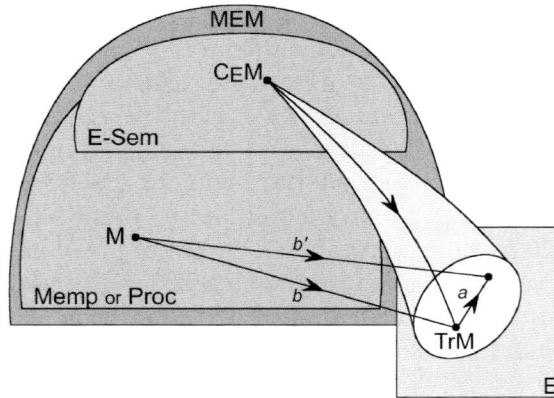

Fig. 8.7 Formation of an E-concept.
E is a co-regulator and M a record in the memory MEM. A subset of agents of E recognizes certain aspects b, b' of M. These agents (indexed by the various aspects b) form the components of a pattern TrM of agents of E, called the E-trace of M; a distinguished link from TrM_b to $\text{TrM}_{b'}$ is defined by a link a in E such as $b' = ba$. Through the operation of a higher level co-regulator, a limit $C_E M$ of TrM is formed, and becomes the E-concept of M. The other instances of this E-concept (not shown) are all those records which form 'similar' E-traces, in the sense that their E-traces also admit $C_E M$ as the limit; with M, they form the E-invariance class of M, or of $C_E M$. Over time, this class may be extended by the formation of new instances. The E-concepts are the components of the E-semantic memory E-**Sem**, an evolutive sub-system of the memory MEM.

7.3.   *Formation of an E-Concept*

The classification described in the above section is only implicit; the fact that the co-regulator E responds to two records M and M′ in a manner which is similar (*i.e.* by pro-homologous patterns) can be apprehended only externally; in particular, if there is a higher level co-regulator which may simultaneously observe the E-traces of M and M′, and over a longer time scale. This is in conformity with: 'meaning arises only for someone else and for a temporality differing from a moment reduced to its instantaneity' (translated from Draï, 1979, p. 82). Such a higher co-regulator will confer a meaning to an E-invariance class, namely what is equivalent among, or what remains invariant between, its various elements (or instances) M. This in-variance will be stored via the formation at a given time of a more complex unit of categorization, called an E-*concept*, which represents the E-invar-iance class of M at this time by a single object, a record of a certain type which we shall denote by $C_E M$ (or simply CM). All the records of the same E-invariance class have the same E-concept. This E-concept will be modelled by the limit of the E-trace of M. Since two pro-homologous patterns have the same limit (Chapter 4), all the records in the E-invariance class of M have the same E-concept, and the E-concept characterizes their E-class. For instance, in a library database, authorship characterizes the class of books written by the same author.

More formally, the E-concept will be formed at the level of a higher level co-regulator, having a longer period, which simultaneously receives information about the records M and their E-traces Tr_EM. One of the objectives of its procedure will be to classify the E-trace by the formation of a limit of Tr_EM which will give a formal representation of the invariance class of M with respect to the attributes classified by E.

Definition. We say that a(n item with a) record M admits an E-*concept* C_EM if the E-trace Tr_EM of M admits a limit C_EM. In this case, M is called an *instance* of C_EM (see Fig. 8.7).

The set of the instances of an E-concept C_EM at a given time forms its E-*invariance class* at this time. The E-concept itself, as a record, is a particular instance of its E-invariance class, and it plays the role of a prototype, in being the closest E-concept linked to M in the following sense:

Theorem 3. *If M has an E-concept C_EM, there exists a link m° from M to C_EM satisfying the 'universal property': any other link g from M to an E-concept C_EN factors uniquely through m°, so that $g = m^\circ g^\circ$ for a unique g° from C_EM to C_EN (see Fig. 8.8).*

Proof. The different aspects of M for E form a distributed link from M to its E-trace TrM, which is classified by a link m° from M to the limit CM of TrM. Let g be a link from M to an E-concept CN; this concept is the limit of the E-trace TrN of a record N. We define a distributed link (g_d) from M to TrN as follows: an index d of TrN is an aspect of N for an agent of E; the composite of g with the corresponding projection link q_d from the limit CN is an aspect of M for E, hence an index for the pattern TrM, and g_d will be the projection link from the limit CM of TrM associated to this index. The distributed link from CM to TrN so constructed is classified by a link g° from CM to the limit CN of TrN satisfying the equation $m^\circ g^\circ = g$. The link m° will be called the E-*universal link* associated to M.

As for any record, an E-concept C_EM is not a fixed entity, but it is an evolving and flexible representation. Thus, the E-invariance class consisting of its instances may vary; in particular records formed while the concept C_EM already exists will become new instances if their E-trace admits C_EM for its limit (and thus is pro-homologous to the trace of M). For example, a librarian classifies books into a finite number of subjects (identified by their names), and the subject-concept of a book will be the subject which most

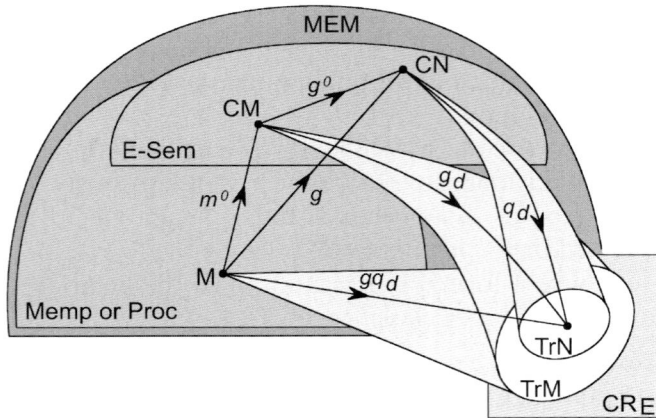

Fig. 8.8 E-universal link from a record to its E-concept.

M is a record having an E-concept CM; the aspects of M recognized by E form a distributed link to its E-trace TrM, which is classified by a link $m°$ from M to the limit CM of TrM. The record M is linked to the E-concept CN of a record N by a link g. We construct a distributed link (g_d) from CM to TrN as follows: for any index d of TrN, the composite gq_d of g with the projection link q_d (from the limit CN of TrN to TrN$_d$), is an aspect of M recognized by E, hence an index of TrM; let g_d be its associated projection link from CM to TrN$_d$. The distributed link (g_d) from CM to TrN is classified by a link $g°$ from CM to CN; and we have $g = m°g°$. Roughly, CM is the 'nearest' E-concept linked to M (by $m°$), since every other E-concept CN linked to M is also linked to CM; it justifies to call $m°$ an E-universal link.

covers the contents of the book; if he receives a new book, he begins by determining its subject.

The recall of an E-concept can potentially be done by any of its instances M via the E-universal link $m°$ from M to C_EM; conversely the recall of C_EM via one of its aspects f for some co-regulator may also recall its instance M via the composite $m°f$; and the same for another instance M′. The succession of these two operations (from M to its E-concept, and then to another instance M′) is called a *shift* from M to M′. It can be used (*e.g.* in the interplay among the procedures) to recall the instance of the concept the most adapted to the context.

By definition, two instances of an E-concept have pro-homologous traces; these traces can be connected by a cluster of links between agents of E, so that their similarity is seen at the level of the co-regulator. However, if the memory evolutive system is complex enough, the traces of M and M′ can be pro-homologous without being connected, and their homology can only be observed externally to E, through their global behaviour (they receive distributed links from the same objects), and their E-concept is a *pro-multifold* object (Chapter 4). In this case, its formation is an emerging property of the whole system. The existence of such pro-multifold objects in a category means that its opposite satisfies the multiplicity principle. Thus, *the condition which allows the system to develop an elaborate* E-*classification, with pro-multifold* E-*concepts is that both the memory evolutive system and its opposite satisfy the multiplicity principle.*

7.4. Links between E-Concepts

The E-concepts form the components of an evolutive sub-system of the memory MEM of the memory evolutive system. Now we are going to characterize the links that model their interactions. We begin by claiming that the links between E-concepts are obtained as the images by a functor 'E-concept' of links between their instances.

Theorem 4. *There exists an evolutive functor* C_E *from the full evolutive sub-system of* MEM *having for components the records admitting an* E-*concept, onto an evolutive sub-system* E-**Sem** *of* MEM *(see Fig. 8.9).*

Proof. Let M and N be two records which have E-concepts CM and CN respectively, and let *f* be a link from M to N in (a configuration category of)

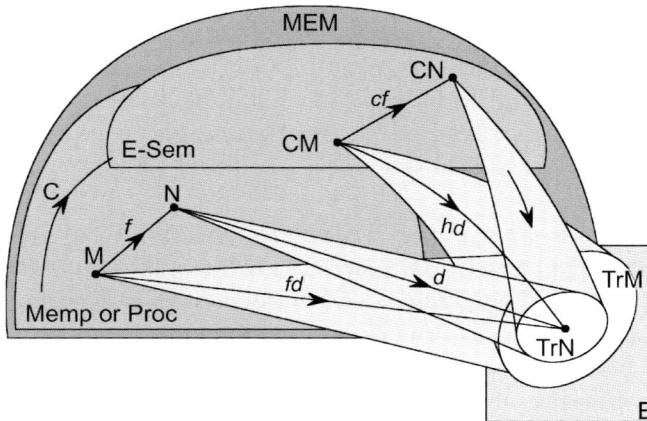

Fig. 8.9 The functor E-concept.
M and N are records which have E-concepts, denoted by CM and CN, and *f* is a link in the memory MEM from M to N. Thus, CM is the limit of the E-trace TrM of M and CN the limit of TrN. We construct a distributed link (h_d) from CM to TrN as follows: if *d* is an aspect of N recognized by E, the composite *fd* is an aspect of M recognized by E, and h_d is the corresponding projection link from the limit CM of TrM to the target TrN$_d$ of *d*. This distributed link is classified by a link C*f* from CM to CN. There is a partial functor C_E from MEM to E-**Sem** which associates to a record M its E-concept, if it exists, and to a link such as *f* the link C*f* constructed above.

the memory MEM. The E-concepts CM and CN are the limits of the E-traces TrM and TrN. We construct a distributed link (h_d) from CM to TrN as follows: an index d of TrN is an aspect of N for an agent of E; the composite fd is an aspect of M which is the index of the component of TrM corresponding to the same agent, and we take for h_d the projection link associated to fd, which is a link from the limit CM of TrM to this agent. The links h_d form a distributed link from CM to TrN, which is classified by the link Cf from CM to CN.

The evolutive sub-system which has for components the E-concepts, and for links the links so defined above, is called the E-*semantic memory*, denoted by E-**Sem**. Its formation requires the existence of the necessary limits.

Theorem 3 can be interpreted in terms of the functor C_E of Theorem 4 above: it means that the E-universal link m^o from M to its E-concept C_EM defines C_EM as a free object generated by M with respect to the insertion functor from E-**Sem** to MEM (so that C_E is a partial adjoint to this functor; see Chapter 1). The uniqueness (up to isomorphism) of a free object leads to another characterization of an E-concept and of its E-class: namely, the E-concept of a record M (if it exists) is the unique E-concept to which M is connected by an E-universal link; and the instances of an E-concept A are the records which are linked to A by an E-universal link.

The following theorem (which could be directly deduced from the general properties of adjoint functors) shows that the E-concept of a complex object is obtained as the colimit in E-**Sem** (which can be different from the colimit in MEM) of the pattern formed by the E-concepts of its elementary components (Fig. 8.10).

Theorem 5. *The partial evolutive functor C_E from* MEM *to* E-**Sem** *preserves colimits: If M is the colimit of a pattern Q of records (M_i) and if M and the various M_i have E-concepts, then C_EM is the colimit in* E-**Sem** *of the pattern C_EQ, image of the pattern Q by the functor C_E.*

Proof. Let (h_i) be a collective link from the pattern C_EQ to an E-concept CN. The composites $m_i^o h_i$ of the universal links m_i^o (from M_i to CM_i) with h_i form a collective link from Q to CN which binds into a link g from the colimit M of Q to the E-concept CN, and (by Theorem 3) this link g factors through the E-universal link m^o from M to CM into a unique link h from CM to CN; the link h binds the collective link (h_i).

The consequence is that, once 'elementary' E-concepts are constructed (or are initially given, as will often be the case in a neural system), more complex E-concepts are obtained as colimits of patterns constructed on them.

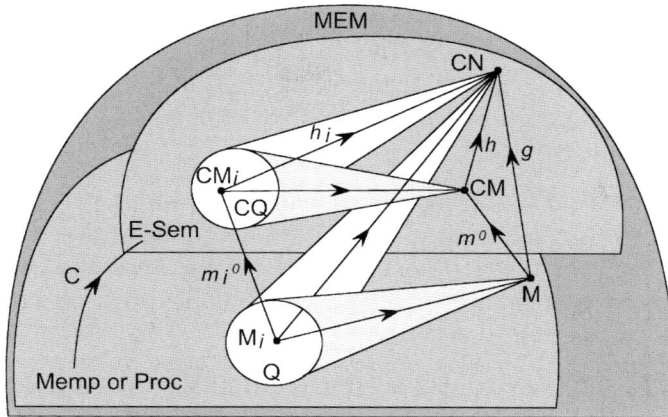

Fig. 8.10 Colimit of E-concepts.

M is a record which is the colimit of a pattern Q of records M_i. Then its E-concept CM is the colimit of the pattern CQ, image of Q by the E-concept functor C. This is proved by taking a collective link (h_i) from CQ to an E-concept CN. The composites $m_i^o h_i$, where m_i^o is the E-universal link from M_i to its E-concept CM_i, form a collective link from Q to CN which binds into a link g from M to CN. This g factors through the E-universal link m^o from M to CM into a link h from CM to CN binding the collective link (h_i). This proves that the functor E-concept preserves the colimits.

7.5. Semantic Memory

Up to now we have considered invariance classes and concepts relating to a particular co-regulator. Generally, the same record M will have also an invariance class with respect to some other co-regulators, corresponding to a classification with respect to different attributes. For example, in the library database mentioned previously, a book by X on systemics and a book by X on physics share the same authorship X; but the subject-concept of the first one is 'systemics', whereas that of the second one is 'physics'; to have a finer classification, the librarian can also classify books by both author and subject. The invariance class of M with respect to more than one co-regulator, say E_1 and E_2, can be obtained simply as the intersection of the invariance classes relating to E_1 and E_2, and the corresponding (E_1, E_2)-concept is the product (Chapter 4) of the concepts with respect to E_1 and E_2 (for example 'author X on physics' is the product of the author-concept X and of the subject-concept 'physics'). More generally, richer concepts are defined as limits of patterns of already existing concepts. On the other hand, more abstract concepts are formed to unite the invariance classes of known concepts, for instance 'Polygon' as the sum of 'Triangle', 'Square' and the various 'n-gons'.

For this, let us consider the evolutive sub-system **USem** of the memory MEM union of the E_k-**Sem** for the different co-regulators E_k with respect to which E_k-concepts can be formed. The formation of richer and more abstract concepts over time is modelled by the extension of **USem** through successive mixed complexifications.

Definition. The *semantic memory* of a memory evolutive system is defined as an evolutive sub-system **Sem** of MEM which contains as sub-systems the different E_k-**Sem** for the various co-regulators E_k, and develops from their union **USem** by a sequence of mixed complexifications. A component of **Sem** is called a *concept.*

Thus, richer and increasingly abstract concepts are constructed in several steps, from concepts relating to a particular co-regulator up, by the successive formation of limits or colimits of patterns of concepts already constructed. However, it remains the problem to define what are the instances of such a concept and characterize its invariance class. This problem did not arise for E-concepts, since they have been defined the other way round, first determining E-invariance classes and then representing such a class by an E-concept. The characterization of the instances of an E-concept by a property of their E-traces (to admit the E-concept as their limit) cannot be generalized to a general concept, because we cannot define something analogous to the E-trace; similarly, their later characterization in terms of E-universal links cannot be generalized because the same record may be an instance of several concepts at different levels of abstraction; for instance a dog is an instance of the concept 'Mammal', but also an instance of the concept 'Animal'. For a general concept, we must use the fact that it is constructed by induction, being at each step a limit or colimit of a pattern of concepts constructed in the preceding steps. The idea is that, if the concept A is the limit of a pattern P of concepts whose instances have already been defined, then an instance of A is an instance of each component of P; and if A is a colimit of a pattern Q, an instance of A is an instance of at least one of the components of Q. More precisely

Definition. We say that a link m^o defines the record M as an *instance of* the concept A in each of the following cases:

(i) A is an E_k-concept and m^o the corresponding E_k-universal link from M to A.
(ii) A is the limit of a pattern P of concepts, and m^o classifies a distributed link (m_i^o) from M to P in which, for each index i of P, the link m_i^o defines M as an instance of P_i.
(iii) A is the colimit of a pattern Q of concepts and m^o has at least one Q-factor m_j^o defining M as an instance of Q_j (*i.e.* $m^o = m_j^o c_j$, where c_j is the

binding link from Q_j to A; *cf.* Chapter 3); moreover m^o is the unique link from M to A with this property.

The set of instances of a concept A is called the *invariance class* of A.

The invariance classes provide a classification of the records having a concept, but, while the E-invariance classes associated to a co-regulator E were disjoint, the invariance classes of abstract concepts are not disjoint: a record may belong to several invariance classes. Thus, the instances of a concept A are well defined, so that the concept 'represents' a unique invariance class, but we cannot speak of 'the' concept of a record M since M may be an instance of several concepts. An abstract concept such as that of 'eternity' may have no real instance.

As already explained for E-concepts, the recall of a concept can be done by any of its instances, with a possible *shift* between its instances to adapt to the context. Over time, a concept (as any component obtained by a mixed complexification process) takes its own identity; it follows that the associated invariance class may also vary. In particular newly formed records may become new instances of the concept; conversely some instances may disappear (*e.g.* if new facts lead to classify an animal in another species then initially). Thus, the classification afforded by the concepts is not fixed once for all, but it adapts to take account of the new situations met by the system.

The development of a semantics in a memory evolutive system gives added flexibility to the choice of procedures, and to the interplay among the procedures. Indeed, the system will recognize invariants in spite of variable circumstances, and will react in a specific way not to a particular situation but to all similar situations. For example, a book by X can typically be identified whatever its format or date. Moreover, since the procedures are particular records (in the procedural memory), they can also be classified into concepts. Then higher level co-regulators (which can observe concepts in their landscape) will be able to select, not a single procedure, but a concept of procedure, without having to specify a particular instance (procedure) of the concept. As different co-regulators cooperate (possibly with conflicting procedures), this latitude confers a new degree of freedom to the interplay among the procedures that leads to the formation of the operative procedure actually carried out on the system. The choice of a concept of procedure instead of a single procedure will make it possible to carry out shifts between the various instances of this concept, in order to select the best adapted procedure among them, within the context of the other procedures selected by other co-regulators and relayed to the system. For example, the commands of a movement to open a box will activate different muscles, depending on the position of the box, its shape and its size.

8. Some Epistemological Remarks

Consideration of the memory of a system, in particular as storage that can later be recalled, raises epistemological questions about the relation between it and the learning process. Likewise, it raises the question of whether the contents of the memory can be interpreted as the knowledge of the system.

8.1. *The Knowledge of the System*

Following Bachelard (1938), we consider that *knowledge* is a ternary relation: knowledge of something attributed to some agent by an interpreter (possibly the knowing agent itself). Here *agent* and *interpreter* can be living organisms, social groups or machines. The attribution of knowledge to the agent by an (internal or external) interpreter is based on any or all of the following:

- direct observation of the behaviour of the agent,
- (partial) reading of the memory of the agent,
- material traces produced by the agent (such as books, files, CD),
- inquiry to the agent (second degree: the agent must already interpret his own knowledge).

In each case the attribution may be incorrect, since the interpreter has only an external and partial view of the agent's memory (even if it is the agent itself). For instance, the principle of charity (Quine, 1960; Dennett, 1990) attributes by default a rational basis to actions; but the agent can act for other reasons unknown to the interpreter, or deliberately deceive him. Two different interpreters (one of which can be the agent itself) can differently attribute knowledge to the same agent: a teacher can judge that a pupil does not know the lesson which the pupil thought he knew.

Approximate knowledge can be used to react quickly to various situations, and this may lead afterwards to better knowledge, through the modification of some parameters. However, if the errors of the approximation are too great, whether as a result of insufficient information, inadequate analysis of the situation (as in *e.g.* optical illusions) or a changed context, the response may fail. For conceptual knowledge, errors can also result from an incorrect or too narrow interpretation of the concepts used. Most errors made by students in standard mathematics are of this type; for example they interpret a mathematical concept (say, a derivative) on the basis of a particular representation, such as a formula learned by rote. They do not integrate the concept into a wider conceptual framework, and so are unable to use it successfully in contexts even slightly different from the original.

In a memory evolutive system, we can interpret the contents of its memory as its knowledge, though the term is not well adapted for simple biological systems, such as a cell. Its basis is a kernel of innate knowledge (stored in the innate memory) which is later developed by learning. This knowledge is distributed among its different co-regulators and their agents, though it remains mostly implicit for them (as in the social comportments analysed by Goffman, 1973), and they only have the capacity to use part of it (in particular their admissible procedures), in appropriate situations. However, as we have seen, there may exist higher level co-regulators which interpret and classify the implicit knowledge of lower level co-regulators, and develop a semantics that allows for a more flexible and deliberate use of knowledge. The flexibility of knowledge is ensured by the fact that records in the memory are multifold objects, and thus can be applied in various contexts by unfolding different ramifications.

8.2. *Acquisition of Knowledge by a Society*

The constructions we have done for a memory evolutive system are suggested by the evolution and development of the knowledge of an individual or of a society. It is natural therefore to try to apply them explicitly in these cases. It would be difficult to explicitly describe the relevant categories, and so the result would be more a metaphor. We will return to the case of the cognitive system of an animal in the next chapter; here we present some ideas about the way a human society extends its knowledge, for instance by developing scientific knowledge.

We consider a memory evolutive system which models some real world system which a group of people (taken as a co-regulator) wants to investigate. Its components are monitored by the group, but only via some of their aspects. The development of the knowledge of the group will be accomplished through its landscape. At each step its objective will be to notice or discover new features of the system under investigation, either by simple observation, or by first perturbing or otherwise changing it. Taking into account their present knowledge, the agents of the co-regulator (the members of the group) agree to select a procedure intended to lead to deeper or more accurate observations. These may be achieved by modifying the context (an instrument may be used, or a controlled perturbation introduced), or changing the system directly (*i.e.* conducting an experiment). The expected effects of the procedure are described in the anticipated landscape. However, the procedure is admissible only if it respects the physical constraints, and it may not have the expected result, since it is based on partial and possibly incorrect information. It will be evaluated at the end of the step, by comparison of the new landscape with the anticipated one. At this juncture, the difference between the expectations and the result is measured by trying to form the comparison functor, which may indicate necessary adjustments to be carried out at the next step.

Let us note that when the agents choose their procedure, they may not know if it is admissible or not. Consequently, their knowledge will develop in two different ways. Negatively, *e.g.* by *falsification*; as asserted by Popper (1972): if a procedure assumed admissible does not succeed, implying it was not admissible, that leads to asking which hypothesis was wrong. Positively, through *verification*: the procedure succeeds, so that the comparison functor exists. This makes it possible to measure the effectiveness of the result, and in particular to discover emerging objects or properties.

8.3.   The Role of the Interplay among Procedures

In the discussion above, we have only considered one group of people (one co-regulator). In fact the development of knowledge generally depends upon the cooperation and/or competition between several groups (*e.g.* inter-disciplinary studies (Lunca, 1993); competition between research teams). The interplay among their procedures plays an essential role, as do the fractures it may cause in some of their landscapes. Let us give two examples.

Construction of an Instrument.   To better understand some part of an object (say, a particular organelle in a cell), a group of scientists (the first co-regulator) may use their knowledge to devise a new instrument that could assist them. Schematically, the process might be as follows. In a first step, the scientists draft the instrument's specifications. Then they ask a team of engineers (the second co-regulator) to build the instrument. For this, the engineers select a procedure to gather adequate materials, and assemble them according to the draft, the purpose of this building proce-dure being to obtain the instrument (modelled by a new colimit). This procedure will succeed only if the material constraints are respected. If it fails, the engineers will determine what the problem is (by trying to form the comparison functor), and they will undertake to correct it at their next step, possibly with the help of the scientists, to modify the draft (interplay among the procedures of the two co-regulators). The errors will also be recorded, and they will help to 'falsify' the theory on which the scientists relied initially.

If the construction finally succeeds, the next procedure of the scientists will be to verify that the instrument allows new knowledge to be obtained of the object under investigation. Its later use may either confirm some the-oretical hypotheses (verification of the theory), falsify others or reveal un-expected properties (measured using the comparison functor) that will be the starting point for a new theory.

This is a version of the process described by Bunge (1967, Vol. 2, p. 142) for the introduction of the electron microscope, which made it possible to see finer structures in a cell, thus increasing knowledge, but also raising new problems to be studied at a later step.

Development of a Theory.   The same considerations apply to the development of a new theory (we consider that concepts and theories form part of a real world, *cf.* the third world of Popper and Eccles, as in Popper, 1972, p. 94). The explicit construction of a theory, through a sequence of complexification processes with respect to operative procedures, models the generation of a theory starting from its core (in the terminology of Bunge, 1967; Lakatos, 1970). It explains how a theory can emerge or be temporarily restrained, because of the interplay among the procedures of competitive co-regulators. It also explains what the repercussions are on the behaviour of the people who participate in the process.

   These ideas are illustrated by a personal example, involving the development of a mathematical theory. In the late 1960s, a small group of young research students (first co-regulator) working with Charles and Andrée Ehresmann developed the theory of sketches (Ehresmann, 1968), which thus became an internal record for it. As a tightly-knit group, they adopted particular concepts and even particular notations that were very different from those used by mainstream category theorists (second co-regulator). Since they had almost no contacts with the mainstream, their work, published in not widely circulated journals, remained unknown for a long period. When they first attended an international conference in 1970, the mainstream academic establishment could not understand them, because they were far removed from the current problems (topos and triple theory) and the notations were unusual. This cold reception caused a fracture to the landscape of the group. However, it had also beneficial effects: contacts were established, especially thanks to the international conferences they organized in France. These meetings allowed for a rich interplay among the procedures of (the co-regulators representing) different schools of category theorists, leading them to harmonize the notations and better explain and understand the motivations. Thus, the theory of sketches became broadly known, and accepted common knowledge. It was widely developed in the 1980s, with important applications in computer science (Barr and Wells, 1984; Gray, 1989; Walters, 1991).

8.4.   Hidden Reality

The process illustrated by the examples given above corresponds to a short-term evolution. Its reiteration over the long-term leads to the development of a coherent corpus of knowledge. Let us consider this long-term development, as assessed by a macro-CR taking what may be termed a historical over-flight, covering a long time period. The successive steps of the co-regulator modelling a group of people are too short to be distinguished separately at this higher level; only their total results are taken into account. Thus, the micro-evolution of the knowledge appears at the macro-CR as if it were smooth, between the fractures at the historical level. In the case of science, this models 'normal science', which is interrupted at long intervals by fractures corresponding to a change of paradigm. Let us note that the epistemologists do not all agree with Kuhn (1970), who speaks of 'revolutions' rather than of consistently regular, progressive evolution. The memory evolutive system model could help to compare and classify various theories of knowledge, in particular according to the chosen landscape.

This analysis gives a new view on the *problem of reality*. The human landscape constitutes a 'thoroughly objective synthesis of the communicable human experience' (d'Espagnat, 1985, p. 31). However, since the temporary internal representation it gives of the system must later be revised, it cannot be confused with the real world. This reality will be known, even in a diachronic way, only via the obstacles which stand in the way of knowledge of it. Thus, solipsism is averted, but there remains a hypothetical 'hidden reality' (in the terminology of d'Espagnat, 1985, p. 219), the disclosure of which will always run up against a transcendental border that man cannot overcome. This framework is admitted, at least implicitly, by most scientists: there is a hidden reality, be it atemporal, or in evolution more or less dependent on the agents; it can be gradually discovered, but never totally unveiled. Only an observer able to capture the entire evolution of the system at once, and at all its levels (God!), could have a true global vision, so that such a vision of complex systems, especially autonomous systems, cannot exist for human beings (Matsuno, 1989).

PART C. APPLICATION TO COGNITION AND CONSCIOUSNESS

Chapter 9 Cognition and Memory Evolutive Neural Systems

1   A Brief Overview of Neurobiology
    1.1   Neurons and Synapses
    1.2   Coordination Neurons and Assemblies of Neurons
    1.3   Synchronous Assemblies of Neurons
    1.4   Binding of a Synchronous Assembly
2   Categories of Cat-Neurons
    2.1   The Evolutive System of Neurons
    2.2   Cat-Neurons as Colimits of Synchronous Assemblies
    2.3   Interactions between Cat-Neurons
3   The Hierarchical Evolutive System of Cat-Neurons
    3.1   Higher Level Cat-Neurons
    3.2   Extended Hebb Rule
    3.3   The Evolutive System of Cat-Neurons
4   The Memory Evolutive Neural System
    4.1   The Memory of an Animal
    4.2   The Memory as a Hierarchical Evolutive System
    4.3   A Modular Organization: The Net of Co-Regulators
5   Development of the Memory via the Co-Regulators
    5.1   Storage and Retrieval by a Co-Regulator
    5.2   Formation of Records
    5.3   Procedures and their Evaluation
6   Applications
    6.1   Physiological Drives and Reflexes
    6.2   Conditioning
    6.3   Evaluating Co-Regulators and Value-Dominated Memory

Chapter 10   Semantics, Archetypal Core and Consciousness

1   Semantic Memory
   1.1   Perceptual Categorization
   1.2   Concept with Respect to an Attribute
   1.3   The Semantic Memory
   1.4   Recall of a Concept
2   Archetypal Core
   2.1   The Archetypal Core and Its Fans
   2.2   Extension of the Archetypal Core: The Experiential Memory
3   Conscious Processes
   3.1   Intentional Co-Regulators
   3.2   Global Landscape
   3.3   Properties of the Global Landscape
   3.4   The Retrospection Process
   3.5   Prospection and Long-Term Planning
4   Some Remarks on Consciousness
   4.1   Evolutionary, Causal and Temporal Aspects of Consciousness
   4.2   Qualia
   4.3   The Role of Quantum Processes
   4.4   Interpretation of Various Problems
   4.5   Self-Consciousness and Language
5   A Brief Summary
   5.1   Basic Properties of Neural Systems
   5.2   Interpretation in Our Model

Chapter 9

Cognition and Memory Evolutive Neural Systems

How can an animal learn to recognize its environment, and develop more and more complex adapted responses? What can be said about the mind–brain problem? These questions will be examined in this chapter. Here the theory of memory evolutive systems developed in the preceding chapters is applied to study the processes by which the neural system of an animal learns, up to the development of higher order sensory–motor and cognitive processes.

For this, we construct a specific memory evolutive system built by successive complexifications of the categories of neurons which model the neural system of an animal. Its components, which we call category-neurons (abbreviated to cat-neurons) comprise formal units that model the components of the neural system, not just at the neuronal level, but also at the levels of neuronal assemblages, super-assemblages and still higher order assemblages. The higher level cat-neurons represent more and more complex mental objects and cognitive processes. We will define this memory evolutive system, called the *memory evolutive neural system*, in greater detail below, after presenting the relevant neurobiology.

1. A Brief Overview of Neurobiology

In this section we recall the neurobiological notions that are the basis for the memory evolutive system by which we model the functioning of the neural system and its cognitive capacities.

1.1. Neurons and Synapses

The key constituents of the neuronal system are neurons and the synapses between them. The state of the system varies over time; at a given time t, its state depends on the activities of the neurons and on the strengths of the synapses (Fig. 9.1). The activity of a neuron N at t is measured by its instantaneous firing (spike) frequency and its threshold potential (*i.e.* the smallest membrane potential difference triggering a spike), we say that N is *activated* at t if its activity is (almost instantaneously) increased at t. A synapse f from N to N′ has N for its pre-synaptic neuron and N′ for its post-synaptic neuron. Its strength at t depends on the probability that an

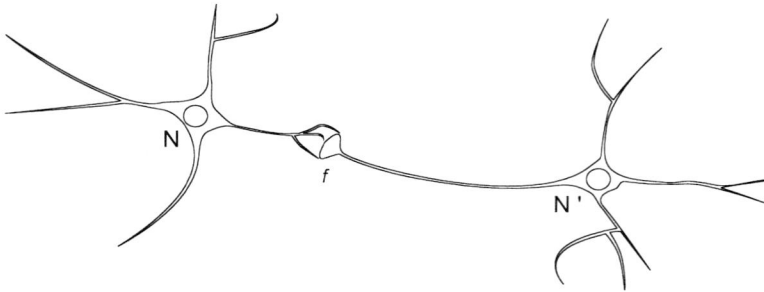

Fig. 9.1   Neurons and synapses.
A neuron N is a cell of the neuronal system. It has two prolongations: an axon and a dendritic tree. Two neurons N and N' are linked by synapses such as *f*, which joins a dendrite of N to the axon of N'. There may exist 0, 1 or several synapses between two given neurons (on the figure, only one). The synapse transmits a spike (action potential) of the pre-synaptic neuron N to the post-synaptic neuron N' if the threshold potential of N' is exceeded.

activation of N propagates across the synapse to activate N'; this probability is the inverse of the average number of spikes of N necessary to activate N' via *f*; it is inversely related to the threshold of N', and also (Carp *et al.*, 2003) to the propagation delay of *f*. The strength of a synapse varies more slowly than the activity of a neuron, and may act as a memory of its past activity. We assume that: *the activity of a neuron* N *is a function of the sum of its spontaneous activity and of the activities of the neurons* N_i *to which it is linked, weighted by the strengths of the synapses from* N_i *to* N (the strength of an inhibitory synapse being computed as negative); this function is upper bounded by the threshold potential of N.

The synapses correspond to the direct links between two neurons; however, a neuron N'' can also be activated indirectly by a neuron N if there is a synapse *f* from N to N', and a synapse *g* from N' to N''. The strength of the synaptic path (*f*, *g*) is defined as the product of the strengths of *f* and *g*. More generally, a synaptic path of length *m* from N to N_m is composed of a sequence of synapses from N to a neuron N_1, from N_1 to N_2,... from N_{m-1} to N_m (Fig. 9.2); its strength is defined as the product of the strengths of its successive synapses, and its propagation delay as the sum of their propagation delays.

1.2.   Coordination Neurons and Assemblies of Neurons

An animal is able to distinguish and learn to recognize various features of its environment, to develop adapted responses and to perform more or less complex operations. What is the neural basis of these functions? For simple stimuli, such as visual features, recordings of individual neurons in visual

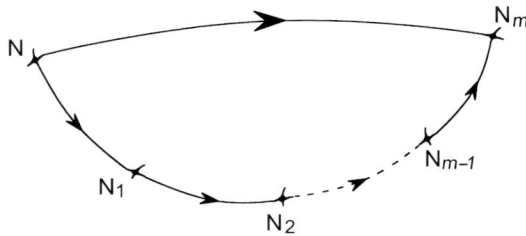

Fig. 9.2 A synaptic path.
A synaptic path is a sequence of consecutive synapses, joining N to N_1, N_1 to N_2, ...,
and N_{m-1} to N_m. If the activity of the first neuron N is strong enough and there is no
inhibition, it may be transmitted to the last neuron N_m via this path.

area 17 of the brain have shown that there exist particular neurons which are
activated by lines having a specific orientation (simple cells), or by an angle
(complex cells; Hubel and Wiesel, 1962). There also exist neurons activated by
more complex stimuli having particular importance for the animal; for
example, a hand holding a banana for the monkey (Gazzaniga, 1985). Some
authors have assumed that any known stimulus would similarly activate a
specific *coordination neuron*: seeing grandmother with her glasses would recruit
such a 'grandmother neuron' (Barlow, 1972). However, in spite of the great
number of neurons, this is not very plausible and seems contrary to experimental facts (Hebb, 1949; Changeux, 1983; Gilinsky, 1984; Edelman, 1989).

A well-known theory in psychology, associationism, supposes that an
animal has a number of representations of elementary objects or processes,
and that learning consists in forming associations of these elements to represent more complex ones. Even if this hypothesis is no longer accepted *per
se*, it has been adapted to the neural setting by several authors, who have
emphasized the role played by groups of neurons operating in synergy
(Hebb, 1949); these are also referred to as neuronal graphs (Changeux,
1983), neuronal groups (Edelman, 1989) or synaptic patterns (von der
Malsburg and Bienenstock, 1986). Brain imagery seems to confirm that a
complex item (an object or a process, such as a sensation, perception,
cognitive process, motor activity) requires the activation of a neuronal assembly, composed of connected neurons more or less widely distributed in
the brain, and that the recognition or execution of the item corresponds to a
coordinated firing of the neurons of this assembly (Stryker, 1989).

1.3. Synchronous Assemblies of Neurons

Learning leads to reinforcement of the assemblies of neurons activated by
complex items, by strengthening of the synapses between their neurons. This

strengthening, and likewise the plasticity of neuronal connections, depends on various complex sub-cellular mechanisms (*e.g.* metabolic changes, formation of vesicles and so on). Of this complexity we will retain only what can be encapsulated by the following general hypothesis, adapted from that given by Hebb in 1949 (and already suggested by Hering in the 19th century), for which we retain the well-known name:

Hebb Rule. If the activities of the pre-synaptic and of the post-synaptic neurons are simultaneously increasing, then the strength of the synapse joining them increases at the same rate; conversely, if they vary in opposite ways, the strength of the synapse decreases.

This rule has been experimentally confirmed for synapses in many areas of the brain (*e.g.* Zhang *et al.*, 1998; Frey and Morris, 1997), and also for groups of synapses (Engert and Bonhoeffer, 1997).

The more its connections are reinforced, the more an assembly of neurons takes on a distinct identity as a unit. Experimental data support the notion that the unit as a whole is activated when the activations of its neurons are tightly synchronized. Equally, we can distinguish the activation of different items by the fact that their respective activities are not synchronized (Singer, 2003). Thus we will refer to one of these units as a *synchronous assembly of neurons*.

The nature and extent of the synchronization requires some further elaboration. Some regions of the brain display certain natural oscillations of neural activity of various frequencies (*e.g.* 40 Hz, 60 Hz, 200 Hz). For example (Fisahn *et al.*, 1998), there are oscillations of 40 Hz in the hippocampus, due to a mechanism of negative feedback: the activation of pyramidal cells activates inhibitory inter-neurons, which in turn inhibit the pyramidal cells. Thus, an assembly of neurons of a given area is described as synchronous if all its neurons are activated during the same cycle of the oscillation for that area. The synchronization lasts only a short time (*cf.* Rodriguez *et al.*, 1999; Miltner *et al.*, 1999; Usher and Donnelly, 1998), and its duration is related to the complexity of the neuronal assemblage (*e.g.* longer in the associative cortex).

1.4. *Binding of a Synchronous Assembly*

The recognition of an item or the development of a process which is not too complex (we will come back to the case of more complex items or processes in Section 3) can be explained as follows: the neurons related to its various features are all activated in synchrony, and learning consists in the strengthening of the synaptic paths which connect them, to form a synchronous assembly of neurons, which becomes an internal representation of the object. The activation of an assembly of neurons P will be transmitted to another neuron N if all the neurons P_i of P are connected to N by synaptic paths f_i, having propagation delays such that the activation transmitted by f_i to N is synchronous with that transmitted by the path (x, f_j), for each synapse x of the assembly from P_i to P_j.

In some cases, there is a *coordination neuron* cP which 'binds' the assembly P, in such a way that cP is activated if and only if the assembly is synchronically activated. In this case, cP transmits to another neuron N the same activation as the whole assembly. Mathematical criteria, related to the form of the pattern, have been given for a pattern to admit such a coordination neuron (Amari and Takeuchi, 1978).

More often, however, such a coordination neuron cannot be found. In these cases learning consists in reinforcing the strengths of the synapses of the assembly according to the Hebb rule, so that the assembly becomes synchronous (since, as said in Section 1.1, the propagation delay of a synapse is inversely related to its strength). In fact, experimental studies (Edelman, 1989, p. 50) prove that more than one particular assembly can lead to the same output, and a particular item or process may activate several more or less different synchronous assemblies depending on the context, with possible changes over time. Thus the item or the process must be internally represented not by a unique assembly but by all these different synchronous assemblies, which are not necessarily inter-connected. This 'degeneracy' (in the terminology of Edelman, 1989, p. 50) is important, because it is at the root of the robustness and flexibility of brain functioning. For instance, it helps to explain perceptual constancies: animals are able to recognize the form of an object despite change in orientation, apparent size or position in the visual field. Likewise, they are able to recognize colours despite changes in the spectral quality of the illumination. It also helps explain how brain functions can be recovered after small lesions.

To model the class of these assemblies activated by a single item, we introduce a conceptual object, which can be thought of as a 'higher order neuron': we call it a *cat-neuron* (abbreviation of *category-neuron*, a name suggested to us by Albert Ducrocq, 1989). A cat-neuron is a multifold dynamic unit, activated by a neuronal event, namely the activation of any one of the assemblies it subsumes. The word multifold relates to the fact that a cat-neuron can be activated by various non-connected assemblies, and in particular that the memory allows for possible changes over time (Edelman, 1989, speaks of re-categorization, p. 110), *e.g.* we recognize a friend not seen for a long time, in spite of aging. A cat-neuron would represent an autonomous 'mental object' (such as a percept, a memory image or a concept) in the terminology of Changeux (1983, p. 179).

2. Categories of Cat-Neurons

The preceding concepts are used to construct a memory evolutive system that, at the lower levels, models the organization of the nervous system of the animal, and at higher levels corresponds to the development of mental objects of increasing complexity. First, we define the evolutive sub-system **Neur** associated to the neuronal system, whose components model the neurons. Then we describe how cat-neurons are formed, and what their interactions are. In turn, these give rise to a hierarchical evolutive system, the evolutive system of cat-neurons. The idea is to consider an assembly of neurons as defining a pattern in **Neur,** and its synchronization as the formation of a cat-neuron which becomes its colimit in a complexification of **Neur**.

2.1. *The Evolutive System of Neurons*

The neuronal system of an animal changes over time, with destruction and creation of neurons, and likewise of synapses. Thus it will be modelled by an evolutive system, **Neur**. This evolutive system has for its time scale the life-time of the animal. Its configuration category at a given time t is the *category of neurons* Neur_t defined as follows:

(i) It has for generators the graph of neurons at t (*cf.* Fig. 9.3). Its vertices model the neurons (and are also called neurons) existing at this time, and its arrows the synapses existing between them around that time; a synapse is directed from the pre-synaptic neuron to the post-synaptic one. Between two neurons there may exist one, zero or several synapses. This graph is labelled in the multiplicative monoid of the real numbers, by taking as the weight (see Chapter 1) of a synapse its strength at t.
(ii) Neur_t is the category associated to this labelled graph. It is constructed in two steps (Chapter 1): First, we label the category of paths of the graph of generators, by taking for the strength of a path the product of the strengths of its factors; by convention the strength of an identity (null path) is taken as 1. We then consider the relation on it which identifies two synaptic paths between two neurons if they have the same strength. (This identification has already been proposed by Zeeman, 1977.) And Neur_t is the resulting quotient category.

Thus the category of neurons Neur_t has for objects the neurons, and for links from N to N' the classes of synaptic paths (these classes in turn still called synaptic paths if no confusion is possible) from N to N' which have the same strength, hence are functionally equivalent in transmitting the activity of N to N'. They are labelled by the strength of the synaptic paths.

In the evolutive system **Neur**, the transition from t to t' is the partial functor from Neur_t to $\text{Neur}_{t'}$ that maps the configuration of a neuron at t

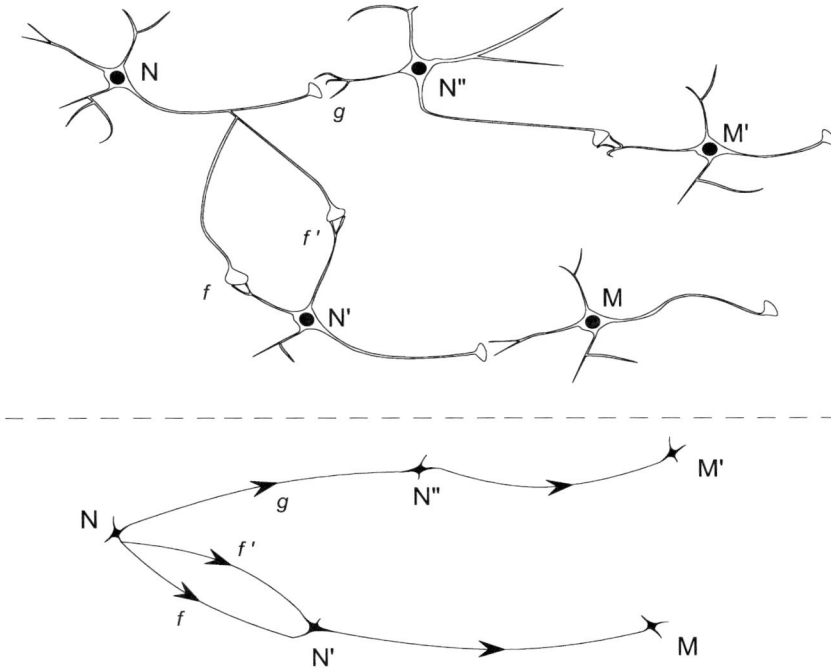

Fig. 9.3 Graph of neurons.
The upper half of the figure depicts a system of neurons and synapses, and the lower half depicts the corresponding graph. Its vertices model neurons N, N′, N″, M, M′ while its edges, oriented from the pre-synaptic to the post-synaptic neurons, model synapses between them. To simplify, the same symbols are used in both (though they represent biological objects in the first case and their conceptual representation in the second). There may exist several parallel synapses between two neurons, such as f and $f′$ from N to N′. The graph is labelled by the strengths of the synapses, but these are not depicted in the figure.

onto the configuration of the same neuron at $t′$, if it still exists; and similarly for synaptic paths. The components correspond to the neurons; not simply as anatomical objects, but with their trajectory as dynamic units, with successive activity configurations. Similarly, the links are dynamic models of the synaptic paths.

2.2. Cat-Neurons as Colimits of Synchronous Assemblies

If we think of a neuron as being a component of **Neur**, we must consider the status of a synchronous assembly of neurons, and give a representation for it. An assembly of neurons will be modelled by a pattern P in **Neur**; that is a pattern in a configuration category, and its successive configurations which model the successive configurations of the assembly at corresponding times. The components of P represent the neurons P_i of the assembly, and its distinguished links represent the classes of synaptic paths which transmit an activation between two neurons of the assembly. The assembly operates as a whole on another neuron N by means of the collective links of the pattern to N. The pattern P admits a colimit in **Neur** if and only if the assembly has a coordination neuron.

If this is not the case, we have seen above that the assembly may nevertheless develop as a unit by strengthening its links, and becoming a synchronous assembly. We are going to model this unit by a colimit cP of the pattern P in a larger evolutive system, called the *evolutive system of cat-neurons* (*cf.* Section 3 below), which contains **Neur** as an evolutive sub-system. The colimit will be the cat-neuron we have associated above to the assembly, or rather to the class of all the synchronous assemblies which are activated by the same item or the same process (Fig. 9.4). Its construction is an application of the complexification process, described in Chapter 4, to form a higher order unit that integrates a class of patterns by becoming the colimit of each of them.

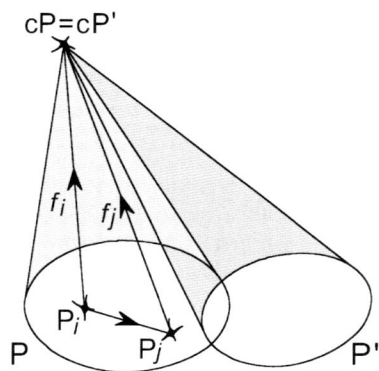

Fig. 9.4 A cat-neuron.
A synchronous assembly of neurons P is formed by a family of neurons which are synchronously activated through certain synaptic paths; it is modelled by a pattern in the evolutive system of neurons **Neur**. It admits a colimit in this category if and only if there exists a coordination neuron whose activity corresponds to that of the assembly acting synchronously. Otherwise, we model the assembly by a formal unit cP, called a category-neuron (abbreviated to cat-neuron), obtained by adding a colimit to P in a complexification of **Neur**. This cat-neuron cP can also be(come) the colimit of other patterns P', if P and P' represent synchronous assemblies of neurons which have the same functional role. Once formed, the cat-neuron becomes a component of the evolutive system of cat-neurons, and represents, not a unique assembly, but the whole class of functionally equivalent synchronous assemblies of neurons which it binds; this class may vary over time.

More specifically, the cat-neuron will emerge at a given time *t* as a formal unit cP, added to become the colimit of the pattern P in a category obtained from the category of neurons Neur, by the complexification process with respect to an option, having the binding of P as one of its objectives. Once formed at *t*, cP becomes a component of the larger evolutive system of cat-neurons, and remains the colimit of P in it, as long as P represents the assembly, though acquiring a complex identity of its own. According to the construction of the complexification, such a cat-neuron cP is, or becomes later, the colimit not only of P, but also of all the patterns P′ which are homologous to P; that is, which represent synchronous assemblies with the same functional role. Thus *the cat-neuron characterizes the invariant that the different assemblies it binds have in common*, and it therefore represents the mental object corresponding to the item activating P. As these assemblies are not necessarily inter-connected (as said in Section 1), the corresponding patterns are not always connected, and the cat-neuron cP which binds them is a multifold object of the complexification (in the sense of Chapter 3). It should be noted that a cat-neuron has a triple aspect:

(i) Formally, it is a component of the evolutive system of cat-neurons; as such, it has a sequence of configurations deduced from the first one (corresponding to its formation at *t*) via transition functors.
(ii) Physically, each of its configurations is an abstract object, but corresponds to a physical event, the local recruitment and strengthening of assemblies of neurons from which the cat-neuron inherits a more or less distributed structure.
(iii) Dynamically, it can be recalled in various ways, through the temporary activation of any one of these assemblies, with possibly a complex switch to another one. Once formed it becomes a stable component, with a possibly long stability span; however it is not a formal rigid object such as an encoded message, but a multifold component, with a large flexibility. Indeed, it can be differently activated depending on the context, and can later integrate a new synchronous assembly that becomes representative of the same item.

2.3. Interactions between Cat-Neurons

Several authors (see for example von der Malsburg and Bienenstock, 1986) have raised the problem of defining the nature of interactions between synchronous assemblies of neurons, in relation with the

Binding Problem. How are the different functional areas coordinated, so that not only their neurons but also their assemblies of neurons (or, for us, cat-neurons) work together, thus forming higher order super-assemblies, which represent more and more complex items?

We can approach this problem by modelling the assemblies by cat-neurons that are objects of a complexification of a category of neurons. Indeed, the construction of the complexification determines what the appropriate links are between cat-neurons, and therefore between the synchronous assemblies they integrate, so that we may define patterns of such assemblies, and bind them into super-assemblies. Let us characterize these links.

First, let P and Q be the patterns modelling two assemblies of neurons. The simple interactions between the assemblies are modelled by the clusters from Q to P in **Neur**. We recall that a cluster is generated by a set of synaptic paths, such that each component Q_k of Q is linked to at least one neuron P_i by a synaptic path; if it is linked to several, they are correlated by a zigzag of distinguished links of P (Fig. 9.5). An example would be what Edelman calls a 'classification couple' connecting two neuronal groups (1989, p. 111). We define the *propagation delay of the cluster* as the maximum of the propagation delays of the links of the cluster. The *strength of the cluster* is an increasing function of the strengths of these links.

If P and Q model synchronous assemblies, a cluster from Q to P binds into a (Q, P)-*simple link* from cQ to cP, where cP and cQ are the cat-neurons which integrate them (Fig. 9.5). We have said that cat-neurons are colimits of various synchronous assemblies; similarly a simple link can bind various

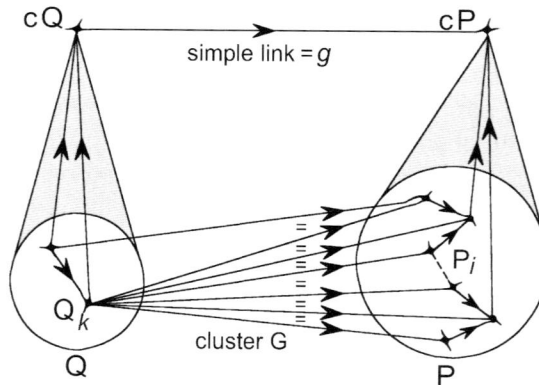

Fig. 9.5 Cluster and simple link between cat-neurons.
The cat-neurons have been obtained as colimits of synchronous assemblies of neurons, added in a complexification of the category of neurons (and later taking their own identity). The links between them are determined by the construction of the complexification. In particular, some of the links are simple: if Q and P are synchronous assemblies of neurons bound respectively by the cat-neurons cQ and cP, a (Q, P)-simple link *g* from cQ to cP binds a cluster G of synaptic paths from the pattern Q to the pattern P. We recall that the 'simplicity' of a link depends on the choice of the decompositions (here, Q and P) of the objects (here the cat-neurons cQ and cP) it joins.

clusters. More precisely, a (Q, P)-simple link f from cQ to cP can also (but not necessarily) be (Q′, P′)-simple for patterns P′ and Q′ having also cP and cQ respectively as their colimits in the complexification. The *propagation delay of f* is then defined as the minimum of the propagation delays of the clusters which it binds, and its strength as the maximum of the strengths of these clusters.

The general construction of the complexification (Chapter 4) shows that its objects are the neurons and the cat-neurons, and that the links are:

- the simple links defined above;
- complex links which are composites of simple links binding non-adjacent clusters; the propagation delay of such a complex link is the sum of the propagation delays of its factors, and its strength is the product of their strengths;
- more generally, complex links obtained by successive applications of the two processes: bind a cluster of (simple or complex) links already constructed, take a composite of them.

The neurons and cat-neurons are the components of an evolutive system admitting **Neur** as an evolutive sub-system, with the links defined above. This evolutive system will be further enlarged to allow for the formation of cat-neurons representing more complex mental objects or cognitive processes. The fact that a cat-neuron is a multifold object (as said above) means that this evolutive system satisfies the multiplicity principle. Here, this principle formalizes what Edelman calls the *degeneracy of the neuronal code*: 'more than one combination of neuronal groups can yield a particular output, and a given single group can participate in more than one signalling function' (1989, p. 50). Let us recall that initially we have introduced the multiplicity principle to generalize this degeneracy property to any memory evolutive system.

3. The Hierarchical Evolutive System of Cat-Neurons

In the preceding section, we have started from the neurons and constructed cat-neurons integrating synchronous assemblies of neurons. Now we define more complex cat-neurons by iteration of the same process, to represent mental objects of increasing complexity.

3.1. Higher Level Cat-Neurons

Once a complexification of a category of neurons is constructed, it affords a larger category on which the complexification process can be iterated. This leads to higher level cat-neurons, obtained as colimits of a pattern of cat-neurons already constructed, and to their simple and complex links. Let us say that:

- a neuron (component of **Neur**) is a cat-neuron of level 0;
- the cat-neurons defined in the preceding section as colimits of synchronous assemblies of neurons are cat-neurons of level 1;
- by induction we define a *cat-neuron of level k* as the colimit of a pattern of lower level cat-neurons, in a complexification of a category of cat-neurons of level strictly less than k.

Using the results on iterated complexifications given in Chapter 4, we can describe such higher level cat-neurons. First, a *cat-neuron A of level 2* is a formal unit, which emerges to become the colimit of a pattern P of cat-neurons P_i of level $\leqslant 1$ in a complexification of a category of cat-neurons, and thus in a 2-iterated complexification of a category of neurons (Fig. 9.6). A then assumes a complex identity as a component of the larger evolutive system of cat-neurons, and is, or later becomes, the colimit of all the patterns homologous to P (such as R in the figure). Physically, the emergence of A necessitates that the distinguished links of the pattern P be strengthened, so that the cat-neurons P_i can be synchronously activated. Since by definition each cat-neuron P_i itself is a colimit of a pattern P^i of neurons, A has ramifications $(P, (P^i))$ down to the neuronal level. Such a ramification can be thought of as a *synchronous super-assembly* (*i.e.* an assembly of assemblies)

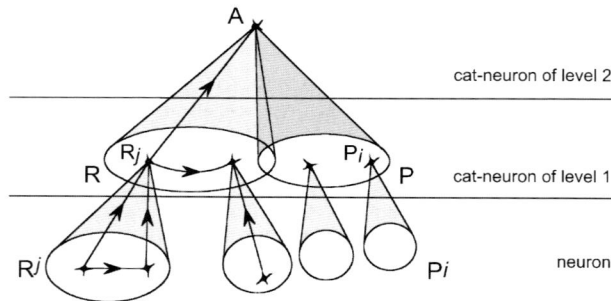

Fig. 9.6   A cat-neuron of level 2.
A cat-neuron of level 2 is a formal unit A obtained by binding together a synchronous assembly P of cat-neurons of level 1. Thus it requires three levels: the level 0 of neurons and synapses, the level 1 of cat-neurons, for instance P_i, binding a synchronous assembly of neurons P^i, and the level 2, where A binds the synchronous assembly of cat-neurons P. The cat-neuron A also binds the other such assemblies R of cat-neurons which are homologous to P. In fact A as a cat-neuron of level 2 models a whole class of synchronous assemblies of cat-neurons, and it has ramifications $(P, (P^i))$, but also $(R, (R^j))$, down to the level of neurons, of which it is an iterated colimit.

of neurons: the links in P correlate the various synchronous assemblies (to be compared to the 'classification *n*-tuples' of Edelman (1989) formed by his 'reentry process', p. 48). Thus A represents a class of such synchronous super-assemblies. Though A is a formal unit, its later recall corresponds to a dynamic physical process, namely the *unfolding* in two steps of one of the synchronous super-assemblies it binds: simultaneous activation of the synchronous assemblies P^i over a sufficiently long time; and synchronization of their activities along their distinguished links, leading to the activation of P. This process has been observed in for example odour encoding (Wehr and Laurent, 1996).

More generally, a cat-neuron of level k emerges to become the colimit of a pattern of cat-neurons of strictly lower levels, acting as a synchronous assembly of cat-neurons, and then takes its own complex identity as a component of an evolutive system of cat-neurons. It follows that, as for any object obtained by iterations of the complexification process, it is also an iterated colimit of at least one ramification of length k, based on the lower level of neurons. A ramification of length k can be thought of as a synchronous hyper-assembly of neurons with k levels, *i.e.* an assembly of assemblies of assemblies ... of neurons (k times). It is activated through the dynamic gradual unfolding of the ramification formed by these successive synchronous assemblies, from the level of the neurons up.

As already said, cat-neurons of level 1 represent some mental objects. By combining cat-neurons of lower levels (through the binding of patterns of such cat-neurons), higher level cat-neurons of increasing levels will represent mental objects corresponding to more and more complex stimuli, processes or behaviours, up to higher mental or cognitive processes of various kinds. Like the cat-neurons of level 1, the higher level cat-neurons are both robust multifold units, with long stability spans, and flexible enough to adapt to contextual changes. This flexibility increases with the level. Indeed, a cat-neuron of level k can be recalled through the unfolding of any of its ramifications down to the neurons, with a possible choice, at each intermediary level, of which lower ramification will be activated. Thus, its degrees of freedom are multiplied, allowing it to adapt to a wealth of situations by selecting an appropriate ramification.

3.2. *Extended Hebb Rule*

The construction of cat-neurons of increasing levels via successive complexifications of a category of neurons shows what the simple and complex links between them are. As we have said, the formation of a cat-neuron of level k as the colimit of a pattern of lower level cat-neurons requires that the distinguished links of this pattern be strengthened, so that the pattern acts as a synchronous assembly of cat-neurons. How can learning lead to the formation of such synchronous assemblies? For cat-neurons of level 1, experimental facts seem to prove that the strengthening of links still follows the Hebb rule (as for neurons). This rule will be extended to higher levels shortly.

The recall of a cat-neuron of level 1 depends on the activation of one of the synchronous assemblies it represents. It follows that the activity of this cat-neuron of level 1 at a given time is an increasing function of the activities of the neurons of the most activated of the assemblies it represents. As a higher level cat-neuron is the colimit of a pattern of cat-neurons of lower levels, by induction, its activity will be defined as an increasing function of the activities of the cat-neurons of the most activated of these patterns. We have defined the propagation delay and the strength of a link between cat-neurons of level 1 just using the delays and strengths of the links between neurons. The same process extends to define the propagation delays and the strengths of links between cat-neurons of any level (since they are iteratively constructed by a similar complexification process). We extend the Hebb rule by supposing that the following rule is valid:

(Extended) Hebb Rule. The strength of a link between two cat-neurons (of any level) increases if the activities of both cat-neurons simultaneously increase, and it decreases if these activities vary in the opposite way.

This rule can be deduced from the usual Hebb rule. We do this here for simple links, say between cat-neurons of level 1; the result extends by induction to higher levels. Let g be a (Q, P)-simple link from N to M binding a cluster G. Let us suppose that the activities of N and M are simultaneously increasing through the activation of the synchronous assemblies P and Q. The usual Hebb rule implies that the strength of a link of the cluster will increase, since the activities of the neurons Q_j and P_i which it links increase. The strength of the link g being an increasing function of the strengths of the links of the cluster, it follows that it also increases.

3.3.   *The Evolutive System of Cat-Neurons*

The cat-neurons with their links are at the basis of a memory evolutive system which will model the functioning of the neural system of an animal such as a higher vertebrate, up to the development of higher cognitive processes. Until this point, we have only considered cat-neurons constructed by a complexification process adding colimits. However, cat-neurons can also be combined as above, but using the limit operation instead of the colimit one. These classifying cat-neurons arise in the construction of procedures in the procedural memory, and of concepts in the semantic memory, and we will return to them later. To include them in the following definition we specify that cat-neurons can be constructed by a mixed complexification process (Chapter 4).

Definition. The *evolutive system of cat-neurons* of an animal is a hierarchical evolutive system over the lifetime of the animal, which contains **Neur** as an evolutive sub-system. Its configuration category at t is a category of cat-neurons deduced from a category of neurons by a sequence of (possibly mixed) complexifications, and the transitions are partial functors from a category of cat-neurons to one of its complexifications.

The evolutive system of cat-neurons (which will be dotted of the structure of a memory evolutive system in the next section) models not only the dynamics of the neuronal system of the animal, but also the development and functioning of higher level organized neural systems that reflect the animal's various sensory, motor, affective or cognitive processes, and shape its behaviour. The operations take also into account the whole body of the animal through the perception of its internal states and its actions on the environment (as in the active perception theory, see *e.g.* Thomas, 1999); we will explain later how the formation, development by adaptation to the context, and later recall of a cat-neuron require active (though possibly automatic) involvement of the animal.

Let us remark that this model is very different from neo-connectionist models of neural systems, which give only a description at the sub-symbolic level, without taking into account the interactions between the different levels. In particular, these models can only describe the formation of what we describe as cat-neurons of level 1 (corresponding, in neo-connectionist models, to attractors of the dynamics), but not their complex links. Thus they cannot account for the higher level cat-neurons modelling complex brain processes. By contrast, the construction of the simple and complex links between cat-neurons can be considered in terms of the synchronous (hyper-)assemblies of neurons by which they are activated, and so can describe the possible interactions between such assemblies; thus solving the binding problem of Section 2. Moreover, it becomes possible to 'compute' with cat-neurons; that is with (hyper-) assemblies of neurons, as if they were simple neurons, and to develop a real *algebra of mental objects*, or rather of mental processes, allowing them to be combined to form more complex ones, and to compare them (following the proposal of Changeux, 1983, p. 181).

More philosophically, the representation of higher order processes by cat-neurons allows another approach to the:

Mind–Brain Problem. What is the correlation between mental states and brain states?

The recall of a simple mental object, represented by a cat-neuron of level 1, corresponds to the activation of a synchronous assembly of neurons, which is a well-described brain event. Now a higher order cat-neuron representing a more complex mental object has no direct neuronal correlate, but several dynamic ways to emerge from the neuronal level, namely its ramifications down to this level. The mental process consisting in its recall requires the stepwise dynamic unfolding, through the various intermediate levels of cat-neurons, of one of these ramifications. At each step, the unfolding can proceed along one or another decomposition of multifold cat-neurons into synchronous (super-)assemblies, with possibly a complex switch between them; this switch might originate at the quantum level (*cf.* Eccles, 1986); from some higher level random influence (neural noise); or it might be internally controlled. Thus the recall of a complex mental object requires a whole active sequence of lower level mental events (the stepwise unfolding of a ramification), with multiple choices at each step, before reaching the neuronal level.

From the preceding discussion, we deduce the correlation between mental states and brain states: the mental states emerge in a dynamic way from brain states, the process requiring a number of steps increasing with their complexity; however the correlation is degenerate, meaning that a single mental state can emerge from many different brain states (*cf.* Edelman, 1989, p. 260). This picture could qualify as an *emergentist monism*, in a sense a little more general than in Bunge (1979, for whom mental processes are identical with brain processes of certain types). The fact that a cat-neuron, though being an abstraction, ultimately operates through physical brain processes (via its ramifications down to the neuronal level) explains how it can cause a physical event, and how mental properties supervene on physical properties with multiple realizability (the various ramifications of a cat-neuron), making *mental causation* (Kim, 1998) possible, while preserving the physical closure of the world (something, incidentally, that is not accepted by Bunge.)

4.   The Memory Evolutive Neural System

The preceding description of the evolutive system of cat-neurons is in a certain way external, since it does not explain how the system is internally organized, in particular how the assemblies of (cat-)neurons which will generate more and more complex cat-neurons are recruited. Neurobiological data support the idea that this system is self-organized, with a net of partial internal regulatory modules (the co-regulators) which modulate its dynamics. These also participate in the development of a central (long-term) memory, formed by cat-neurons of increasing orders. Equipped with these co-regulators and the memory, the evolutive system becomes a memory evolutive system, which will be called the *memory evolutive neural system* of the animal.

4.1. The Memory of an Animal

What we call the memory represents a system of long-term, adaptable internal representations, or mental objects, of the items (external objects, signals, past events or internal states) that the animal can recognize, and of the sensory–motor or cognitive processes it can recall. It is a long-term memory, and can be compared with what Edelman calls a 'recategorical memory', which he describes as 'a process involving facilitated pathways, not a fixed replica or code' (Edelman, 1989, p. 111).

Animals, including insects, have an innate store of information necessary for their survival, such as characteristic features of their environment, internal states (hunger, orientation reflex, avoidance responses and so on) and instinctual sensory–motor processes. For instance, a moth recognizes light and automatically orients towards it, thanks to neural feedback from the eyes to the wings. This store of information forms a sub-system of the memory called the *innate memory*. Animals with more advanced neural systems later extend this innate memory; they learn to recognize and associate together more complex internal states or external circumstances, such as often encountered objects or situations, and to develop adapted responses to cope with them, such as habits, learned skills, other miscellaneous sensory–motor programs or cognitive processes. Their memory is flexible, content addressable and widely distributed within the brain, implicating various areas, in particular the sensory and motor cortex, cerebellum, basal ganglia and hippocampus. Generalizations are possible, the animal learning to interpolate from a set of examples (in particular in vision and motor control; Poggio and Bizzi, 2004), and to extract regularities. For example, honeybees have been trained to recognize some complex patterns sharing a common layout with four-edge orientations, and to generalize to other patterns presenting the same common layout (Stach *et al.*, 2004).

A new object, unknown to the animal, can temporarily recruit and activate an assembly of neurons, possibly forming a *short-term memory*. This does not lead to the strengthening of the links of the assembly to become a synchronous assembly, except if the object is repeatedly encountered or has some emotional value. In that case, the assembly strengthens and a cat-neuron emerges, which has a longer life; it is said that the short-term memory is *consolidated* into a long-term memory (Dupréel, 1931). Both short- and long-term memory rely on the phosphorylation of certain proteins, but the first one uses only pre-existing proteins and fades with their depletion, while the second one requires synthesis of new proteins (see Sweatt and Kandel, 1989; and for a review of the variety of molecules on which the consolidation depends, see for example Wright and Harding, 2004).

For us, the memory corresponds to a long-term memory, and does not contain non-consolidated short-term memories. It contains the innate memory (as described above), and develops by storing the experiences of the animal (perceptions, movements, choices of procedures and their results, cognitive processes of any kind) in a flexible way, under the form of neurons (for simple stimuli), of cat-neurons representing (classes of) synchronous assemblies of neurons, or of more complex higher order cat-neurons. They will be recalled in similar situations later on, thus leading to more adapted behaviours.

4.2.   *The Memory as a Hierarchical Evolutive System*

Formally, in the evolutive system of cat-neurons, the (long-term) memory of the animal is modelled by an evolutive sub-system, called the *memory* and denoted by MEM. Its components are cat-neurons called *records* (from the general name given to components of the memory in a memory evolutive system), the activation of which leads to the recognition of external items that the animal has already met or of internal states it has already felt, or to the recall of various processes: sensory–motor procedures, skills, cognitive processes, possibly with their emotional or affective undertones.

The formation of a record of an item C (which is not necessarily by intention) corresponds to its assimilation (following Piaget, 1940). Afterwards it will be consolidated, yet remain adaptable to gradual enough temporal modifications of the system and of the environment (Piaget's accommodation). Its domain of application can be later extended, for instance, by the formation of new ramifications, leading to a flexible multifold record. In this way, the record assumes a distinct identity as a component of the system, with its successive states becoming more and more independent from the particular decomposition used in its formation, and it can be recalled in different ways appropriate to the context. Since the memory relates to a long-term memory, the records have a long enough lifetime. It is not a replicative memory as in a computer, with the recognition depending on an exact encoding and decoding. Recall by activation of a cat-neuron is more flexible, and depends on contextual influences; perception and action are intermingled. Thus a record is progressively updated to take new circumstances into account. It can be generalized or, on the contrary, more finely tuned. If it is no longer adapted, it will be modified or even destroyed.

The memory will develop over time by the formation of higher order cat-neurons, obtained by binding or classifying patterns of already existing records (through mixed complexification processes), thus allowing for more complex processes to be remembered. These complex records interact through simple and complex links, as described by the complexification process. A complex link from M to M″ relates not only the two records, but also the intermediate multifold records which occur in its formation, through switches between two of their decompositions (change of parameters). Thus the cohesion it creates between M and M″ reflects more than a local cohesion between lower level decompositions of M and M″; it reflects something of the overall structure of the lower level memory (containing the decompositions), emerging at the level of the link (*cf.* the discussion on emergence in Chapter 4). For example, the formation of complex links allows for chains of inferences using metaphors (Paton, 1997, 2002) that reveal new overall perspectives. Indeed, a metaphor can be interpreted as a switch between two decompositions of the same record; *e.g.* the genome can be considered as a chemical structure, or looked at as a text.

We distinguish different kinds of memory, forming evolutive sub-systems of the memory MEM.

(i) The *empirical memory* (or recollection memory) **Memp** covers both what is often called the perceptual memory (memory images in the terminology of Changeux, 1983) and the episodic memory. It allows for the recognition of items often met, such as environmental cues, internal states, facts and events with survival value, but also, through a longer mental time travel, relevant past episodes, even possibly encountered only once.
(ii) The *procedural memory* **Proc** allows for the retrieval of sensory–motor processes and plans, or procedures to command effectors. Its records have the form of procedures in the sense of Chapter 8.
(iii) For higher animals, the *semantic memory* **Sem** (often called categorical memory) is formed by concepts that classify items and their records. It will be discussed in the next chapter, along with the *archetypal core* **AC**, in relation with self and consciousness.

4.3. A Modular Organization: The Net of Co-Regulators

How are the dynamic brain processes directed; and in particular, how is a higher animal able to develop its memory and learn more adapted behaviours? Several authors have proposed a modular theory of brain function (*e.g.* Fodor, 1983), wherein distinct modules are postulated to be responsible for specific operations—for instance, a colour module to treat colours. Even if such a theory is not accepted neat, it is well recognized that there exist specialized parts of the brain. Some are easily distinguishable by their gross anatomy, such as the brain stem, the cerebellum, the different large cortical

areas, the limbic system and so on. Even within them, we find more specialized areas; for instance, the visual cortex and the auditory cortex, or the
hypothalamus and the amygdala, in the limbic system, which have a specific
role related respectively to vital needs and to affective states (Stefanacci,
2003). These specialized areas in turn contain further areas of sub-special-
ization. Edelman speaks of 'multiple functionally segregated areas' (Edelman, 1989, p. 70), of which he distinguishes more than 20 in the visual
cortex, each one responding best to a specific attribute (*e.g.* orientation,
motion or disparity); the integration between them comes from a 'process of
reentrant signaling along inter-areal connections' (*cf.* our interplay among
the procedures of the co-regulators). In the spinal cord, there are small
neuronal networks called central pattern generators which generate vertebrate movements such as walking and swimming (Gosgnach *et al.*, 2006).

With respect to vision, Crick (1994) assumes the existence of a hierarchy of
'*treatment units*', some non-conscious (as in the V1 area), others 'conscious'
(pp. 336–339). He supposes that each unit has its own form of representation, its own delays for the treatment and its short-term memory, in which
the thalamo–cortico–thalamic circuit acts as a reverberating circuit. Higher
units may exercise a global control over lower ones. These properties are
exactly those we have attributed to the co-regulators of a memory evolutive
system. Indeed, let us recall that a co-regulator is an evolutive sub-system of
a memory evolutive system with a discrete time scale (extracted from the
continuous time scale of the system); it has its own treatment delays (period
and time lag) which must satisfy specific structural temporal constraints, and
its own form of representation corresponding to its admissible procedures;
the landscape it forms at each step acts as a short-term working memory.
Each co-regulator has a differential access to the memory through its landscape, thus allowing it to recognize recorded items and to access its admissible procedures via afferent links to its agents. It can activate (some of) the
effectors of these procedures via efferent links to these effectors, the composite of an afferent and an efferent link being a command of the procedure.
The co-regulators also participate to the development of the memory,
directly and/or indirectly, via the interplay among their procedures.

Thus we feel justified in modelling the modular organization of the brain
described above by a net of co-regulators of a memory evolutive system,
which will also have more abstract higher level co-regulators, meaning that
their agents are not neurons but higher order cat-neurons; nevertheless they
are doubly based on the neuronal level: vertically through the multiple
ramifications of these cat-neurons, and temporally through the successive
activities of these cat-neurons, which correspond to the activation of synchronous hyper-assemblies of neurons (as explained in Section 3). Thus we
can give the following:

Definition. We define the *memory evolutive neural system* (MENS) of an animal as the memory evolutive system obtained by equipping its evolutive system of cat-neurons with a hierarchical evolutive sub-system called the *memory* (or MEM) modelling the memory of the animal, and with a net of co-regulators, which are evolutive sub-systems based on the neuronal level of various brain areas.

We could add to this system lower sub-cellular levels, so that it would have as sub-systems the memory evolutive systems associated to the various neurons (as to any cell). This would be necessary to explain the underlying biological mechanisms allowing for the activity of neurons, and the formation and strengthening of synapses.

5. Development of the Memory via the Co-Regulators

As in any memory evolutive system, the dynamics of the memory evolutive neural system are internally controlled by the co-regulators and the interplay among their procedures. Let us recall from Chapter 8 how this is done, and how it contributes to the emergence of higher order cat-neurons, allowing for the development of the memory.

5.1. Storage and Retrieval by a Co-Regulator

Here we consider a particular co-regulator, say E, and analyse one of its steps. To be more concrete, we suppose that the agents of E are neurons and that the event starting its step at t is the presentation of an item S (external stimulus or internal state), of which E can perceive some attributes (*e.g.* its shape if E is a shape-CR). S activates a pattern R of (cat-)neurons acting as receptors.

The first phase of the step consists in forming the actual landscape L_E of E. The item S will be observable in this landscape only if it activates some of the agents of E; the synaptic paths which transmit this activation are modelled in L_E by a pattern r_E of perspectives coming from a sub-pattern of R.

The second phase of the step consists in the selection of a procedure through the landscape, which will be carried out in the third phase. The pattern R_E that is the image of r_E by the difference functor from the landscape to the system has its links strengthened. There are several cases:

• If S has not yet been encountered, the procedure will be to try to remember it by transforming R_E into a synchronous assembly of neurons, leading to the formation of a cat-neuron M_E colimit of R_E (see Fig. 9.7), which is the *partial record* of S for E (as defined in Chapter 8); it is an E-*internal record* of S if it has a perspective which is the colimit of r_E in the

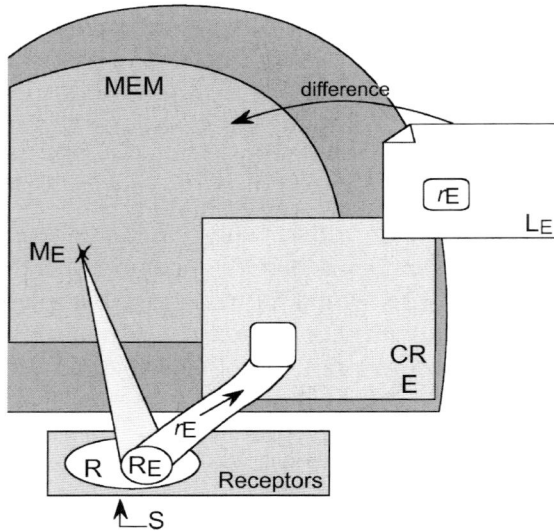

Fig. 9.7 Partial record.
An item S activates a pattern of receptors R. A particular co-regulator, say E, receives some aspects of R, the perspectives of which form a pattern r_E in its landscape L_E. This pattern is depicted twice: formally in the landscape, shown on the right; and in the system, by the aspects of its different perspectives. The image of r_E by the difference functor, from L_E to the system, is a pattern R_E which has for components some of those of R. Let us remark that a component of R may have several perspectives for E, in which case it will correspond to several components of R_E with different corresponding indices. S has a partial record for E if R_E admits a colimit M_E. However, this partial record M_E may have no aspect for E (as in this figure).

landscape (Fig. 9.8). Once formed, the cat-neuron M_E becomes a component of the memory and takes its own complex identity.

• If S is already known by the animal and has a partial record M_E, the activation of the synchronous assembly R_E activates its colimit M_E, so that E participates in the recognition of the item, and the procedure will be to strengthen this partial record. If this partial record is an internal E-record M_E, it will be recognized in the landscape of E through the activation of its perspective m_E. This results from the (extended) Hebb rule: indeed, since M_E has a perspective m_E that is the colimit of r_E, there exists an agent A such that both m_E and each perspective in r_E have some aspects arriving at A. The presentation of the item will activate A both via the aspect of r_E, and directly via M_E. The activities of A and of M_E being simultaneously increasing, the Hebb rule implies that the strength of a link from M_E to A increases; thus the perspective m_E becomes strong enough to be visible in the actual landscape of E, in turn allowing for the internal recognition of the attributes of S detected by E.

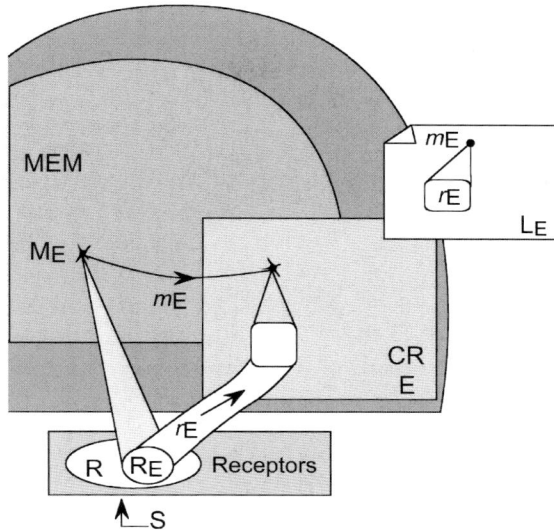

Fig. 9.8 Internal E-record.
Same situation as in Fig. 9.7, except that the partial record M_E of S has a perspective m_E for E, which is the colimit of r_E in the landscape of E. Then M_E is called an internal E-record of S; in this case S can be later recognized directly by E in its landscape, through the recall of its record M_E via the perspective m_E.

- If moreover S has an internal E-record which is connected to an admissible procedure Pr_E for E by an activator link h_E observable in the landscape of E, the objective of E may then be to activate this response procedure (*e.g.* for a food item, to eat it) through the activation of h_E (Fig. 9.9). In this case the complete step of the co-regulator, from the recognition of the item to the commands of the procedure, would correspond to the activation of what Edelman calls a 'global mapping' (Edelman, 1989, p. 54). The result is evaluated at the next step, and remembered by strengthening the activator link if the procedure has succeeded, otherwise by decreasing its strength.

In any case, the step can be interrupted, either by a fracture imposed by the situation or the system (such as by the impossibility of recognizing the item as a result of conflicting data), or by the impossibility to respond with an appropriate procedure.

5.2. *Formation of Records*

Each co-regulator proceeds in its landscape; however its selected procedure is relayed to the system, and the operative procedure Pr^o (Chapter 6) carried out at a given time will be obtained after the interplay among the procedures

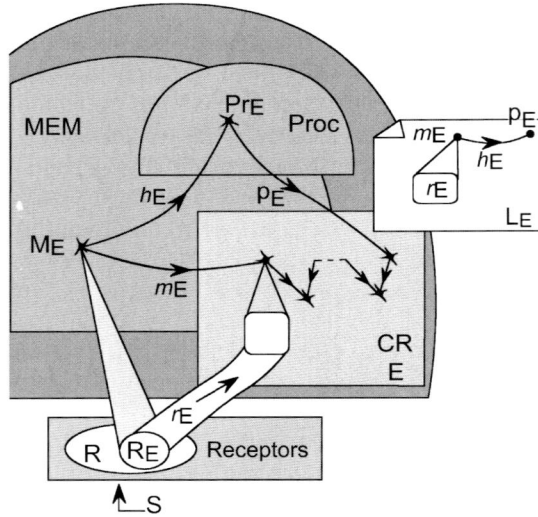

Fig. 9.9 Activation of a procedure.
The item S has an internal E-record M_E with an activator link h_E to a procedure Pr_E, which has a perspective p_E in the landscape of E. This link defines a link in the landscape L_E, from the perspective m_E of M_E to p_E. This is depicted by the fact that m_E and $h_E p_E$ are connected by a zigzag of links between agents. Then E can recall S by its internal E-record, the activation of which will activate the procedure Pr_E through (the perspective of) h_E in the landscape of E.

relayed by the various co-regulators, and possible other procedures independent from them.

For the formation of a record of the item S (as defined in Chapter 8), its various partial records M_E must be synchronously activated, giving rise to a synchronous assembly of cat-neurons, which is modelled by a cat-neuron M, the record of the item. Thus S has a *record* M if the pattern Q of the partial records of S has M for its colimit (Fig. 9.10).

Generally, the whole record M will only emerge from the interplay among the procedures of the different co-regulators that bind the various partial records into M, so that the recognition or recall of M necessitates a synergetic action of these co-regulators. This is corroborated by experimental data. For example, Mingolla (2003) shows that 'perceptual unit formation is a highly context-sensitive and dynamical process in which complexes of form, colour and depth emerge from network interactions' (2003, p. 115). Once formed, the record (as any cat-neuron) assumes a distinct identity over time, and will be consolidated and/or updated depending on the context of later experiences.

The recall of a record M is an active process, which will be more or less easy depending on the number of co-regulators participating in it through

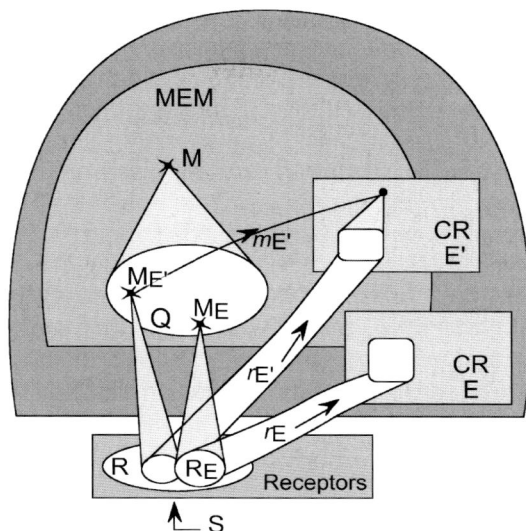

Fig. 9.10 Record of S.
Different co-regulators have partial records of the item S; in particular, M_E for E and an internal E'-record $M_{E'}$. These various partial records form the components of a pattern Q. The record M of S will be the colimit of this pattern Q.

partial records: if it contains simultaneously visual, auditory and proprioceptive data, the cat-neuron M will be retrieved through multiple channels. Conversely, a record with only a few referents, or which must recall partial records from damaged co-regulators, can become difficult to form or retrieve. In some cases a higher co-regulator, say E', may entirely recognize the item S, and later recall it directly; for this, E' must admit the (complete) record M of S as its internal E'-record of S; that is, each receptor activated by S must activate some agent of E'. We should be aware that, even in these cases, E' accesses not the record M itself, but only its perspective in the actual landscape of E'; *i.e.* it accesses a subjective representation of M (in the case of higher co-regulators, this difference could be at the root of *qualia*; see Chapter 10). This difference between the record M and its perspective accounts for certain memory dysfunctions which come not from a destruction of records, but from the severing of their links to the appropriate co-regulator(s). For example, when we meet somebody we may not come up with his name immediately, though we recall it later on.

To summarize, the co-regulators participate in the development and updating of the memory, directly by their choice of procedures, but also indirectly through the interplay among their procedures. In particular, without forming a partial record itself, a co-regulator modelling areas in relation with intentional processes may still participate, by a choice of procedures that facilitate the storage or later recall of (partial) records, through the strengthening of the patterns of perspectives received from an item. For example, we can make an effort to memorize a phone number.

5.3. *Procedures and their Evaluation*

At birth an animal can perform some basic inborn or instinctual actions required for survival: control of bodily functions (respiration, cardiac rhythm, emotional responses), reflexes and primitive responses to some aspects of its environment (eating, movement, escape and so on). Later it may learn or develop more complex sensory–motor responses, habits, skills, behaviours, action schemes of various types, and for higher animals cognitive and intentional processes. We model them as specific records, called *procedures,* and with their links, they form an evolutive sub-system of the memory MEM called the *procedural memory* **Proc**. As explained in Chapter 8, a procedure consists of a formal unit Pr in **Proc**, and of the pattern OPr of its commands. OPr has for indices the commands of Pr, represented by links from Pr to effectors (*e.g.* muscles); this pattern is stored in the empirical memory by its colimit effPr, a cat-neuron called the *effector record* of Pr, which can be thought of as an internal predictive model of the result of the procedure (see Fig. 9.11).

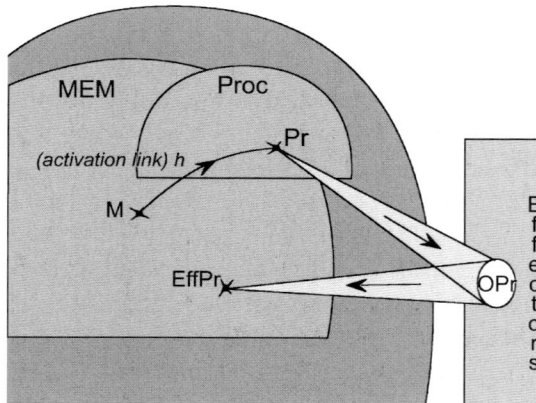

Fig. 9.11 Activator link.
A procedure Pr has a pattern OPr of commands to effectors. OPr is stored in memory by its effector record EffPr, obtained as the colimit of OPr. If there exists an activator link *h* from a record M to Pr, the recall of M activates Pr via *h*, and Pr activates its effectors via its different commands.

Procedures are constructed as iterated limits of patterns of basic procedures. **Proc** is generated from the initial sub-system of basic innate procedures by the formation of iterated limits during successive mixed complexifications. Given a pattern P of procedures, the procedure Pr that is the limit of the pattern is constructed through the synergetic action of the co-regulators and their interplay, by the same process we have described above for other records (except that the preceding colimits are now replaced by limits). As proved in Theorem 3 of Chapter 8, the commands of Pr unite the commands of the various procedures P_i of the pattern, and its effector record effPr is the colimit of the pattern formed by their effector records. Thus its formation consists in the strengthening of the links of this pattern, so that it becomes a synchronous assembly of cat-neurons. And the recall of Pr physically consists in the activation of one of the homologous synchronous assemblies of cat-neurons that its effector record effPr binds.

For the animal to survive, it is important that the chosen procedures be well adapted to the situations it encounters. It should be able to evaluate these procedures and to remember their result for later use, so that the selection of a procedure (by the co-regulators, or more globally through the interplay among their procedures) takes account of its previous experiences. A procedure succeeds if it realizes its objectives (*e.g.* the animal being hungry, prey is captured), satisfies primitive drives (no more hunger), fulfils homoeostatic requirements and possibly increases fitness and adaptation (*e.g.* a new kind of prey or a more efficient behaviour is discovered); this is recognized by the fact that the effector record is activated. It fails if this effector record is not activated, which may cause a fracture to one of the co-regulators—for instance to a co-regulator evaluating internal states (the hunger persists).

When a procedure Pr selected to respond to a situation S has succeeded, an *activator link h* is formed from the record M of S to Pr (Fig. 9.11). Each time Pr is selected in the presence of M, and Pr succeeds, the simultaneous activation of M and Pr increases the strength of h (by the Hebb rule), up to some threshold. Conversely, if Pr fails, the strength of the activator link is decreased. For basic procedures, the activator link can be inborn (examples are given below). For more complex procedures, the activator link will be formed or strengthened during the mixed complexification processes leading to the development of the memory. A later presentation of the situation will activate its record M and, if the strength of the activator link h is great enough, Pr will be recalled, so that the commands of Pr are transmitted to the effectors. Thus the activator link plays the role of a policy function in reinforcement learning theory (Sutton, 1998).

The formation and recall of a procedure in response to the situation S depends on the synergetic action of the various co-regulators that form partial records M_E of S. The record M of S is the colimit of the pattern Q of these partial records. Let us consider the following cases:

(i) If there exists an activator link h from M to a procedure Pr, then there is also a 'partial' activator link h_E from M_E to Pr, obtained by composing with h the binding link from M_E to the colimit M. The links h_E form a collective link which h binds. Each co-regulator recognizing a partial record M_E will activate Pr through the corresponding h_E. Though the strength of one of these links may not be sufficient to activate Pr, the simultaneous activation of the collective link they form is transmitted to its binding h, and allows it to recall Pr. For instance, a complex movement requires the coordination of several muscles. The recall of Pr can fail if some of the partial records cannot be recognized.

(ii) If each co-regulator forming a partial record M_E is forced to select a procedure Pr_E through an activator link k_E from M_E to Pr_E, and if these procedures are compatible, the operative procedure Pr^o arising from the interplay among the procedures will be the limit of a pattern having these procedures for components, and an activator link k (Fig. 9.12) will be formed from M to Pr^o to combine the different k_E (as explained in Chapter 8, Section 6.2).

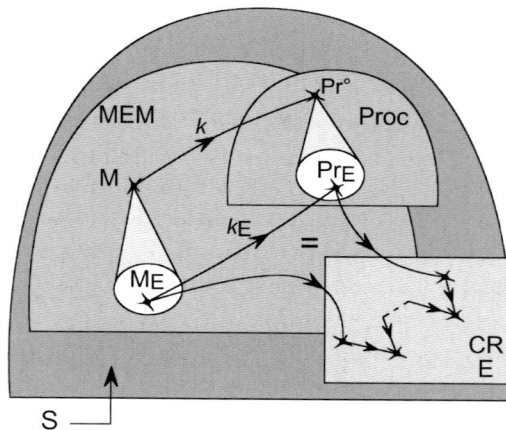

Fig. 9.12 Formation of a procedure.
An item S has different partial records M_E, which are bound by its record M. Each M_E has an activator link k_E to a procedure Pr_E. If these procedures are compatible, the operative procedure arising from the interplay among the procedures will consist in adding a limit Pr^o to the pattern they form; Pr^o becomes a new procedure (added to the procedural memory **Proc**), and the various links k_E are glued into an activator link k from M to Pr^o.

A particular record may have activator links to several procedures; which one is chosen (locally by a co-regulator or globally by the operative procedure) depends on the respective strengths of these links in the present context. For example, the sight of prey will have no effect for a satiated predator. If the predator is hungry, it starts a procedure to capture the prey; this operation requires coordination between the visual co-regulators which evaluate the position and the size of the prey, and the motor co-regulators which control the movements to catch it. However, if the prey moves unexpectedly quickly, or if, in an unforeseen way, the position information coming from visual co-regulators arrives too late for the motor co-regulators to compensate, the prey will escape.

We insist on the different roles played by the co-regulators and the procedures (which are not well distinguished in most models, *e.g.* as in Josephson, 1998): the procedures are just records, and their recall necessitates the cooperative operation of the co-regulators which select the procedures (directly or via their interplay) and send their commands to effectors.

6. Applications

Here we show how some general kinds of behaviours and of learning are integrated in our model.

6.1. *Physiological Drives and Reflexes*

At birth an animal (specially a primitive one) has automatic or instinctual responses to a variety of situations important for its survival. For instance, it has homoeostatic regulatory mechanisms controlling its internal environment (such as respiration, cardiac rhythm, temperature), innate reactions to preservative drives related to hunger or thirst, and to protective or defensive drives related to possibly harmful situations (*e.g.* orientation reflex, targeting reflexes, escape) and so on. Such an automatic response is modelled by a procedure commanding the appropriate effectors, which is activated via an activator link coming from the record of the motivating situation (internal state and/or stimulus).

For example, in the marine snail *Aplysia,* a touch S on the siphon causes a withdrawal of the gill into the mantle cavity; this defensive siphon-gill reflex depends on a synaptic path from the sensory neurons of the siphon (which detect S) to motor neurons controlling the gill and its withdrawal. This reflex has been studied by Kandel and Schwartz (1982), who have described its molecular basis. In particular they have shown that the sensitivity of the reflex is heightened (sensitization) if the touch is forceful, and thus causes an adverse effect. Inversely, if the touch is repeated and mild, the sensitivity of the reflex diminishes (habituation). Both processes are regulated by the same neuronal locus, having synapses with the sensory neurons.

The reflex will be modelled by an activator link h from the record M of the touch to the procedure Pr which commands the withdrawal of the gill (the effectors being the motor neurons of the gill). The touch has a partial record M_E for a co-regulator modelling the regulating locus which measures the adverse effects on the internal state. As this reflex is a defensive reflex, for a forceful touch, M_E is activated, increasing the activation of M (via the binding link from M_E to M), and the activator link h is strengthened (sensitization). If the touch is mild, and so not followed by an adverse effect, M_E is not activated, and the strength of the activator link lessens with the number of repetitions (habituation). This process generalizes for simple behaviours in response to vital drives, for acquired reflexes or even more complex behaviours, as the following examples show.

6.2.   Conditioning

Conditioning is one of the forms of learning which have been studied most often, and this has led to several models being proposed (for a recent review of these models, see Vogel *et al.*, 2004). It allows a response to be generalized to more or less different stimuli; for instance lures, which preserve only the main features of the prey, can produce in the animal the same capture behaviour as the prey itself.

Classical conditioning, discovered by Pavlov (1927), is a process wherein behaviours are acquired by the association of two stimuli. A stimulus, called the *unconditioned stimulus,* activates a strong autonomic response, called the *unconditioned reflex.* If a different stimulus, the *conditioned stimulus*, is presented several times just before the unconditioned stimulus then the animal learns to respond to the conditioned stimulus by activating the response unconditioned reflex (before the unconditioned stimulus, and even when the unconditioned stimulus is not presented). This is especially the case when the unconditioned stimulus is *pregnant* (in the terminology of Thom, 1988), meaning that it is associated to an essential drive or urgent need. In Pavlov's original study, the unconditioned stimulus was food presented to a hungry dog, and the unconditioned reflex was the salivation which precedes the eating of the food and the extinction of hunger. If over several repetitions the dog hears the sound of a bell (the conditioning stimulus) just before he sees the food, then he learns to associate the bell to the food, so that the sound of the bell alone makes him salivate (Fig. 9.13).

This is easily explained in our model. There exists a record M of the situation consisting of the unconditioned stimulus (or US), and the internal state I, as well as a strong activator link h from M to the procedure Pr

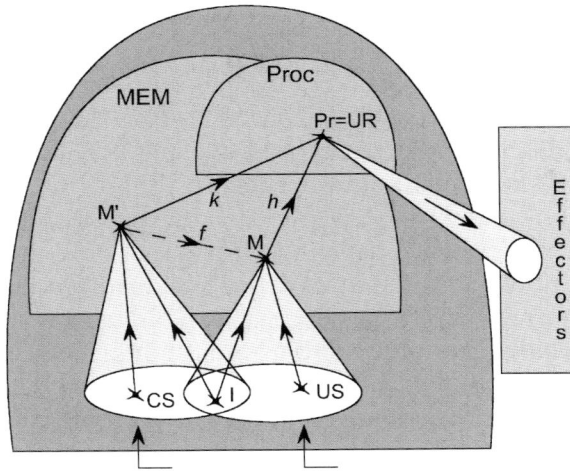

Fig. 9.13 Conditioning.
Initially an unconditioned stimulus US (*e.g.* food) presented in the context of a specific internal configuration I (hunger) activates an unconditioned reflex UR (salivation). This is modelled by an activator link *h* from the record M of the situation US + I to UR. The conditioning consists in preceding the US by the repeated presentation of a conditioned stimulus CS (*e.g.* the sound of a bell). This leads to the formation of a record M' of the situation CS + I, and of a link *f* in the memory from M' to M, modelling the fact that the situation recorded by M' precedes that recorded by M. By composing this link with the activator link *h*, an activator link *k* is formed from M' to UR, so that the presentation of CS in the context of the internal state I activates the UR. However, this link *k* disappears later if the presentation of CS is no longer followed by the presence of US (not shown). In the figure, the records of US and CS are also denoted US et CS, and the different times are not distinguished (except that the link from M' to M is hatched).

commanding the unconditioned reflex (or UR). If a conditioned stimulus (say, CS) is presented several times just before the US, a record M' of both the CS and the I is formed, and a link *f* from M' to M is created in the memory MEM; the simultaneous activation of M' and M strengthens this link *f* (by the Hebb rule). The composite of *f* with the activator link *h* forms an activator link *k* from M' to Pr, and the CS alone (in the state I) may cause the activation of the UR.

The connection between conditioned stimulus and unconditioned reflex can be severed if several later presentations of the conditioned stimulus are not followed by the unconditioned stimulus. For example, with previously conditioned dogs, if the sound of the bell is no longer followed by the presentation of food, so that the animal remains hungry, he will eventually cease to salivate in response to the bell. In the model, M records both the US (food) and the I (hunger) in which US activates UR. If the CS is presented alone, US does not appear while I remains, and thus M is not activated while M′ is still activated, so that the strength of the link *f* weakens (by the Hebb rule); since the activator link *k* from M′ to Pr is the composite of *f* with *h*, it leads to a reduction in the strength of this link, so that M′ may become unable to activate the unconditioned reflex.

This conditioning allows a response to be generalized to more or less different stimuli. For instance lures, which preserve only the main features of the prey, can produce in the animal the same capture behaviour as the prey itself.

Operant conditioning is another kind of conditioning described by Skinner (1938). The resulting acquired behaviours are of the general response-stimulus type, to which the behaviourists thought all learning could be reduced. During operant conditioning, an animal learns to associate a behaviour (the response) with a specific consequence, and repeats or suppresses the behaviour depending on the desirability or undesirability of the consequence. Examples are provided by the rat who learns to press a lever to obtain food, or to avoid a branch of a maze that provokes an electric shock.

Gilinsky (1984) considers this conditioning to be basically similar to the classical one, insofar as the consequence is assimilated to an unconditioned stimulus for the behaviour. However the setting is different: in the classical case, the effect does not depend on the animal but on the experimenter (food is presented or not after the conditioned stimulus), while now the effect is under the control of the animal itself (it repeats or not the triggering behaviour). Thus, we prefer to model it in the framework of reinforcement learning (Sutton and Barto, 1998), in which the behaviour is influenced by the expectation of a reward, or more generally, the evaluation of a procedure.

6.3.   *Evaluating Co-Regulators and Value-Dominated Memory*

Among the co-regulators, there are those which play a particular role in the adaptation of the animal to its environment, namely the co-regulators evaluating homoeostatic drives (such as hunger or fear) or hedonic states (pleasure, pain and so on). These co-regulators are related to the emotive brain, essentially based in the brain stem and the limbic system; for instance, Stefanacci (2003) indicates how the amygdala plays a main part in affective states, in particular in the recognition of frightening objects and escape from them. They will be called *evaluating co-regulators*, because they allow evaluation of procedures as a function of their consequences on these internal states of the animal. Indeed, let Ev be an evaluating co-regulator. For a procedure Pr to be evaluated by Ev, it should have aspects to some agents of Ev; for instance, an aspect v to an agent + modelling a state of well-being of the animal. If Pr is activated and its result is beneficial for the animal, the simultaneous activation of Pr and + increases the strength of v (by the Hebb rule); on the other hand this strength is decreased if the activation of Pr does not have a beneficial result (+ is not activated). Let us suppose that Pr is activated via an activator link h from a record M to Pr. If Pr succeeds, the strength of v increases (as said above), and h defines a link in the landscape of Ev from the perspective of hv to that of v. Since these two perspectives are simultaneously activated, the strength of h is increased in the landscape (still by the Hebb rule). If Pr fails, v weakens, as well as h (this change of h seems dependent on dopamine transmission, *cf.* Berridge and Robinson, 1998). In particular, if Pr causes a fracture in an evaluating co-regulator (some need is not met), the strength of h will be decreased.

Another interpretation arises if we consider that an evaluating co-regulator plays the role of an effector, and that an aspect v of Pr represents a command added to the procedure Pr. Thus the result of the procedure, in terms of the desirability or lack thereof of this command, will be included in the effector record effPr of Pr, which operates as a predictive internal model. In particular, if there exists a higher co-regulator which perceives effPr in its landscape, it will be able to 'intentionally' select or not the procedure, depending on the expected result. This agrees with the model of intention described by Eagleman (2004). Operant conditioning enters into this general framework. Let us consider the example of the rat which learns to associate one branch of a maze with an electric shock, and to avoid it. The different co-regulators recognize various aspects of the maze, and the co-regulator evaluating pain recognizes the shock. The global record of the maze will contain, as a partial record, the pain record of this co-regulator, which is linked to an avoidance procedure. Thus the procedure associated to the maze will acquire a new command to avoid the branch, and this inhibits the other commands to take it.

Records of items which have partial records for some evaluating co-regulators form a sub-system of the memory which can be compared with what Edelman calls the *value-dominated memory system* (1989, p. 99); and he explains its functioning by the existence of loops between the emotive brain and the thalamo-cortical system. Stefanacci calls this the *stimulus-reward memory* (2003, p. 21), and shows that lesions of the amygdala impair it. The role of the evaluating co-regulators, and in particular the amygdala, is confirmed by these experimental results. Indeed the amygdala corresponds to one of the evaluating co-regulators. If a stimulus has a partial record for it, and if the amygdala is damaged, its partial record of the stimulus cannot be formed or recalled, thus preventing the formation or access of the whole record. Conversely, lesions of the amygdala do not impair the recognition memory of inputs that lack emotional undertones (Stefanacci, 2003), which have no partial record for evaluating co-regulators. It could also explain the conscious inhibition of painful memories (those hidden in what is classically referred to as the Freudian unconscious), by the inhibition of the recall of their partial records for evaluating co-regulators.

Semantics, Archetypal Core and Consciousness

Higher animals are able to not only recognize specific stimuli and learn more or less complicated behaviours, but also to extract invariants out of the wealth of information they receive externally and internally. In this way animals with a developed associative cortex will classify their records into invariance (equivalence) classes with respect to a particular attribute, and associate to each class a formal unit, called a concept for this attribute. More general concepts are formed by combining concepts for several attributes. These concepts form an animal's *semantic memory*. In higher animals, the semantic memory allows for the formation of an intricate part of the memory, called the *archetypal core*, at the root of the notions of self and of consciousness, which we shall try to characterize.

The chapter ends with a brief summary of the main characteristics of neural systems allowing for the development of higher cognitive processes, and of the manner they are modelled in memory evolutive neural systems.

1. Semantic Memory

The components of the memory are records representing various items (objects, signals or features of the environment, past events or situations, behaviours, sensory-motor programs and so on). A record corresponds to a specific item, but in a flexible way, so that the correspondence holds despite variations in the item's contextual attributes, such as its location or apparent size. A more complex operation consists in forming classes of records, based on similarity: although the member records have intrinsic differences, they nevertheless share one or more common features or attributes, and thus form invariance classes; for instance we recognize a chair we has never seen before, for it has the features we have learned to associate to chairs. An invariance class is internalized by the formation of a formal unit which we call a concept for the attribute(s). Then these concepts with respect to attributes are combined to form more abstract concepts, which lead to a more or less fine classification, depending on the number of already formed concepts which are combined: we can form the class of all animals, or the class of mammals, or the class of dogs, or the class of spaniels and so on. Edelman

(1989, p. 141) names this process 'perceptual categorization', and he uses also the name concept. However, he thinks that the word semantics should be reserved for the case where there is a language. We follow his lead in using the term concept (also used by Changeux, 1983), but we consider that the concepts are the components of a semantic memory even in the absence of language. This is justified by recent results which prove that semantic distinctions are understood by 5-month-old children, well before they have a language (Hespos and Spelke, 2004).

1.1. *Perceptual Categorization*

An animal can classify objects with respect to basic attributes such as colour, orientation, size, tonality, motion and so on. Neurobiological data demonstrate the existence of specialized areas of the brain which treat these attributes. They act as distinct modules, discriminating objects according to specific attribute(s), without taking into account their resemblances or differences as determined by other such modules. An object displaying a certain attribute activates an assembly of neurons within the module more or less specific for this attribute, and two objects are distinguished by the module if they activate non-similar assemblies. For instance a square will be distinguished from a circle by the shape-module, while if they both have the same colour, the colour module will not distinguish between them. This comparison does not imply that the module performs a real perceptual categorization: the module acts in the same way in the presence of two similar objects, but this pragmatic comparison is not remembered and does not entail an internal characterization of the similarity by the module. The colour module acts similarly if this square and that circle are both red, but it does not give a characterization of a red object.

Only sentient beings of a higher order, those with a developed associative neo-cortex, can recognize this similarity at a higher level, and remember it in a formal unit, representing an invariance class and taking its own identity. We call this unit a *concept* for this attribute. Such concepts can be combined to form concepts classifying records with respect to several attributes (*e.g.* a blue square). More complex concepts are formed by combining (binding or classifying) together more elementary ones. Concepts, and the appropriate links between them, form the semantic memory. It allows for partial similarity-based categorization and for generalizations, so that in particular, a procedure can be adapted to similar but not identical situations.

As explained in Chapter 8 for general memory evolutive systems, the construction of the semantic memory is modelled in three steps:

(i) A pragmatic classification with respect to particular attributes is implemented by lower level co-regulators (corresponding to the specialized areas of the brain and modules previously referred to).

(ii) This classification by a co-regulator, say E, is internally reflected at a higher level by the formation of abstract units, called E-concepts, which formally represent the invariance classes so distinguished. Once formed, an E-concept evolves, as well as its invariance class, to take account of the successive experiences of the animal. Links between E-concepts are constructed.

(iii) More and more complex concepts are formed by combining existing concepts. The concepts and their links form the semantic memory.

1.2. Concept with Respect to an Attribute

In the memory evolutive neural system of the animal, the modules treating specific attributes are modelled by some of its co-regulators. Let E be a specific co-regulator, such as a colour-CR modelling a colour module. Let us suppose that an item S, say a stimulus, which has a record M is presented to the animal at time t. First, it activates a pattern of receptors (*e.g.* retinal cells) of which the record is the colimit, thus activating this record via the binding links to this colimit. The record is transmitted to E by synaptic paths, which are interconnected by the operating links between these agents. Thus, we get the pattern $Tr_E M$ of (cat-)neurons activated by M: its components are indexed by the different aspects of M for E (Fig. 10.1). In Chapter 8 we have called this pattern the *E-trace of* M, or of the item S. For a colour-CR, the trace will consist of all the paths along which the colour characteristics of the stimulus (hue, intensity, and so on) are transmitted to the units of the colour areas, disregarding all other information treated by other co-regulators.

The E-trace represents the manner in which E responds to the presentation of the item, or directly of its record M. It follows that two records M and M' may induce similar E-traces, while they induce differing traces on other co-regulators; in this case, we say that M and M' are in the *same E-invariance class*. For two records to have 'similar' traces means that the patterns of agents they activate are sufficiently close to decode the same kind of information, translated in the model by requiring their E-traces to be prohomologous (Chapter 4). For instance, a red circle and a red square are in the same invariance class for a colour-CR but not for a shape-CR, since they have different geometric shapes (Fig. 10.2). Or if we consider the two patterns of retinal cells activated by two vertical bars in different positions in the visual field, they activate two different simple cells, but they may activate the same complex cell, signifying that they have the same invariance class for an orientation-CR.

Locally, E reacts similarly in the presence of two items (or their records) having the same E-invariance class. However, at its own level there is no

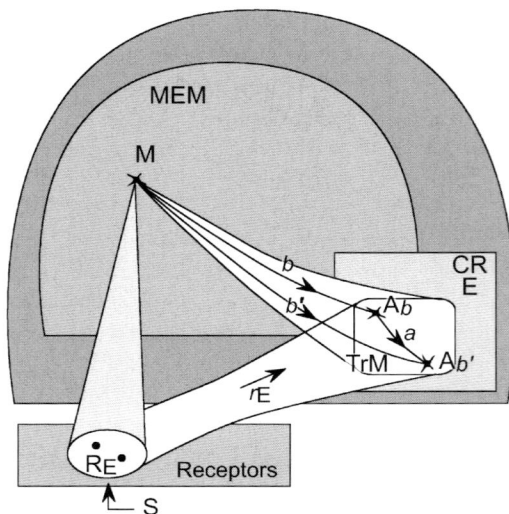

Fig. 10.1 E-trace of a record.
An item S has a record M, and E is a co-regulator for which the record M has several aspects (such as b, b'). The trace TrM of M for E is the pattern having for indices these aspects, and for components the agents A_b to which these aspects arrive, indexed by the corresponding aspects b. The distinguished links are defined by links a in E from A_b to $A_{b'}$ commuting with the corresponding aspects, so that $b' = ba$. The trace must not be confused with the pattern of perspectives r_E of the stimulus for E, which is used to define a partial record of S. This last pattern r_E has for components perspectives (not aspects, but classes of aspects with the same effect on the agents), hence is the pattern internally perceived by E; on the other hand the trace TrM is the pattern of agents activated by the record M, and thus represents the action of the record on this co-regulator.

explicit comparison between items. Higher animals may pursue the discrimination task further, and recognize that two items have the same E-invariance class, for instance recognizing 'the class of blue objects'; moreover, the classification is not fixed but is adapted to the successive experiences of the animal. This can be done by a higher level co-regulator, acting on a longer time scale, which can apprehend externally to E the fact that two items are similarly treated by E, by perceiving that their E-traces decode the same kind of information. The invariance so demonstrated will be remembered by the formation of a cat-neuron, called an E-*concept*, which may be thought of as an abstract *categorization unit* associated to an E-invariance class.

Applying the definitions given for any memory evolutive system in Chapter 8, we define the E-*concept* C_EM of M (or of an item that M remembers) as the limit of the E-trace of M; its *instances* are all the records (or items)

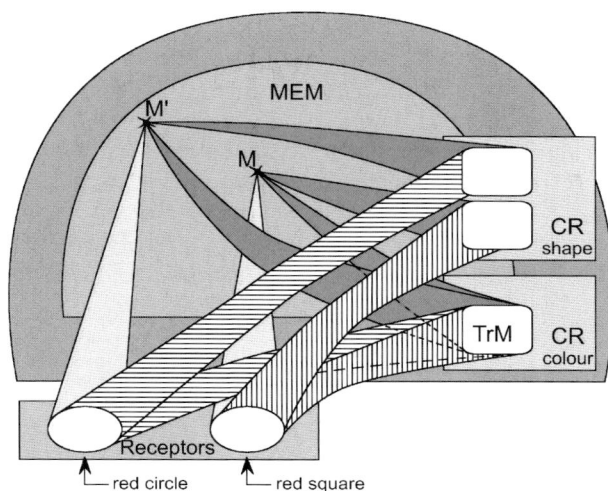

Fig. 10.2 Different traces of a record.

M and M′ are respectively the records of a red circle and of a red square, and they have traces for a colour-CR and for a shape-CR. Both have the same trace for the colour-CR (both are red), but different traces for the shape-CR, since one is the shape-trace of a circle, the other the shape-trace of a square.

which have an E-trace admitting $C_E M$ as their limit, and they form its E-invariance class (Fig. 10.3); in particular among these instances figures the E-concept itself (looked at as a record). As any cat-neuron, once formed the E-concept evolves depending on the experiences of the animal, and so does its E-invariance class. Indeed, the E-concept $C_E M$ of M emerges at a given time t as the limit of the E-trace of M (through a mixed complexification process with respect to an option whose objective is to classify this E-trace). At this time, its E-invariance class is formed of the items actually experienced so far by the animal which have an E-trace pro-homologous to TrM. At a later time t', the instances of the configuration at t' of the E-concept will be all the records (or items) whose E-trace at t' admits $C_E M$ for its limit; thus, the E-invariance class may then also include items that the animal had not yet experienced at t but has encountered later. Roughly, first there is a classification in E-invariance classes; then a categorization unit is associated to each class; and finally these units, the E-concepts, take the precedence, and act as prototypes (in the terminology of Rosch, 1973) to which an item may be compared to characterize its E-invariance class. For example, though initially formed to classify the blue objects already known, the colour-concept 'Blue' virtually determines the class of all blue objects the animal will encounter (Fig. 10.4) (or, for man, even imagine).

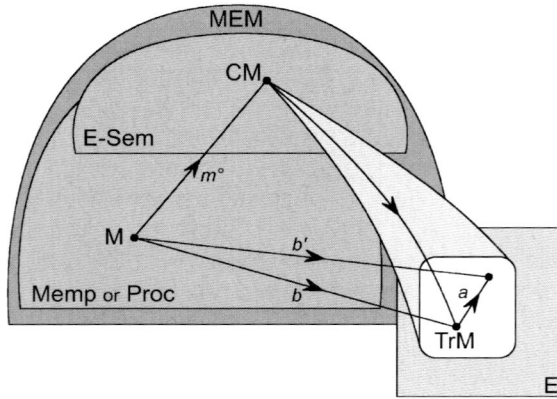

Fig. 10.3 E-concept.
The record M has an E-concept if its E-trace TrM has a limit CM. There is then an E-universal link m° from M to CM, through which any link from M to an E-concept factors. The instances of the E-concept are all the records such that CM is the limit of their E-traces.

Higher order animals are able to make elaborate classifications, in which two items are in the same E-invariance class though their E-traces are not physically connected; in this case their E-concept will be a pro-multifold object (Chapter 4). For this to be possible, we suppose that there exist pro-multifold components, *i.e. both the memory evolutive neural system and its opposite satisfy the multiplicity principle.*

The E-concepts are the components of an evolutive sub-system of the memory MEM, called the E-*semantic memory,* denoted by E-**Sem**, and there is a functor C_E from a full sub-evolutive system of MEM onto E-**Sem** which preserves the colimits (Theorems 4 and 5 of Chapter 8). The E-trace of M determines a distributed link from M to the E-trace of M, so that it is classified into a link m° from M to C_EM which can transmit the activation of M to its E-concept (Fig. 10.3). Theorem 3 of Chapter 8, shows that m° is an E-universal link, meaning that it defines C_EM as the best approximation of M in E-**Sem**, and as the best representative of the E-invariance class of M. Thus, the instances of an E-concept can also be characterized as the records linked to it by such an E-universal link.

1.3.   The Semantic Memory

A record M may belong to several invariance classes, each the result of classification according to a particular attribute, and thus each corresponding to a different co-regulator. For example, a blue triangle and a blue circle have the same colour-concept 'Blue'; but the shape-concept 'Triangle' of the

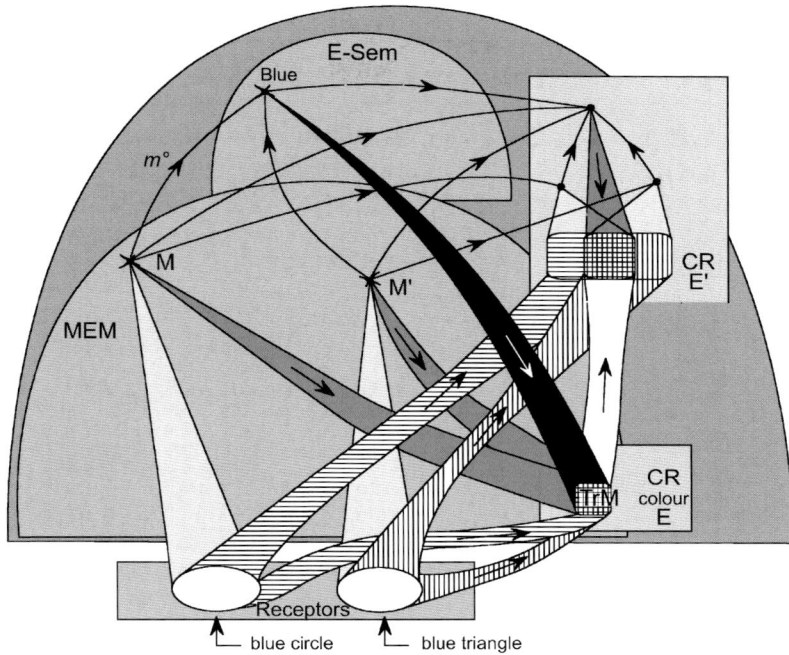

Fig. 10.4 Instances of an E-concept.

E is a colour-CR. The records M of a blue circle and M′ of a blue triangle have pro-homologous E-traces; this is observed by E′, a higher level co-regulator, which receives aspects of M, M′ and of their E-traces in its landscape. One of the objectives of E′ will be to add a limit to these E-traces, which becomes the colour-concept of both M and M′ (the E-traces being pro-homologous, they have the same limit), in this case the colour-concept Blue. Both M and M′ are instances of 'Blue' so that there exist an E-universal link m^o from M to Blue, classifying the distributed link from M to Blue; and the same for M′. The top-most link represents an aspect of Blue for E′. More generally the instances of the E-concept Blue can also be defined as all the records with an E-universal link to Blue.

blue triangle represents the class of triangles, while the shape-concept 'Circle' of the blue circle represents the class of circles. Concepts with respect to several attributes can be stored as products of the family of concepts with respect to these different attributes (the concept 'Blue-circle' is the product of 'Blue' and 'Circle', Fig. 10.5).

More general concepts are obtained by combining together concepts with respect to various attributes, through the successive formation of limits or colimits of patterns of already formed concepts. More precisely a *concept* is a component of the *semantic memory* **Sem,** which we have defined as an evolutive sub-system of the memory MEM obtained through successive

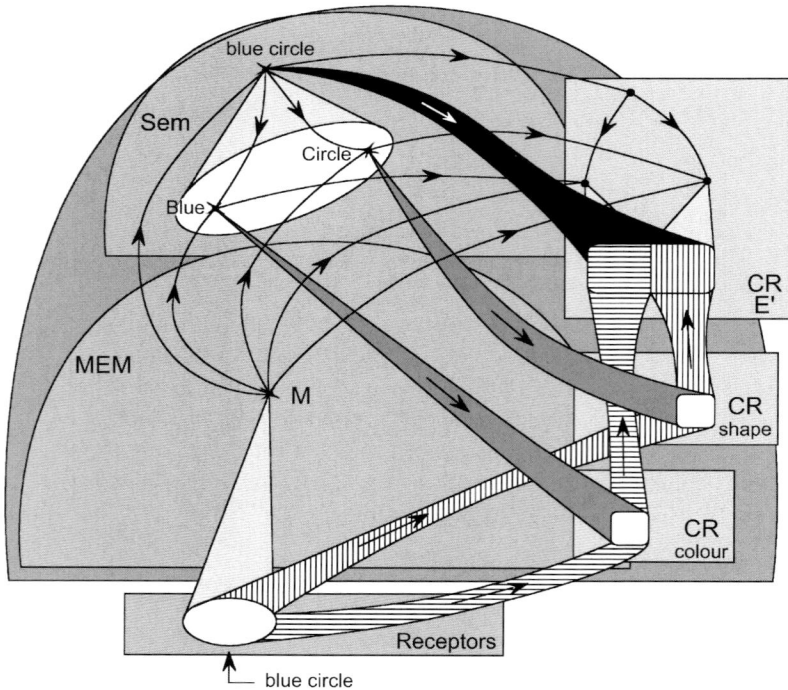

Fig. 10.5 Concept for several attributes.
The record M of a blue circle has the colour-concept Blue and the shape-concept
Circle. E' is a higher co-regulator which receives aspects from the traces of M for the
colour-CR and for the shape-CR. Then it may form the (colour, shape)-concept Blue-
circle, which is the product of Blue and Circle in the memory MEM.

mixed complexifications of the union of the E_k-**Sem** for the different co-
regulators E_k associated to attributes.

    As said above, the E-concepts have been initially constructed to formalize
the E-invariance classes the animal can distinguish at a given time; then they
evolve and take their own identity, allowing to extrapolate the E-classification to new experiences of the animal; thus, we go from the instances, to the
E-concept, and back to a more flexible E-classification. For general concepts,
the situation is different: they emerge as formally combining already known
concepts, with no mention of a classification; the associated classification in
invariance classes must be deduced, by defining their instances. We have
given two different characterizations of an instance M of an E-concept C: the
E-trace of M admits C for its limit; or there exists an E-universal link from M
to C. These characterizations cannot be extended to general concepts. Thus,
we must define the instances M of a concept A and their defining link to A by
induction on their construction as explained in Chapter 8

(i) If M is an instance of an E_k-concept A, its defining link is the E_k-universal link from M to A.

(ii) If A is the limit of a pattern P of concepts, M is an instance of A if it is an instance of each component P_i of P and if the defining links m_i^o from M to P_i commute with the distinguished links of P, so that they form a distributive link from M to P; then the link classifying this distributive link is the defining link from M to A.

(iii) If A is the colimit of a pattern Q of concepts, M is an instance of A is it is an instance of at least one component Q_j, and the defining link is the composite of the defining link m_j^o from M to Q_j with the binding link from Q_j to the colimit A; this link must be uniquely characterized by this property.

The instances of a concept form its well-defined *invariance class*. These invariance classes provide a classification of the records having a concept: two records are in the same invariance class if and only if they are instances of the same concept. However, these invariance classes are not disjoint, so that we cannot speak of 'the' concept of a record M; for instance an apple is an instance of the concept 'Fruit', but also of the concept 'Food'. It is the reason for which we do not call the defining link m^o of M to A a universal link (as in the case of concepts related to an attribute). However, in some cases m^o is universal for **Sem** (meaning that it defines A as a free object generated by M with respect to the insertion functor from **Sem** to MEM); in this case, M (and also A) has an attribute associated to a co-regulator E if m^o factors through an E-universal link.

Once formed a concept takes its own identity; the definition of its instances applies to any of its configurations, so that its invariance class may vary to take account of the new experiences of the animal. The classification also evolves by the formation of new concepts (*e.g.* different kinds of preys are better distinguished; for man, new words are defined). The formation of limits leads to a finer classification, the invariance class of the limit of P being included in the intersection of the invariance classes of the components of P (there are less small blue triangles than triangles). Conversely, the formation of a colimit leads to a larger class, uniting the invariance classes of its components; for example the invariance class of 'Mammal' contains those of 'Rat', 'Dog', 'Man' and so on.

Within particular contexts, the formation of more specific concepts is forced by the occurrence of fractures to some higher co-regulators. For instance, in the development of language by a child, terms take on a more precise meaning as experiences accumulate. Initially the child may call all wheeled vehicles 'Cars', but then he will eventually distinguish trains from cars. Conversely, this might explain several types of neural degeneracy, such as aphasias or apraxias, as depending on the severing of communications between specific co-regulators, and so disrupting the transmission of activation between a record and its concept, or vice versa.

1.4. Recall of a Concept

A concept A can be recalled through the selection of any one M of its instances, indeed the activation of M is transmitted to A via the defining link m^{o} from M to A. Conversely, the activation of A can recall one of its instances. Indeed, suppose that, at a time t, there is one of the instances M′ of A which is independently activated via a diffuse activation of the memory (*e.g.* if the item remembered by M′ has been observed a few instants before, it remains an awareness of it, whence a 'priming effect'). If A is activated at t, the simultaneous activation of M′ and of A strengthens the defining link m'^{o} between them, according to the (extended) Hebb rule. If A has an aspect f for a co-regulator, the composite $m'^{o}f$ is an aspect of M′, which is also strengthened, leading to the recall of M′. If initially A had been activated via another instance M, we have a passage from M to M′ which we call a *shift*.

The possibility of such a shift, or even of a sequence of shifts between different instances of a concept, gives more flexibility to the animal's interactions with the context, in particular to the interplays among procedures of the various co-regulators: there are two degrees of freedom in the activation of a concept: selection of the instance which is activated, possibly after a shift between instances; and then activation of this instance (as for any cat-neuron) by the unfolding of any one of its ramifications, with possible complex switches among the (patterns intervening in these) ramifications (Chapter 3). Note that the two operations, 'shift' and 'switch', are of a different nature:

(i) A complex switch between two patterns (or between two ramifications) corresponds to a change from one decomposition (or ramification) of a single higher order object to another one, the two being non-connected.
(ii) The shift between two instances of a concept corresponds to a change from one of the presentations of the concept to another.

In the first case, the change is between two alternatives which merge at the higher level; in the second, it appears by dissociation of a common base (the concept).

The development of a semantic memory allows for more adapted behaviours, thanks to the formation of concepts whose instances are procedures, called *procedure-concept*. The choice of a procedure-concept instead of a unique procedure by a higher co-regulator gives more latitude, since it can be realized by anyone of its instances, and then by anyone ramification of this instance, depending on the context. As the different co-regulators all cooperate (eventually with conflicting procedures) in the dynamics of the system, the choice of procedure-concepts by some of them will produce different results according to the choices of the other co-regulators, their choices being made coherent by using shifts between instances of the procedure-concepts and switches between ramifications of these instances. For example, the procedure of seizing an object will be abstractly represented by a concept, but its activation will consist in the selection of one of its instances, which will be activated via the successive activations of synchronous assemblies of cat-neurons based on lower motor areas, down to simple synchronous assemblies of neurons activating the muscles. The selection of the instance and of the assemblies activated at each level depend on the size and shape of the object to seize, and may be modified during the motion to ensure that the object is in fact seized.

The semantic memory also makes the activation of a procedure via an activator link more efficient. Indeed, when an item is recognized by its concept and this concept A has an activator link h to a procedure, each instance M of A is then connected to that procedure by an activator link (composite of the defining link m^o from M to A, with h), so that all instances lead to the same procedure, allowing for generalizations. A lion may chase any zebra, not just the ones it has already met.

In general, a concept is a pure abstraction, though it acts through its instances. As explained, simpler concepts (the concepts with respect to some attributes) are formed in relation to real-world references: the animal meets an item in its environment, remembers it, and then forms its concept of the record with respect to specific attributes, later the concept acquires more instances. However, more complex concepts are formed more abstractly, by combining simpler concepts, using the specific links between them already formed in the semantic memory, and their instances are only determined afterward. Thus, semantics is not always related to reference, but relies on the connections between records and concepts, in particular their links to other concepts, and to procedures. In human beings, language allows concepts to be named, and thus for them to be referred to in a more efficient way, permitting more complex operations and combinations.

Works of experimental psychology have well underlined the process of semantic emergence described above in relation to the cognitive development of children. For instance Houdé (1992) proposes a polymorphic system of calculation and meaning, making it possible to form a synthesis of the epistemic constructivism of Piaget (1964) and the pragmatic constructivism of Varela (1989). He describes the idea as follows.

• Initially, at the pragmatic stage, there is assembly and remembering of groups of contiguous elements; this is revealed by success with the inclusion problems of Piaget (formation of records).

- At the pragmatico-functional stage, there is a beginning of classification in terms of substitutable elements, very influenced by the context (the invariance class is formed, but without being still internalized in a concept).
- Finally, at the functional stage, the classification emerges from the context (emergence of the concepts), and logical symbolization develops (extension of the semantic memory).

2. Archetypal Core

We have distinguished three evolutive sub-systems of the memory MEM, with links between them: the empirical memory **Memp**, the procedural memory **Proc**, and lastly the semantic memory **Sem**. Now we further distinguish records according to their significance for the animal, thus displaying two other sub-systems of the memory: the *archetypal core* **AC** and the *experiential memory* **Exp**.

2.1. The Archetypal Core and Its Fans

The mental liberty afforded by concepts allows records of various types (in particular sensory, motor, emotive) to intermingle and activate one another. Records and concepts having greater and more lasting importance for an animal (for instance stable aspects of the environment in contrast with more variable ones, deep feelings, and so on) are, over the lifetime of the animal, activated more often, and for longer periods. They are connected into quickly activated patterns, which integrate the major experiences, sensory, cognitive, motor, or whatever, with their emotional overtones (this role of emotions has been well emphasized by Damasio, 1999), and with the main procedures associated to them. For example, the recall of blue sky is linked to records of perceptions, motor processes, sensations and emotions (*e.g.* sun, swimming, heat, well-being).

These records form what we call the *archetypal core*, which associates sensory-motor and perceptual categories with value states, and thus represents a personal, emotive memory for the animal: of its body, its experiences and its acquired knowledge—be it pragmatic, social or conceptual. The archetypal core can be compared with what Dehaene *et al.* (1998, p. 14529) call the *global workspace*, which they describe as a computational space 'composed of distributed and heavily interconnected neurons with long-range axons'. It corresponds to what Edelman calls the '*value-category memory*', which he puts at the basis of a 'primary consciousness' (Edelman, 1989, p. 152).

In the memory evolutive neural system, the archetypal core is modelled by an evolutive sub-system of the memory MEM, still called the archetypal core, and denoted by **AC**. Its components, or *archetypal records*, can be neurons or higher order cat-neurons (for example remembering a complex situation with its affective value). Edelman (1989, p. 152) explains that the 'linkage between category and value is based on two very different kinds of nervous structures and functions: the limbic and brain-stem system and the thalamo-cortical system'. Following him, we suppose that **AC** is based on these systems, meaning that it is obtained by successive mixed complexifications of the evolutive sub-system of the memory evolutive neural system modelling these systems with their links (in particular, those in the thalamo-cortical loops connecting different specialized areas of these sub-systems). In other terms, archetypal records are mental objects or concepts of increasing complexity, obtained by combining (binding or classifying) assemblies of neurons more or less distributed in these two sub-systems.

One of the characteristics of the archetypal core is that the activation of an archetypal record is quickly propagated to other archetypal records, so that a whole sub-system is autonomously activated (above the background noise) as soon as a small part of it is stimulated; and this activation can be sustained, whatever the variations of the stimuli, through a sequence of loops along specific channels. This is modelled by associating, to each archetypal record A, a bundle of particular links of **AC** (based on the loops of the neural system referred to above), along which the self-activation will preferentially propagate from A to other records; we call such a bundle a *fan*. The links of a fan have the property that they can be composed with links of other fans to form easily activated circuits (Fig. 10.6). More formally:

Definition. To each record A in the archetypal core **AC** is associated a set of complex links in **AC**, called the *fan associated to* A, with the following properties:

(i) Each link *f* of the fan is the first factor of a path of links belonging to fans which have for their composite a loop from A back to itself.
(ii) The propagation delay of the links of the fan decreases over time, while the strength of the links increases, up to a threshold.

Formally, the fans equip **AC** with a *Grothendieck topology* (Grothendieck and Verdier, 1963–1964).

The fans act as channels (to be compared to the chreods of Waddington (1940) who uses the word in embryology to mean a critical development path) through which the activation of the archetypal record A is relayed and amplified, as it propagates first to the target concepts, and then oscillates through a sequence of loops based on shifts between various instances of the concepts, and switches among their ramifications. It leads to a process of stochastic resonance in the brain which has been experimentally observed

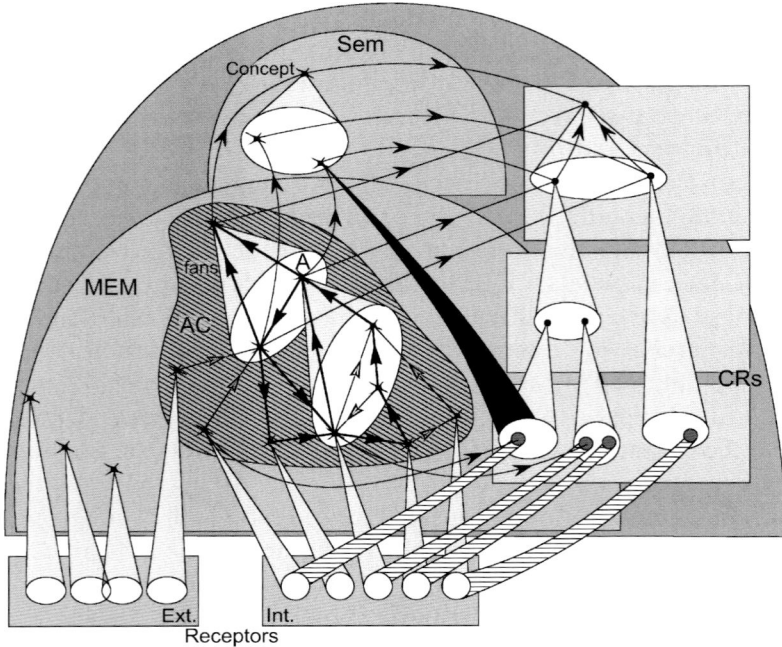

Fig. 10.6   Archetypal core and its fans.
The archetypal core (modelled by **AC**) is the part of the memory containing the records most important for the animal, connected via strong links, quickly activated. To each record A in **AC** is associated a bundle of such links (darker in the figure), called a fan, which transmit its activation to other archetypal records; and there is a loop from A to A formed by links belonging to successive fans. When an archetypal record is activated, this activation will propagate to other records through links belonging to fans; and then to decompositions of these records (with possible switches between decompositions). This activity of the archetypal core remains self-sustaining for a long period and may extend to other parts of the system, in particular to co-regulators, as shown by the various links in the figure.

(*cf.* Wiesenfeld and Moss, 1995; Levin and Miller, 1996; Collins *et al.*, 1996). Over time the fans acquire more links, and these links are gradually strengthened, up to a threshold, leading to more and more integration of the whole archetypal core.

2.2.   *Extension of the Archetypal Core: The Experiential Memory*

The archetypal core, as a persistent internal representation of the animal, its phenomenal experiences, and its acquired knowledge, could be the basis of the inherent notion of self. As any record, an archetypal record evolves in time, and an archetypal concept may acquire new instances. The archetypal core itself evolves, and expands over time. More complex archetypal records are formed by combining (binding or classifying patterns of) simpler ones. Moreover, a record N which initially is not archetypal can later become archetypal if it is strongly connected to a particular archetypal record through a preferential link, say g from N to A_n, through which the activation of N is propagated first to A_n and then diffuses to the other archetypal records linked to A_n via the links in the fan of A_n. Roughly N admits A_n as the closest archetypal record which it can activate. We call a record such as N an *experiential* record. These records may represent external stimuli as well as internal states, behaviours or procedures, or an association of such; and they form a kind of 'halo' around the archetypal core.

In our model, the records admitting such a closest archetypal record, with the links between them, form an evolutive sub-system of MEM, the *experiential memory*, denoted by **Exp**, which contains **AC** (Fig. 10.7). The

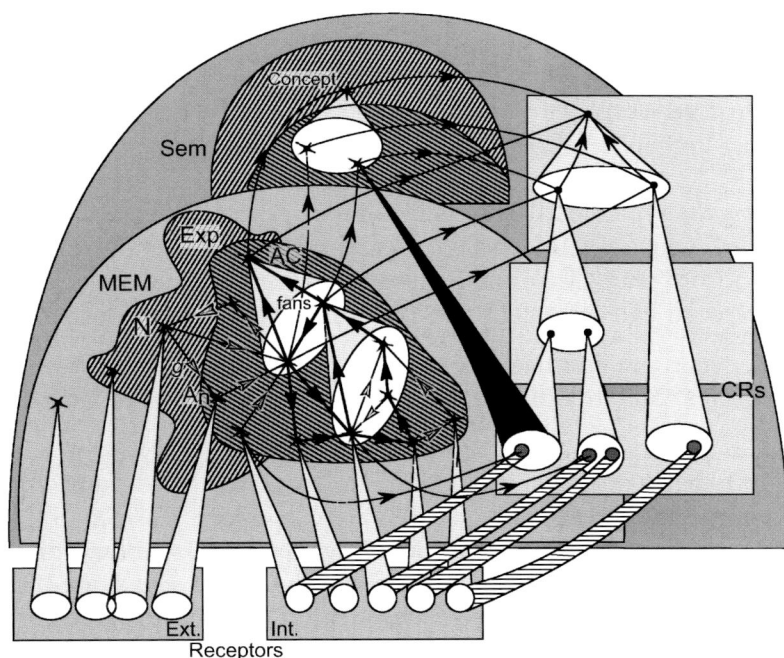

Fig. 10.7 **AC** and **Exp**.
Representation of a part of the memory MEM containing the experiential memory **Exp**, and the archetypal core **AC** (included in **Exp**). The semantic memory **Sem** is entirely contained in **Exp**, and only partially contained in **AC**. A record N in **Exp** is linked (by g) to a 'closest' record A_n in **AC**, meaning that any other link from N to an archetypal record factors through g. On the figure, N also receives a link from an archetypal record.

characteristics of an experiential record to have a closest record A_n in **AC** is modelled by the existence of a link g from N to A_n such that any other link from N to an archetypal record uniquely factors through g; or, in other terms, A_n is a free object generated by N with respect to the insertion functor from **AC** to **Exp** (*cf.* Chapter 1). Thus, the following definition:

Definition. Exp is the largest evolutive sub-system of the memory MEM which contains **AC** and with the property: Each record N in **Exp** generates a free object A_n with respect to the insertion functor from **AC** to **Exp**.

A record N can become experiential if it has a distributed link (g_i) toward a pattern P in **AC** whose links strengthen over time. Then the pattern is classified into a new archetypal record A_n (the limit of P, added in a mixed complexification process), and (g_i) is classified by a link g from N to A_n, so that A_n becomes the archetypal record 'closest' to N. For instance, a newly heard melody N evokes memories of known melodies, and then of experiences related to them. If it is heard several times, this pattern of archetypal records can take a unique identity A_n, and become the closest archetypal record of N; later the melody itself can become archetypal. As this melody, a record N in **Exp** may take more significance for the animal over time, and later become archetypal. For that, it must also receive links from archetypal records to N, so that a loop from N back to itself can be formed; such a loop will begin by the preferential link g, then links in fans, and finally a link to N. The repeated activation of this loop strengthens g. Then N becomes an archetypal record (Fig. 10.8), and g one of the links of the fan associated to N.

Thus, a record can become archetypal through a two-step process:

• first it forms a preferential link to one archetypal record;
• and then this link strengthens sufficiently for the record to be integrated in **AC**.

While only parts of the empirical memory and of the procedural memory become experiential over time, the semantic memory, formed by records which are more pregnant for the animal (the concepts), will develop into an evolutive sub-system of **Exp**. And, at least at the beginning of the life of the animal, the majority of the experiential memory is progressively integrated in the archetypal core, in priority the records coming from internal stimuli. In the terminology of Thom (1988), the archetypal records would be pregnant and the experiential ones salient, and the integration into **AC** of a record in **Exp** corresponds to what he calls the 'encircling of a salient form by a pregnant one' (translated from Thom, 1988, p. 55).

As an illustration of these ideas, consider the following example: an infant has an innate holding reflex in its archetypal core. Progressively, it will acquire a well directed sensory-motor procedure for seizing an object.

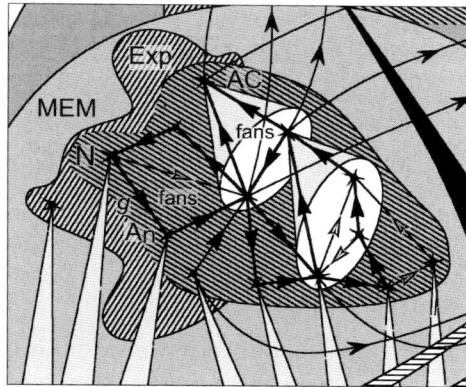

Fig. 10.8 Extension of the archetypal core.
Same situation as in Fig. 10.7 (only the central part is shown), except that here the record N in **Exp** has been integrated into **AC**, following a sufficient strengthening of the link g from N to A_n, and of the link arriving to N. Then g becomes a link of the fan associated to the, now archetypal, record N; and it is incorporated into a loop from N to N formed by links in fans.

Indeed, a visual co-regulator will learn to recognize some objects (in **Memp**), an emotional co-regulator will recall the pleasure (in **AC**) of seizing these objects, and a motor co-regulator will recall how an object has been seized (procedure in **Proc**). When the baby recognizes an object, the links between these different records are activated, with shifts between the different instances of the concepts associated to them, so that these co-regulators will cooperate, by the interplay among the procedures, to try to seize the object. The baby will learn to adjust its motions to the size and shape of the object, and this will be remembered via the formation first of a record representing this particular action, and then of a concept of the sensory-motor behaviour *seizing an object*. This concept admits the holding reflex as a closest record in **AC**, so that it pertains to **Exp**, and it will become archetypal through its connection to other archetypal records.

3. Conscious Processes

In recent years, consciousness has become a much-studied topic, with often diverging viewpoints, although what has been termed a 'global neural network space paradigm' (Wallace, 2004, p. 2) seems to be emerging. Several models (in particular, the primary consciousness of Edelman [1989], and that of Dehaene *et al.* [1998] for effortful tasks) rely on two key sub-systems: one for 'long-term memory', and another, a 'modular control system', which briefly or intermittently acts upon this memory. In the model we have proposed (Ehresmann and Vanbremeersch, 1992b, 2002, 2003), we also have the analogues of these two main sub-systems:

• the memory MEM of the memory evolutive neural system, and more specifically the archetypal core, which play the role of the long-term memory;
• the net of co-regulators; and more particularly some higher level ones, called intentional co-regulators, which play the role of a modular control system.

3.1. *Intentional Co-Regulators*

A lower-level co-regulator may respond in an automatic manner to a given situation, or have its procedure imposed by another (generally higher level) co-regulator; thus, it operates non-consciously, as in the 'treatment units' of Crick (1994, p. 336). A higher level co-regulator has a larger choice of admissible procedures, which can be recalled through activator links, or even formed for the occasion by combining other procedures, and it can evaluate the results of its procedures. In particular, we distinguish certain higher level co-regulators, which will be the main agents of conscious processes. They would correspond to the 'conscious units' of Crick (1994, p. 336); they can also be compared to the intentional systems of Dennett (1990), which act as if they make rational choices. We call them *intentional co-regulators* (the word intentional being taken in the sense of Brentano, 1874, meaning passively oriented toward something, and not actively pursuing an objective).

An intentional co-regulator is a higher level co-regulator which is based on associative cortical areas, so that its agents are cat-neurons representing synchronous hyper-assemblies of neurons in these areas. It has some latitude to select its procedures, and to receive feedback about their effects on the well-being of the animal, the feedback being produced via links coming from the archetypal core and from evaluating co-regulators (Chapter 9), which measure the consequences of the procedures on the homeostatic balance and hedonic states of the animal.

Conscious processes are triggered by a relevant event without an automatic response, which exceeds the empirical background noise. Such an event might originate either externally (unknown or threatening stimuli, complex situation), or internally (related to feelings, activities or procedures in progress). As Edelman explains, the first response is a general increase of attention (Edelman, 1989, p. 205) through the activation of several areas, in particular the reticular formation, which diffuses to other areas on which is based the archetypal core, and persists for a long enough time, via the autonomous activation of the archetypal core; Edelman (1989, the title of its book) speaks of the 'remembered present'; following James (1890), we prefer to call this period the *specious present*. It allows the formation of a transient working memory (which we model by the 'global landscape') on which a coherent action of the various intentional co-regulators will be developed (von der Malsburg, 1995, has emphasized the role of coherence in consciousness). This global landscape can be compared with a 'theatre' (in the title of Baars, 1997); and also with the activity (denoted by C') of the 're-entrant dynamic core' of Edelman (2003, p. 5522; but we do not consider that consciousness [his C] reduces to the variations of C'). We model *consciousness* as a 3-part process:

(i) Formation, using the subjective experiences stored in the archetypal core, of a holistic global landscape; it acts as a transient working memory which persists during the actual presents of intentional co-regulators, and it extends and correlates their landscapes.
(ii) Retrospection (looking toward the past) on this global landscape to find the nature of an arousing event.
(iii) Prospection (looking toward the future), to program a sequence of adapted procedures, either retrieved from the procedural memory, or constructed anew by combining known procedures.

The more developed the archetypal core is, the more extensive the global landscape will be, thus allowing for retrospection further back in time, and for the selection of more long-term procedures. How we model these processes is explained in more detail below.

3.2. Global Landscape

An arousing event at a given time t is perceived in the actual landscape of one or several intentional co-regulators, possibly under the form of a fracture. It has for consequence an activation of part of the archetypal core, either as the result of procedures chosen by some co-regulators after a more detailed analysis of the present situation, or it can be automatic, in the case of a fracture. Due to the properties of the archetypal core, this activation diffuses through the links in fans and with processes of shifting among instances of concepts and switching among the ramifications of these instances; and it is autonomously sustained for an extended period (the specious present), say from t to $t+s$. From there, it may spread to cat-neurons in other parts, such as records which, though not in **AC**, are linked to it; and also to agents of various intentional co-regulators. The whole system D_t of cat-neurons which is so activated and remains activated at least up to $t+s$ will be called the *t-activated domain*. In particular it may contain links between agents of different intentional co-regulators. The pattern I_t in D_t having for components activated agents of these intentional co-regulators and for distinguished links the links between them is called the *intentional net at t*. What we call the global landscape at t corresponds to the landscape of the intentional net in D_t; it will represent a transient working memory, persisting up to $t + s$, which not only gives a coherence to the actions of intentional co-regulators but also allows the interplay among their procedures to become more efficient.

Let us give more formal definitions:

• At a given time t, we define the *t-activated domain* D_t as the category whose objects are the cat-neurons M which are activated at t via a loop of links from M to M connecting them to **AC** and containing some links of fans, and whose activation remains auto-sustained for a specious present.
• D_t is the configuration category of an evolutive sub-system **D** of the memory evolutive neural system, which we call the *activated domain*. This domain may have very few components at a given time; in the case of an unforeseen event, D_t extends as explained above.
• We define the *intentional net* **I** as the evolutive sub-system of **D** whose configuration category I_t has for objects the (configurations at t) of the agents of intentional co-regulators which are in D_t as well as the links between them.

Definition. The *global landscape* **GL** is an evolutive system whose configuration category GL_t at t is the field in D_t of I_t (considered as a pattern).

By definition of the field of a pattern (Chapter 3), the objects of the global landscape at t are the perspectives of the cat-neurons in the t-activated domain D_t for the intentional net I_t; and the links between two perspectives are defined by the links in D_t which correlate the two perspectives.

As the activation of a component of **D** is sustained for a long time (thanks to its connections with the archetypal core), the activated domain changes slowly, and idem for the global landscape. However, when an unforeseen event occurs at t, the increased activation due to the surge of attention extends the activated set. For instance a record which had been recalled just before t can be re-activated and become integrated to **D** up to $t + s$. Thus, the global landscape also extends by keeping track of more or less ephemeral events from the recent past (*cf.* the retention process described by Husserl, 1904). For example, if a clock strikes four times at t, our attention might be only aroused at the last stroke; however, we may then recover the hour it has indicated; indeed, if each sound only briefly activates auditory receptors, the record of the sounds lasts longer, and, through the attention increase, remains in the activated domain; thus, during a short period after the clock stops, we can retrieve from the global landscape not the direct effect of the individual sounds, but the sequence of their perspectives persisting in it.

3.3. *Properties of the Global Landscape*

The global landscape allows for the landscapes of the intentional co-regulators to be extended retroactively, and connected. For example, let E be an intentional co-regulator whose agents are in the intentional net at a given time t. Then from the definitions of the global landscape GL_t and of the actual landscape L_t of E at t (both correspond to perspectives for the agents), it follows that L_t is contained in GL_t. We have said that, in any memory evolutive system, the actual landscape of a particular co-regulator retains only the perspectives of components of about the same complexity level and with a long enough stability span (Chapter 6). For E, the extension of its landscape will consist in the recovery of information initially overlooked by it, because it concerned records of objects only briefly activated and/or of a lower level (such as lower level components of the perspectives); some of this information may figure in the global landscape. We define the *extended landscape* of E at t as the sub-category of GL_t whose objects are the perspectives arriving to agents of E. In particular, it will contain the perspective of a cat-neuron B for E that is too weak to be retained in the actual landscape, but is strengthened (as a consequence of the Hebb rule) if the activities of B and E are simultaneously amplified by an attention increase, so that B becomes included in the activated domain, and its perspective in the global landscape.

Thus the global landscape at t is a reflection, not only of the present state at t, but also of recent past states, as recorded in the various landscapes it extends, which were formed during the actual presents of the corresponding intentional co-regulators, and which take into account the specific experiences of the subject as recorded in the archetypal core. It integrates these extended landscapes and connects them, allowing for increased cooperation among the agents of the intentional net. To illustrate, let us suppose that an intentional co-regulator, say E_1, receives a perspective b_1 from a record M of a present stimulus, while another, E_2, receives a perspective b_2 from a procedure Pr in **AC**. If M is linked to Pr by an activator link h, this link cannot be seen by either of the co-regulators separately; but if h is in **D**, it will appear in the global landscape, and, by acting together, the two co-regulators can retrieve this link.

In the evolutive global landscape, the transitions are such that the system keeps its complex identity, in the sense that there is a representative sub-pattern of the global landscape at t which remains a representative sub-pattern of the global landscapes at near enough times (Chapter 5). Indeed, its perspectives last during the specious present, thanks to the stability of **D** (generated by the auto-sustaining activity of the archetypal core), and, in spite of the changes induced by the context, they are maintained in subsequent global landscapes. This could explain the sense of unity of the conscious being.

The formation and persistence of the global landscape during the specious present are exemplified by the confusing experience of the total blank one sometimes experiences after a brusque awakening. Indeed, the preceding global landscapes are almost void, and the formation of the new global landscape occurs more or less slowly. Aspects coming from the higher levels of the memory take especially long to be activated, because the rhythm at those levels is slower. The archetypal core, hence also the activated domain, is activated gradually, through new perceptions which reactivate items in the memory, themselves reactivating other items to which they are linked, up to the restoration of a clear vision of the situation.

Note that the overlapping of successive global landscapes seems lacking in some neural disorders: during migraines, some patients see a sequence of fixed scenes, instead of a continuity of vision and motion (Sacks, 2004). More recently, Gepner and Mestre (2002) have shown that a similar malfunction occurs early in the development of autistic children, and could explain their difficulty communicating, and their introversion.

3.4. *The Retrospection Process*

We have said that the global landscape keeps traces of events of the recent past. It will allow for the intentional co-regulators to conduct a search through it to try to recognize the nature, or possibly the causes, of an arousing event, taking into account information stored in memory, and more specially in the archetypal core, regarding similar situations. For Nietzsche (1888), to become conscious of something is to uncover motivations for it. Metaphorically, this will be what the retrospection process does. Depending on the context, it is directed either by one particular intentional co-regulator in its extended landscape (what is illuminated by the flashlight of Baars's theatre, 1997), or simultaneously, or successively, by several of them, acting cooperatively through the global landscape. The process is based on a series of loops in the global landscape (*e.g.* along the fans of the archetypal core, shifts between instances of concepts and switches between the several ramifications of these instances); the aim is to recover the different records which possess links toward a record of a similar situation, and to evaluate which one might be the most plausible one. For example, if the identified consequence of the event is the activation of a procedure Pr, and if there exists an activator link h from a record M to Pr, this activator link can become included in an activated loop of $\mathbf{D}$ and thus seen in the extended landscape, allowing to find that the event was the activation of Pr by M. In our earlier papers on the subject (Ehresmann and Vanbremeersch, 1992b), we have spoken of an *abduction* process (in the sense of Pierce, 1903) to identify the nature of the original event.

For instance, the sentence 'the fish attacked the man' creates a fracture in a language-CR. Through a process of abduction, it is found that the fracture comes from the fact that a typical fish is not aggressive towards man; subsequently, a process of analogical reasoning, coupled with trial-and-error (through loops in the semantic memory), evokes dangerous animals related to fish, whence the recall of a shark, which makes the sentence meaningful. Another example is the presentation of separate objects to each eye; the receptors activated by the two objects will activate corresponding concepts, say A and B, in the semantic memory, and both can be recovered in the global landscape; as both have the same strength, they will be perceived alternately.

3.5. *Prospection and Long-Term Planning*

One of the roles most generally attributed to consciousness is 'sequencing complex learning tasks [...] permitting planning or 'modelling the world' free of real time' (Edelman, 1989, p. 92). This relies on an anticipation of the results (the importance of anticipation in consciousness is stressed by Goguen, 2003) taking into account previous experiences stored in memory, including the risks of fractures. It is the aim we assign to the prospection process, oriented toward the future. Intermingled with the retrospection process, it leads to the selection and longer term planning (possibly covering several steps ahead) of procedures by the different intentional co-regulators, so that they may respond in the best adapted way. This is accomplished through the formation of virtual landscapes inside the global landscape, with the help of the whole memory, although in particular the archetypal core. In this way, successive procedures (selected as procedure-concepts) can be tested without material cost for the system, by anticipating their possible effects on the future course of events. The construction relies on a comparison between the strengths of activator links to procedures, as they are retrieved from the memory through the global landscape. Let us examine several possibilities.

The prospection may lead to the simultaneous selection of procedures by different intentional co-regulators, which will independently carry them out. Alternatively, a sequence of procedures (Pr_i) can be selected for several steps ahead; each procedure depending on the realization of the preceding ones (by the same or another co-regulator), so that the procedure Pr_{i+1} will be realized only once Pr_i has been realized. The successive commands of these procedures to effectors may be set forth either directly on the landscape of one or several co-regulators, or indirectly by imposing procedures on lower

co-regulators. For instance, it will be possible to program the dynamic formation of a cat-neuron N of level 2 in two steps: the first procedure will
activate cat-neurons N_i of level 1 binding synchronous assemblies of neurons, and connect them by simple or complex links; then a second procedure
will integrate the pattern so formed into a second-order cat-neuron N.

   A third possibility is that a procedure may be selected for several steps
ahead, but allowing for temporary interruptions during which other procedures will take over. (In object-oriented programming, this is the function
served by a thread; see Niemeyer and Peck, 1997; Karmiloff-Smith, 1992;
Josephson, 1998.) This is the case if the interruptions consist in waiting for
some operations imposed by the procedures to be realized by a lower level
co-regulator. It has been proved (Koechlin *et al.*, 1999) that the fronto-polar
prefrontal cortex (which is especially well developed in humans relative to
other primates) selectively mediates the ability, necessary for such multiple
planning, to keep goals in mind while exploring and processing secondary
goals.

   Lastly, it is also possible to simultaneously select several long-term
procedures that preserve a continuous command of lower level effectors, but
which are perceived at the higher level only in an alternating fashion, via
peaks of consciousness corresponding to a quick switch of attention from one
task to another. For example, it is possible to talk while driving a car, each
of these activities asking for more or less attention at a given time. The
conscious reactivation of one of these procedures at a given time can be
planned by an intentional co-regulator, or imposed on it by a fracture
caused by an external stimulus to one of the lower co-regulators involved in
the procedure (a road obstacle requires the driver to pay attention and slow
down). The reactivation of the procedure is rapid and does not cause a
fracture at the intentional level (the conversation is resumed at the same
point).

4.   Some Remarks on Consciousness

Conscious processes will be more or less well developed depending on the
complexity of the neural system of the animal. They can range from a primary
consciousness (in the terminology of Edelman, 1989), to the self-consciousness of some primates, to those involved in language and culture for human
beings. The development of consciousness raises several questions.

4.1.   Evolutionary, Causal and Temporal Aspects of Consciousness

Why has consciousness arisen? Consciousness gives a selective advantage to
an animal and increases its fitness; that explains how it might have arisen by

natural selection. Indeed, retrospection allows the causes of an event to be uncovered, instead of just neutralizing its effects in a purely symptomatic way, via trial-and-error (compare a drug which cures a disease by eliminating its causes to a drug which only alleviates its symptoms: find the causes of a fever instead of just suppressing it). And by planning for the long-term, more efficient procedures and behaviours may be devised, less influenced by immediate concerns.

In what sense can we say that conscious processes cause behaviours? These processes depend upon the activity of a large number of cat-neurons directed by intentional co-regulators. We have already discussed how cat-neurons, though abstract units, can cause physical effects through the unfolding of one of their ramifications (Chapter 9). This extends to consciousness as the cumulative action of intentional co-regulators, and so explains how conscious processes have a causal effect, without negating that the world is physically closed.

How is consciousness related to time? The oscillation between the retrospection and prospection processes in the global landscape contrasts with the more stable image of the world assumed by the memory, and so helps internalize the notion of time. Past and future are experienced through their repercussions in the specious present of the global landscape. A past event is not processed as such, nor even by its more or less stable record M, but by the perspective of M in the global landscape; thus, it acquires a colouration, loosely speaking, that is dependent on the context (*e.g.* our mood influences our view of a situation). The future is anticipated from the construction of virtual extended landscapes, onto which we project our desires and fears. This is to be compared with the views of Kant, for whom time is the form of our internal state; and with Merleau-Ponty, for whom 'consciousness unfolds or constitutes time', and for whom time 'is not an object of our knowledge, but a dimension of our being' (translated from Merleau-Ponty, 1945, p. 474–475).

4.2. *Qualia*

How is consciousness internally experienced, as distinctly first-person knowledge? Or, as Nagel (1974) asks, 'What is it like to be a bat?' (p. 435). This question, extensively and contradictorily discussed by many authors, asks for connecting the neural correlates of consciousness to the subjective experience of *qualia* (*e.g.* the so-called 'redness of red'). It has been called the *hard problem* (Chalmers, 1996), and is one aspect of the mind-brain problem. Some authors consider it to be an ill-posed problem; for instance Edelman (2001) thinks that phenomenal experience is just the manner in which the discriminations made possible by neural events appear.

What of qualia in our model? The global landscape we construct is not formed by records and cat-neurons, but by perspectives of these for various intentional co-regulators. In particular, a perspective of the record of an object is a formal unit, representing a class of particular aspects of this record. Aspect here has the specific meaning of a link from the record to the corresponding co-regulator, which transmits information about the activation of this record. In this way the perspective gives only an internal reflection of the record, and *a fortiori* of the object, which could be related to the phenomenological aspect. Thus, qualia could correspond to these perspectives, which represent the manner by which the activation of a record (hence of one of its ramifications) is observed, internally, through the intentional co-regulators. This is distinct from what an external observer would observe.

4.3. The Role of Quantum Processes

Several authors have proposed that quantum processes are responsible for certain aspects of cognitive processes, but without describing exactly how this is so. In fact, their theories can be seen as relying on the existence of complex switches at the molecular or sub-cellular level: for Pribram (2000), through teledendrons and dendrites; for Hameroff and Penrose (1995), through micro-tubules; for Eccles (1986), at a synaptic level. The general results we have presented show in which precise sense cognitive processes of any level are 'based' on quantum processes: these processes emerge through successive complexifications of the neuronal level, which itself emerges by complexifications of a sub-system of the quantum evolutive system (Chapter 5) which models particles and atoms; it means that they have ramifications down to the quantum level. Now, we have proved that successive complexifications of a category lead to the emergence of a hierarchy of objects with strictly increasing complexity only if the category satisfies the multiplicity principle (Theorem 3 of Chapter 4). Thus:

Theorem 1. *The emergence of cognitive processes is possible only because the quantum evolutive system* **QES** *satisfies the multiplicity principle, and this is a consequence of the quantum laws.*

Let us remark that this explanation leads to a reconciliation of the above recalled dendrite theory of Pribram with his earlier holographic brain model (Pribram, 1971): his holograms correspond to emergent structures formed through successive complexifications of **QES**, and thus can only exist because the laws of quantum physics imply this evolutive system satisfies the multiplicity principle.

4.4. Interpretation of Various Problems

Let us consider the implications of our model for some problems frequently raised in papers related to consciousness.

Blind-sight. The activation generated by an object presented to the subject is maintained through the archetypal core and is functional, but its perspective is not formed (the links forming its aspects are severed), so that it appears in the activated domain, but it does not appear in the global landscape. Thus, the subject is not conscious of seeing the object, though able to unconsciously operate with it.

Dreams. Are we conscious during dreams? As the external stimulations are reduced to a minimum and the commands to the effectors suppressed, the formation of the activated domain and of the global landscape relies on the autonomous internal activation of the archetypal core. Thus, this landscape is reduced, and the retrospection and prospection processes are almost non-existent. It follows that there is less overlap between successive global landscapes, whence the possible feeling of discontinuity that remains when the dream is narrated after awakening.

Mary. Mary lives in a world without colour, but has formally learnt everything about how the brain functions in presence of colour; then she is presented a red object (Jackson, 1986). Will she learn something new? And in this case, does it mean that physical properties are not sufficient to account for the subjective feeling? In our model, she cannot anticipate the redness of red; she knows externally that there will be a perspective of red, but not what it is to experience red internally. It does not mean that there is some non-physical ingredient, but instead that what is reflected is not a passive configuration of the brain, but the dynamical activity of this configuration as it unfolds.

Zombies. Zombies (Dennett, 1991, p. 73) would develop the archetypal core and the distributed activation it generates in the activated domain, but could they form the global landscape and access it?

4.5. Self-Consciousness and Language

The conscious processes we have modelled do not need language, and we admit that higher animals may have a non-linguistic consciousness, which is somewhere between Edelman's primary and higher order consciousnesses. A higher animal may probably develop a *self-consciousness*. For example, chimpanzees recognize themselves in a mirror, as shown by experiments wherein they try to remove a mark on their cheek, made by an experimenter while they were sleeping. Self-consciousness requires an internal apprehension of the changes in the global landscape, while a kind of continuous identity is maintained, via the partial overlapping of successive global landscapes, and their access to the more stable memory. It arises as a construction, initiated by the occurrence of fractures which reveal the existence of what might be called a non-self, and which, by opposition, lead to the differentiation of the self. Whence an interpretation of Descartes' (1637) famous axiom, 'Cogito ergo sum': thought, as the conscious perception of a sequence of fractures, makes sense of one's existence by contrast with one's limitations.

Language allows for more efficient processes, because complex information (higher order cat-neurons) can be stored and retrieved in the more compact form of a word, and thus more operations can be effected on it. The consequence is the development of a richer algebra of mental objects, and of higher order cognitive processes, depending heavily on social interactions. Though we have said that consciousness alone gives some apprehension of time (on this point we disagree with Köhler (1947), and Edelman (1989), who consider that language is necessary for this), language allows for the formation of narratives, which are at the root of human communication, education and cultural development. Language leads to the integration of the various memory evolutive neural systems K_i of the members of a society, into a larger memory evolutive system representing the society with its whole knowledge and culture. (It can be formally defined as a colimit of the different K_i in the category of evolutive systems; *cf.* Chapter 5). This memory evolutive system itself has an analogue of the archetypal core, integrating the various archetypal cores of its members. Conversely, this larger system interacts with these individual archetypal cores to extend them, in particular through education and the diffusion of culture. Some applications of this system in psychiatry are being developed by Marchais (2003, 2004).

5.   A Brief Summary

Here we sum up the main characteristics of neural systems which lead to the development of higher cognitive processes, and recall how we have modelled them in memory evolutive neural systems. In particular we insist on those that we consider to be responsible for conscious processes.

5.1.   *Basic Properties of Neural Systems*

Binding Process with Multiple Realizability (Degeneracy).   Groups of neurons correlate their activities via their synaptic connections, to form synchronous assemblies of neurons having certain functional states. The same functional state can be produced by any member of an entire class of different assemblies of neurons; we call such a class a category-neuron (abbreviated in cat-neuron). By iteration, the binding process is extended to integrate assemblies of neurons and/or of cat-neurons into synchronous hyper-assemblies of increasing complexity, leading to higher order cat-neurons. They represent mental objects, physically activated through the unfolding of one of their hyper-assemblies down to the neuronal level, with possibly a switch to one of the other members of the class.

Hierarchy and Dynamics. The cat-neurons form a hierarchical evolutive system, the dynamics of which are modulated by the (possibly conflicting) interactions between more or less specialized and complex modules (*e.g.* visual areas, amygdala, associative cortical areas). These modules operate separately, at their own pace and on their own referent (their landscape), but must finally produce a coherent global behaviour.

Flexible Memory. An animal develops a memory in which the records (memories of images, behaviours, experiences and so on) are not rigid units but, as cat-neurons, are adaptable to the context, and which consolidate (in the sense of Wright and Harding, 2004) and evolve over its lifetime. These cat-neurons generate an expanding algebra of mental objects. Higher animals are able to classify the records into concepts (*e.g.* perceptual categorization), which form the semantic memory. The activation of a concept has a double degree of freedom: shifts between its various instances, and then switches between hyper-assemblies of neurons activating these instances.

Values and Self. Combinations of concepts integrate the deep experiences of the animal, that is, those having adaptive and/or emotional value, be they sensory, motor, related to basic needs, or otherwise. They form a basic part of the memory that we call the archetypal core. In it, the records are strongly connected; their activation diffuses along specific channels, and is self-sustained over a long time. This archetypal core is the basis of the unity of the animal; its permanence, or at least its very gradual modification, plays a central role in global processes at the basis of self and consciousness.

Consciousness. Higher (that is to say, intentional) modules are able to plan their actions and evaluate the results, in terms of fulfilment of the animal's needs, and any possible rewards. Along with the archetypal core, these modules are responsible for conscious processes. The attention aroused by an unexpected event activates the archetypal core, whose autonomous activation leads to the retention, extension and integration of the information separately gathered by these higher modules. A global landscape is thus formed, which acts as a transient working memory lasting for the specious present. A retrospection process, looking toward the near past, and a prospection process, toward the future, are carried out on the global landscape, in order to better apprehend the current situation, and to facilitate longer term planning.

5.2. *Interpretation in Our Model*

The neuronal system of an animal is modelled by the evolutive system of neurons, and a synchronous assembly of neurons by a pattern in it. In general, this pattern does not have a coordination neuron (what might be termed a 'grandmother neuron'). A colimit of this pattern, called a cat-neuron, is formed in a complexification of the system. More complex cat-neurons are formed by successive complexifications, leading to the evolutive system of cat-neurons. A cat-neuron can be the colimit of several non-connected patterns, *i.e.* it is a multifold object. Thus, the evolutive system of cat-neurons satisfies the multiplicity principle, which in this setting means that it has multiple realizability: a cat-neuron is a conceptual unit which represents a class of functionally equivalent (though some are physically non-connected) synchronous assemblies of neurons and/or cat-neurons, and it may be activated through any of these assemblies, with possible switches between them. General results on complexifications describe the nature of the adequate links between cat-neurons, namely the complex links that take into account the possibility of such switches. And they explain how successive complexifications lead to the formation of cat-neurons of strictly increasing orders, activated by the unfolding of one among several ramifications down to the level of neurons, with a possible switch from one to another.

The cat-neurons of different levels are the components of the memory evolutive neural system which provides not only a model of the neuronal system (as in more usual models), but also encompasses a hierarchy of cat-neurons representing mental objects of increasing complexity, and the manner in which they interact and combine. Its co-regulators are based on neuronal-level modules. Each co-regulator performs a trial-and-error learning process, having its own time scale and complexity: at each step, it forms its landscape which represents the partial information it can gather, and on which a procedure is selected and the results evaluated. The procedures of the various co-regulators can conflict, and must be made coherent through an interplay among them. This process is not centrally directed, but instead similar to Darwinian selection (Changeux, 1983; Edelman, 1989). The interplay can suppress or modify some procedures, thus causing fractures to the corresponding co-regulators.

The co-regulators participate in the development of the memory (both empirical and procedural) formed by specific cat-neurons, called records, of various orders. Its robustness and adaptability to the context come from the fact that these records are multifold objects, and as such have a large plasticity. The semantic memory developed by higher animals is obtained by a mixed complexification process, leading first to the formation of concepts with respect to one attribute, which represent an equivalence class of records (their instances) and are constructed as the limit of the pattern of agents activated by one of their instances. Then more abstracts concepts are obtained by recursively combining concepts with respect to several attributes.

The semantic memory allows for the formation of an evolutive sub-system of the memory, the archetypal core **AC**, formed by complex records representing and blending strong records of various functionalities. Activation of part of **AC** diffuses to its other parts through specific channels, the fans, and is autonomously sustained for a long time. **AC** is connected to other parts of the memory (in particular to the experiential memory) and is differentially accessed by higher level 'intentional' co-regulators which have some latitude in the selection of their procedures and can evaluate the results in terms of reward.

Consciousness. We describe consciousness as a three-part process: at the time an unforeseen event is met, an increase of attention starts circuits of activation in the archetypal core that lead to the formation of a global landscape, which extends and correlates the actual landscapes of the intentional co-regulators. The global landscape allows for a retrospection process, looking toward the recent past, to search for the nature of the initiating event; and for a prospection process, to select a long-term procedure by evaluating its possible effects for several steps to come. These processes rely essentially on the long-term auto-activation of the archetypal core, which acts as a referent for the intentional co-regulators, and coordinates their operations.

Appendix

Many results given in this book can be extended to cases where the binding process is modelled using notions more general than that of a colimit, and we introduce such generalizations here. They can be useful to model some specific situations where the notion of a colimit seems too restrictive, for instance, to represent certain visual illusions, or to model processes which allow for different alternate realizations depending on the context. However, the interest of colimits is that with them the complexification process is well defined, and allows for an explicit computation of the adequate links to and among added colimits, while no such theory exists for the more general notions considered here.

Two main extensions are given:

- the local colimits introduced by one of the authors (Ehresmann, 2002), still in the categorical framework;
- and the more general notion of the hyperstructures of Baas (1976, 1994).

1. Local Colimits

The existence of a colimit C for a pattern P imposes strict global conditions, since any collective action of P on any object is then integrated into an action of C on this object. However, one can imagine situations where the pattern P has different types of collective actions on different objects, each type being integrated into an action of a specific object.

1.1. Multi-Colimit

Before studying the situation in general, let us illustrate it with the transport network described at the end of Chapter 2 In this example, each region tries to construct a central node through which all its selected transport nets to different towns can transit. However, we have seen that the construction of such central nodes is not always possible, at least if we want to respect the rate constraints that were imposed in that example. If no central node can be constructed, it is possible to try to construct not a unique central node, but a central *multi-node*, formed with the fewest possible local nodes C_k: each town A connected to the region by a selected net is directly linked to a unique local node C_k, so that a traveller to A can transit through C_k for the same rate. Thus, each local node C_k acts as a central node, but only for part of the set of all towns connected to the region by a selected net. For example, all the lines towards towns in the south could transit by a local node C_s, those towards the east by a local node C_e and so on. This is the case in Paris, where each railway station serves a specific part of the country.

A central multi-node is an example of the following general situation: for a pattern P in a category, there exists a classification of the various collective links of P, so that to each one of its classes there is associated an object C_k, through which any collective link in this class factors. Naturally there can be several such classifications. The coarsest one is that where each class has only one collective link, and C_k is its target. At the other extreme is the case where P admits a colimit, and so there is a classification with only one class. A central multi-node is an example of the important intermediate case where there is a classification into a smallest number of classes, each one uniquely factoring through a particular object C_k; then the set formed by the objects C_k is called a *multi-colimit* of P.

More formally, we have the following definition (Diers, 1971):

Definition. A *multi-colimit* of a pattern P in a category K is a family (C_k) of objects of K with the following properties (Fig. A.1):

(i) Each C_k is connected to P by a *locally binding* collective link (c_i^k).
(ii) To each collective link (f_i) from P to an object A, there is associated one and only one C_k, such that (f_i) uniquely factors through (c_i^k).

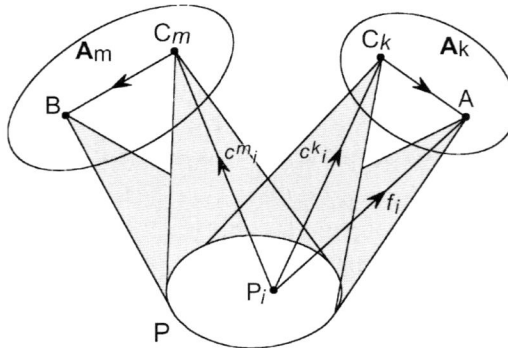

Fig. A.1   A multi-colimit.
The pattern P admits a multi-colimit (C_k) if for each C_k there is a binding collective link (c_i^k) from P to C_k; and each collective link (f_i) from P to an object A factors through one and only one of these binding collective links. The multi-colimit then determines a partition (A_k) of the set of collective links: in the figure (f_i) belongs to A_k because it factors through C_k; on the other hand a collective link from P to B belongs to A_m because it factors through C_m.

A multi-colimit in which there is only one C_k reduces to a colimit. A multi-colimit (C_k) of P determines a classification of the collective links from P to objects of K: each C_k is associated to the class A_k of all the collective links which factor through (c_i^k); the locally binding collective link (c_i^k) is an initial object of the class A_k, which means that each element of A_k uniquely factors through it. Thus, the definition can be formulated in terms of the operating field of P (defined in Chapter 2): P has a multi-colimit if, and only if, its operating field admits an initial set, namely the set formed by the locally binding collective links (c_i^k).

1.2. Local Colimit

The notion of a multi-colimit can be still further relaxed, to that of a *local colimit*. Let us return to the example of the transport network. With a central multi-node, its different local nodes C_k are not connected; however if we think of Paris, the different railway stations are not isolated, but connected by fast metro lines, which can be added to the network; in this way, a southern town is also connected to the *gare du Nord*, through the metro line coming from the *gare de Lyon*. Thus, in some cases a central multi-node could be replaced by a *central net* formed by local nodes connected by specific lines, so that all the selected nets towards a particular town A can transit through a specific 'nearest' local node C_k, the fare from C_k to A being minimal. The difference with the case of a central multi-node is that now A can be connected to several local nodes instead of a unique one; but in this case, among them there is one which is nearest to A in the sense that the fare for the connection to A is lowest.

A central net can be modelled by the notion of a *local colimit* (introduced by Ehresmann, 2002), of which a multi-colimit is a particular case. In a multi-colimit, the different C_k's are not interconnected, so that an object A is linked to only one of them. In a local colimit, this condition is relaxed to admit the existence of distinguished links between the C_k's. Consequently, an object A could be linked to several objects C_k, but a condition is imposed such that, among them, there is a 'nearest' one, in the sense that there is a specific link from it to A through which any link from a C_k to A factors. More formally:

Definition. A *local colimit* of a pattern P in a category K is a pattern LC in K with the following properties:

(i) There is a *locally binding* collective link (c_i^k) from P to each component LC_k of LC, and these collective links are correlated by the distinguished links of LC.

(ii) If f_i is a collective link from P to an object A, there exists a 'universal' link f from one of the C_k to A correlating c_i^k and f_i, and such that any link

from another component C_m of LC to A having this property is the composite of a distinguished link of LC with f.
(iii) There is no smaller pattern having these properties.

In the operating field of P, this means that the pattern having for components the locally binding collective links (c_i^k) and for distinguished links those of LC generates a *co-reflective sub-category* (Mac Lane, 1971), and that this sub-category is a minimal co-reflective sub-category. If it exists, the local colimit is unique (up to an isomorphism). A multi-colimit is a particular local colimit in which LC has no distinguished links, a colimit is a local colimit in which LC is reduced to one component.

1.3. Examples

An *ambiguous image* is one having two or more different interpretations. A person (viewer) can quickly pass from one interpretation to the other, or, once the ambiguity is known, recognize the image as admitting these multiple interpretations. A well-known example is the image which can be interpreted as a vase or the profiles of two faces facing each other. In memory evolutive neural systems, such an image with two interpretations may be represented by three records in the memory: the two interpretations separately, and their union as such. This could be modelled categorically by a local colimit of the pattern of receptors activated by the image. The local colimit is formed by three objects, say C_0, C_1 and C_2, modelling, respectively, the image as a whole and its two interpretations, with a distinguished link from C_1 to C_0 and another from C_2 to C_0. This local colimit is computed in a higher landscape, where the two interpretations can be interchanged during its actual present, and the complete representation exists only after this possibility of interchange is already known.

An ambiguous image could also be represented by a multifold record which has two non-connected decompositions (Chapter 3), each one corresponding to the pattern of receptors activated by one of the two possible interpretations of the image. This representation in term of multifold record is philosophically different from that we have given above in term of local colimit:

(i) In the local colimit, we consider that at the lower levels, a single pattern is activated, but that it is perceived differently depending on the context.
(ii) By contrast, in the multifold interpretation, the ambiguous image would activate two different patterns depending on the context, both having the same colimit, namely, the total record of the image.

The multifold interpretation is more natural when one thinks of the selection of procedures and their realization: a cat-neuron is activated and, according to the context, one or the other of its decompositions is activated. On the other hand, in the recognition phase, the local colimit interpretation seems more natural, the same lower level pattern acting differently on higher agents.

Ambiguities occur also in language, such as the fact that two different concepts can have the same name (*e.g.* race as an ethnic group, or as a contest). One could consider that the sound activates an auditory pattern having the word for its colimit in the auditory memory, but having a multi-colimit in the semantic memory, with two objects corresponding to the two meanings of the word.

As an other example of local colimits, they can be used to model *peaks of consciousness* (Chapter 10) corresponding to a quick switch of attention from one task to another (*e.g.* driving a car and talking at the same time). Both activities remain continuous at lower levels, making it possible, if need be, to have an immediate conscious reaction to an external stimulus relating to the covert activity. The two tasks could be represented by a multi-colimit with two objects, one or the other being activated depending on the context.

2. Hyperstructures

Hyperstructures were introduced by Baas (1992) as a general framework for studying emergence and higher order structures.

2.1. *Construction of Hyperstructures*

Hyperstructures are built up recursively from a given set of objects, abstract or physical. These primary objects are called *structures*, or first-order hyperstructures.

Definition. A *second-order hyperstructure* S^2 is obtained, via a construction process R, from:

(i) a family $S = (S_i)$ of structures,
(ii) an *observational mechanism* Obs^1 able to observe and evaluate the structures of this family, and
(iii) a set Int of *interactions* between the structures.

In equation: $S^2 = R (S, Obs^1, Int)$.

A notion of an *emergent property,* dependent on the observational mechanism, can be defined for hyperstructures: a property of second-order hyperstructures is *emergent* if the observational mechanism Obs^1 associated to them detects this property (directly or by its consequences) for them, but not for the first-order structures on which they are constructed.

The interactions between two hyperstructures with the same observational mechanism Obs, say

$$S^2 = R(S, \text{Obs}, \text{Int}) \quad \text{and} \quad S'^2 = R'(S', \text{Obs}, \text{Int}')$$

are of two types:

(i) Some are directly deduced by computation from the interactions between the elements of S and S' observable by Obs; these are called *deducible interactions*.

(ii) However, there may also exist more complex ones, called *observational interactions*, coming from the fact that the same hyperstructure can be the result of two different constructions and can act as an intermediary in the same interaction successively via one of its constructions and the other (generalizing the role of multifold objects in the formation of complex links, see Chapter 3).

The construction of higher order hyperstructures and their interactions is similar. For instance a third-order structure is constructed on a family of second-order hyperstructures, to which are associated an observational mechanism and related interactions. In this manner an extended hierarchy of hyperstructures can be formed.

2.2. Comparison with Memory Evolutive Systems

In the categorical setting, the colimit of a pattern can be considered as a construction R leading to a second-order hyperstructure, where:

(i) the family of structures is given by the components of the pattern,
(ii) the interactions are their distinguished links,
(iii) various observational mechanisms can be considered, the more trivial corresponding to the set of collective links of the pattern.

In memory evolutive systems, the co-regulators may play the part of these mechanisms.

Then, iterated colimits give rise to higher order hyperstructures, and the deducible interactions between these hyperstructures correspond to what we have called simple links, while the observational ones correspond to our complex links.

There is no general theory of hyperstructures. In a recent paper, Baas *et al.* (2004) have presented some ideas for generalizing the memory evolutive systems setting by replacing the colimit operation with the formation of hyperstructures. The idea is to model the complex components of an evolutionary system at a given time by hyperstructures, the change of configuration being partially directed by observational mechanisms Obs, which play the role of the co-regulators. There may be several Obs acting independently at the same time, and eventually new ones are formed out of emergent hyperstructures.

At a given time t, a given Obs determines which (hyper)structures it observes and what interactions between them it observes; we will say that these form its *landscape at time t.* Then the complexification process is extended as follows: Obs selects (in its landscape) a procedure consisting of some families of (hyper)structures on which new hyperstructures will be constructed. The procedure may also add or remove some existing hyperstructures. If this procedure succeeds, it leads to the construction of a hyperstructure for each chosen family, which will be observed in the new landscape at a later time t'. In the construction, delays may be necessary in order to synchronize the components. So each Obs is subjected to specific structural and temporal constraints which regulate the interactions it may observe and the procedures it can choose, depending on their delays. It could be interesting to generalize the theory of memory evolutive systems in this framework, for example, by studying the interplay among the procedures of the various Obs, and the development of a memory.

Examples. A local colimit of a pattern could also be considered as a particular hyperstructure, the observational mechanism consisting then of the different classes of collective links which the local colimit distinguishes. Another type of hyperstructure could be obtained from the suggestion of Brown and Porter (2003) to replace the colimit construction by consideration of higher order algebraic structures, such as n-simplices or n-categories.

Bibliography

Abramsky, S. and Coecke, B., 2004, A categorical semantics of quantum protocols. In: *Proceedings of the 19th Annual IEEE Symposium on Logic in Computer Science* (LiCS'04), IEEE Computer Science Press, Los Alamos, Calif.

Adelman, R.C., 1979, Loss of adaptive mechanisms during aging, *Fed. Proc.* 38, 1968–1971.

Aerts, D., D'Hondt, E., Gabora, L., 2000, Why the disjunction in quantum logic is not classical? *Found. Phys.* 30,10.

Akin, E., 1979, *The Geometry of Population Genetics*, Springer-Verlag, Berlin.

Albert, M.L., 1987, Caractéristiques cliniques du vieillissement cérébral normal. In: *Le vieillissement cérébral normal et pathologique*, Maloine, Paris, 22–28.

Amari, S. and Takeuchi, A., 1978, Mathematical theory on formation of category detecting nerve cells, *Biol. Cybernet* 29, 127–136.

Arbib, M.A., 1966, Categories of (M, R)-systems, *Bull. Math. Biol.* 28, 511–517.

Arbib, M.A. and Manes, E.G., 1974, Machines in a category: an expository introduction, *SIAM Rev.* 16, 163–192.

Aries, P., 1986, *Le temps de l'histoire*, Seuil, Paris.

Aristotle, Metaphysica, Z. In: Aristote, Physique et Métaphysique (S. et M. Dayan, Eds.), P.U.F. 1966, Paris.

Arnol'd, V.I., 1984, *Catastrophe Theory*, Springer, Berlin.

Aspinwall, P.S. and Karp, R.L., 2003, Solitons in Seiberg-Witten theory and D-Branes in the derived category, *J. High Energy Phys.* 304.

Aspinwall, P.S. and Lawrence, A., 2001, Derived categories and 0-Brane stability, *J. High Energy Phys.* 108.

Aston-Jones, G., Rogers, J., Shaver, R.D., Dinan, T.G. and Mass, D.E., 1985, Age-impaired impulse flow from nucleus basalis to cortex, *Nature* 318, 462–465.

Atlan, H., 1975, Organisation en niveaux hiérarchiques et information dans les systèmes vivants. In: *Réflexions sur de nouvelles approches dans l'étude des systèmes*, E.N.S.T.A., Paris.

Atlan, H., 1979, *L'Organisation Biologique et la Théorie de l'Information*, Hermann, Paris.

Auger, P., 1989, *Dynamics and Thermodynamics in Hierarchically Organized Systems*, Pergamon Press, Oxford.

Baars, B.J., 1997, *In the Theatre of Consciousness: The Workspace of the Mind*, Oxford University Press, Oxford.

Baas, N.A., 1976, Hierarchical systems. Foundations of a mathematical theory and applications, *Report Math.* 2/76, Mate. Inst., Univ. Trondheim.

Baas, N.A., 1992, Hyperstructures—a framework for emergence, hierarchies and complexity. In: *Proceedings Congrès sur l'émergence dans les modèles de la cognition* (A. Grumbach *et al.*, Eds), Télécom, Paris, 67–93.

Baas, N.A., 1994, Emergence, hierarchies and hyperstructures. In: *Artificial Life* III (C.G. Langton, Ed.), *SFI Studies in the Science of complexity, Proc. Vol. XVII*, Addison-Wesley, Redwood City, Calif., 515–537.

Baas, N.A., Ehresmann, A.C. and Vanbremeersch, J.-P., 2004, Hyperstructures and memory evolutive systems, *Int. J. Gen. Syst.* 33 (5), 553–568.

Bachelard, G., 1938, *La Formation De L'esprit Scientifique*, Vrin, Paris.

Bachelard, G., 1950, *La Dialectique De La Durée*, P.U.F., Paris.

Baez, J., 2004, Quantum quandaries: a category-theoretic perspective. In: *Structural Foundations of Quantum Gravity* (S. French *et al.*, Eds.), Oxford University Press, Oxford.

Baianu, I.C., 1970, Organismic supercategories: II. On multistable systems, *Bull. Math. Biophys.* 32, 539–561.

Baianu, I.C., 1971, Organismic supercategories and qualitative dynamics of systems, *Bull. Math. Biophys.* 33, 339–354.

Banaschewski, B. and Bruns, G., 1967, Categorical characterization of the MacNeille completion, *Archiv. Math.* 18, 369–377.

Barlow, H.B., 1972, Single units and sensation: A neuron doctrine for perceptual psychology? *Perception* 1, 371–394.

Barr, M. and Wells, C., 1984, *Toposes, Triples and Theories*, Springer, Berlin.

Barr, M. and Wells, C., 1999, *Category Theory for Computing Science*, 3rd ed.,Les Publications CRM, Montréal.

Bartosz, G., 1991, Erythrocyte aging: physical and chemical membrane changes, *Gerontology* 37 (1–3), 33–67.

Bastiani(-Ehresmann), A., 1967, Sur le problème général d'optimisation, dans *Identification, Optimalisation et stabilité des systèmes automatiques*, Dunod, Paris, 125–135.

Bastiani(-Ehresmann), A. and Ehresmann, C., 1972, Categories of sketched structures (*Cahiers Top. et Géom. Dif.* XIII-2). In: *Charles Ehresmann, Œuvres complètes et commentées* (A.C. Ehresmann, Ed.), Part IV-1, 1982, CTGD, Amiens (reprint).

Bateson, G., 1985, *La Nature et la Pensée*, Editions Seuil, Paris.

Bazerga, R.L., 1977, Cell division and the cell cycle. In: *Handbook of the Biology of Aging*, Vol. 1 (C.B. Finch and L. Hayflick, Eds.), Von Nostrand, New York, 101–121.

Beaber, J.W., Hochhut, B. and Waldor, M.K., 2004, SOS response promotes horizontal dissemination of antibiotic resistance genes, *Nature* 427, 72–78.

Bell, J.L., 1981, Category theory and the foundations of mathematics, *Brit. J. Phil. Sci.* 32, 349–358.

Bellman, R., 1961, *Adaptive control processes*, Princeton University Press, Princeton.

Benabou, J., 1968, Structures algébriques dans les catégories, *Cahiers de Top. et Géom. Dif.* X-1.

Benzecri, J.-P., 1995, Convergence des processus…, *Cahiers de l'Analyse des données* 20 (4), 473–482.

Bergson, H., 1889, Essai sur les données immédiates de la conscience, reproduit dans *Œuvres*, 1959, P.U.F., Paris.

Bernard-Weil, E., 1988, *Précis systémique ago-antagoniste, Introduction aux stratégies bilatérales*, L'Interdisciplinaire, Limonest.

Bernays, P., 1937, A system of axiomatic set theory, *J. Symbolic Logic* 2, 89–106.

Bernays, P., 1941, A system of axiomatic set theory, II, *J. Symbolic Logic* 6, 1–17.

Berridge, K.C., Robinson, T.E., 1998, What is the role of dopamine in reward: hedonic impact, reward learning, or incentive salience? *Brain Res. Rev.* 28, 309–369.

Bertalanffy (von), L., 1926, Untersuchungen über die Gesetzlichkeit des Wachstums, *Roux'Archiv* 108.

Bertalanffy (von), L., 1956, Introduction. In: *General Systems Yearbook*, Vol. 1 (J. Dillon, Ed.), International Society for General Systems Research, Berkeley, 1–10.

Bertalanffy (von), L., 1973, *General System Theory*, Harmondsworth, Penguin.

Bohm, J-P.D., 1983, *Wholeness and the Implicate Order,* Ark Edition, London.

Borsuk, K., 1975, *Theory of Shape,* Monografie Mat. 59, PWN, Warsow.

Bourbaki, N., 1958, *Eléments de Mathématique,* I, Livre I, Chapitre 4 Structures, A.S.I. 1258, Hermann, Paris.

Brandt, H., 1926, Ueber ein Verallgemeinerung der Gruppenbegriffes, *Math. Ann.* 96, 360–366.

Braudel, F., 1969, *Ecrits sur l'histoire*, Flammarion, Paris.

Brentano, F., 1874, *Psychology from an Empirical Standpoint*, International Library of Philosophy, London.

Brown, R. and Porter, T., 2003, Category theory and higher dimensional algebra: potential descriptive tools in Neuroscience. In: *Proceeding of the International Conference on Theoretical Neurobiology* (A. Singh, Ed.), NBRC, New Delhi, 62–79.

Bunge, M., 1967, *Scientific Research*, 1 and 2, Springer, Berlin.

Bunge, M., 1979, *Treatise on Basic Philosophy*, Vol. 4, Reidel, Dordrecht.

Bunge, M.A., 2003, *Emergence and Convergence*, University of Toronto Press, Toronto.

Bunge, S.A., Klingberg, T., Jacobsen, R.B. and Gabrieli, J.D.E., 2000, A resource model of the neural basis of executive working memory, *Proc. Nat. Acad. Sci. USA* 97 (7), 3573–3578.

Burnet, M., 1982, *Le programme et l'erreur*, Albin Michel, Paris.

Burns, E.A. and Goodwin, J.S., 1991, Immunologie et vieillissement. Travaux récents. In: *L'Année Gérontologique* (J.-L. Albarède and P. Vellas, Eds.), Serdi, Paris, 19–29.

Burroni, A., 1970, Esquisses des catégories à limites et des quasi-topologies, *Esquisses Mathématiques* 3, TAC, Paris.

Cantor, G. and Dedekind, R., 1937, *Briefwechsel* (J. Cavailles and E. Noether, Eds.), A.S.I. 518, Hermann, Paris.

Carnap, R., 1928, *Der logische Aufbau der Welt*, Weitkreis-Verlag, Berlin.

Carp, J.S., Tennissen, A.M. and Wolpaw J.R., 2003, Conduction velocity is inversely related to action potential threshold in rat motoneuron axons. *PMID* 12715118 [PubMed—indexed for MEDLINE].

Cavailles, J., 1938a, *Méthode axiomatique et formalisme*, Hermann, Paris.

Cavailles, J., 1938b, *Remarques sur la formation de la théorie abstraite des ensembles*, Hermann, Paris.

Chalmers, D., 1996, *The Conscious Mind*, Oxford University Press, Oxford.

Chandler, J.L.R., 1991, Complexity: a phenomenological and semantic analysis of dynamical classes of natural systems, *WESS Comm.* 1 (2), 34–42.

Chandler, J.L.R., 1992, Complexity II: logical constraints on the structures of scientific languages, *WESS Comm.* 2 (1), 34–40.

Changeux, J.-P., 1983, *L'homme neuronal*, Fayard, Paris.

Changeux, J.-P., Dehaene, S. and Toulouse, G., 1986, Spin-glass model of learning by selection, *Proc. Natl. Acad. Sci. USA* 83, 1695–1698.

Collins, J., Imhoff, T. and Grigg, P., 1996, Noise-enhanced tactile sensation, Nature 383, 770.

Corberand, J.X., Laharrague, P.F. and Fillola, G., 1986, Polymorphonuclear functions in healthy aged humans selected according to the senior protocol recommendations. In: *Modern Trends in Aging Research*, Coll. INSERM-EURAGE, John Libbey Eurotex LTD, Vol. 147, 235–241.

Cordier, J.-M. and Porter, T., 1989, *Shape Theory*, Wiley, New York.

Crick, F., 1994, *The Astonishing Hypothesis*, Macmillan Publishing Company, New York.

Cutler, R.G., 1985, Antioxydants and longevity of mammalian species, *Basic Life Sci.* 35, 15–73.

Cyrulnik, B., 1983, *Mémoire de singe et paroles d'hommes*, Hachette, Paris.

Damasio, A., 1999, *The Feeling of What Happens: Body and Emotion in the Making of Consciousness*, Harcourt Brace, New York.

Danner, D.B. and Holbrook, N.J., 1990, Alterations in gene expression with aging. In: *Handbook of the biology of aging*. 3rd ed., Academic Press, New York, 97–115.

Dehaene, S., Kerszberg, M., Changeux, J.-P., 1998, A neuronal model of a global workspace in effortful cognitive tasks, *Proc. Natl. Acad. Sci. USA* 95, 14529.

Deleanu, A. and Hilton, P., 1976, Borsuk shape and Grothendieck categories of pro-objects, *Math. Proc. Cambridge* 79, 473–482.

Dennett, D., 1990, *La stratégie de l'interprète*, NRF Gallimard, Paris.

Dennett, D., 1991, *Consciousness Explained*, Little, Brown and Co, Boston.

Descartes, R., 1637, *Discours de la méthode*, Garnier-Flammarion, Paris (reprint 1979).

d'Espagnat, B., 1985, *Une certaine réalité*, Gauthiers-Villars, Paris.

Di Cera, E., 1990, Statistical thermodynamics of ligand binding to giant macromolecules, *Il Nuovo Cimento* 12D, 61.

Diers, Y., 1971, *Catégories localisables*, Thèse, Université Paris VI.

Dobzhansky, T., 1970, *Genetics of the Evolutionary Process*, Columbia University Press, New York.

Draï, R., 1979, *La politique de l'inconscient*, Payot, Paris.

Dubois, D., 1998, Introduction to computing anticipatory systems. In: *Computing Anticipatory Systems* (D. Dubois, Ed.), CASYS First International Conference, AIP Conference Proceedings 437, New York.

Ducrocq, A., 1989, *L'objet vivant*, Stock, Paris.

Duncan, G. and Hightower, K.R., 1986, Age-related changes in human lens membrane physiology. In: *Modern Trends in Aging Research*, Coll. INSERM-EURAGE, John Libbey Eurotex LTD, Vol. 147, Paris, 341–348.

Dupréel, M., 1931, *Théorie de la consolidation*, Presses Universitaires de Bruxelles, Bruxelles.

Durbin, R. and Willshaw, D., 1987, An analogue approach to the traveling salesman problem using an elastic net method, *Nature* 326, 689–691.

Duskin, J., 1966, Pro-objects (d'après Verdier), *Séminaire Heidelberg-Strasbourg*, Exposé 6.

Duyckaerts, C., Delaere, P., Brion, J.-P., Fiente, F. and Hauw, J.-J., 1987, Les plaques séniles. In: *Le vieillissement cérébral normal et pathologique*, Maloine, Paris, 157–163.

Eagleman, D.M., 2004, The where and when of intention, *Science* 303, 1144–1146.

Eccles, J.C., 1986, Do mental events cause neural events? *Proc. R. Soc. London, Ser. B* 227, 411–428.

Edelman, G.M., 1989, *The Remembered Present*, Basic Books, New York.

Edelman, G.M., 2003, Naturalizing consciousness: a theoretical framework, *Proc. Natl. Acad. Sci. USA* 100, 5520–5524.

Edelman, G.M. and Gally, J.A., 2001, Degeneracy and complexity in biological systems, *Proc. Natl. Acad. Sci. USA* 98, 13763–13768.

Ehresmann, A.C., 1981, Comments. In: *Charles Ehresmann: Oeuvres complètes et commentées,* Part IV-1 (A.C. Ehresmann, Ed.), CTGD, Amiens.

Ehresmann, A.C., 1996, Colimits in free categories, *Diagrammes* 37, 1–10.

Ehresmann, A.C., 2002, Localization of universal problems. Local colimits, *Appl. Categ. Struct.* 10, 157–172.

Ehresmann, C., 1954, Structures locales (*Ann. di Mat.*, 133). In: *Charles Ehresmann: Œuvres complètes et commentées* (A.C. Ehresmann, Ed.), Part I, 1981, CTGD, Amiens, 411–420 (reprint).

Ehresmann, C., 1960, Catégorie des foncteurs type (*Rev. Un. Mat. Argentina* 19). In: *Charles Ehresmann: Œuvres complètes et commentées*, (A.C. Ehresmann, Ed.), Part IV, 1982, Amiens, 101–116 (reprint).

Ehresmann, C., 1965, *Catégories et Structures*, Dunod, Paris.

Ehresmann, C., 1967, Trends toward unity in Mathematics (*Cahiers Top. et Géom. Dif.* VIII). In: *Charles Ehresmann: Œuvres complètes et commentées* (A.C. Ehresmann, Ed.), Part III, 1980, CTGD, Amiens, 759–766 (reprint).

Ehresmann, C., 1968, Esquisses et type des structures algébriques (*Bul. Inst. Poli. Iasi* XIV). In: *Charles Ehresmann: Œuvres complètes et commentées* (A.C. Ehresmann, Ed.), Part IV-1, 1982, Amiens, CTGD, 19–33 (reprint).

Ehresmann, C., 1980–1982, *Charles Ehresmann: Œuvres complètes et commentées* (A.C. Ehresmann, Ed.), 4 Parts, CTGD, Amiens.

Ehresmann, A.C. and Vanbremeersch, J.-P., 1987, Hierarchical evolutive systems: a mathematical model for complex systems, *Bull. Math. Bio.* 49 (1), 13–50.

Ehresmann, A.C. and Vanbremeersch, J.-P., 1989, Modèle d'interaction dynamique entre un système complexe et des agents, *Revue Intern. de Systémique* 3 (3), 315–341.

Ehresmann, A.C. and Vanbremeersch, J.-P., 1990, Hierarchical evolutive systems. In: *Proceeding of the 8th International Conference of Cybernetics and Systems*, Vol. 1 (A. Manikopoulos, Ed.), The NIJT Press, Newark, New York, 320–327.

Ehresmann, A.C. and Vanbremeersch, J.-P., 1991, Un modèle pour des systèmes évolutifs avec mémoire, basé sur la théorie des categories, *Revue Intern. de Systémique* 5 (1), 5–25.

Ehresmann, A.C. and Vanbremeersch, J.-P., 1992a, Outils mathématiques utilisés pour modéliser les systèmes complexes, *Cahiers Top. Géom. Dif. Cat.* XXXIII-3, 225–236.

Ehresmann, A.C. and Vanbremeersch, J.-P., 1992b, Semantics and communication for memory evolutive systems. In: *Proceeding of the 6th International Conference on Systems Research* (G. Lasker, Ed.), University of Windsor.

Ehresmann, A.C. and Vanbremeersch, J.-P., 1993, Memory evolutive systems: an application to an aging theory. In: *Cybernetics and Systems* (A. Ghosal and P.N. Murthy, Eds.), Tata McGraw Hill, New Delhi, 190–192.

Ehresmann, A.C. and Vanbremeersch, J.-P., 1994, *Emergence et Téléologie*, Publication 14-1, Université de Picardie, Amiens.

Ehresmann, A.C. and Vanbremeersch, J.-P., 1996, Multiplicity principle and emergence in MES, *J. Sys. Anal. Model. Simul.* 26, 81–117.

Ehresmann, A.C. and Vanbremeersch, J.-P., 1997, Information processing and symmetry-breaking in MES, *Biosystems* 43, 25–40.

Ehresmann, A.C. and Vanbremeersch, J.-P., 1999, Online URL: http://perso.wanadoo.fr/vbm-ehr.

Ehresmann, A.C. and Vanbremeersch, J.-P., 2002, Emergence processes up to consciousness using the multiplicity principle and quantum physics, *A.I.P. Conference Proceedings* 627 (CASYS, 2001; D. Dubois, Ed.), 221–233.

Ehresmann, A.C. and Vanbremeersch, J.-P., 2003, A categorical model for cognitive systems up to consciousness. In: *Proceeding of the International Conference on Theoretical Neurobiology* (A. Singh, Ed.), NBRC, New Delhi, 50–61.

Eigen, M. and Schuster, P., 1979, *The Hypercycle: A Principle of Natural Organization*, Springer, Berlin.

Eilenberg, S., 1974, *Automata, Languages and Machines*, Academic Press, New York, London.

Eilenberg, S. and Mac Lane, S., 1945, General theory of natural equivalences, *Trans. Am. Math. Soc.* 58, 231–294.

Eliade, M., 1965, *Le sacré et le profane*, Editions Gallimard, Paris.

Engert, F. and Bonhoeffer, T., 1997, Synapse specificity of long-term potentiation breaks down at short distances, *Nature* 388, 279–282.

Esser, K., 1985, Genetic control of aging: the mobile intron model. In: *The Thresholds of Aging* (M. Bergener, M. Ermini and H.B. Stahelin, Eds.), Academic Press, London, 4–20.

Farre, G.L., 1994, Reflections on the question of emergence. In: *Advances in Synergetics,* Vol. I (G. Farre, Ed.), The International Institute for Advanced Studies in Systems Research and Cybernetics, Windsor.

Farre, G., 1997, The energetic structure of observation, *Am. Behav. Scientist* 40 (6), 717–728.

Finkel, T. and Holbrook, N.J., 2000, Oxidants, oxidative stress and the biology of ageing, *Nature* 408, 239–247.

Finkelkraut, A., 1987, *La défaite de la pensée*, NRF, Gallimard, Paris.

Fisahn, A., Pike, F.G., Buhl, E.H. and Paulsen, O., 1998, Cholinergic induction of network oscillations at 40Hz in the hippocampus in vitro, *Nature* 394, 186–189.

Fodor, J.A., 1983, *The Modularity of Mind*, MIT Press, Cambridge.

Frey, U. and Morris, R., 1997, Synaptic tagging and long-term potentiation, *Nature* 385, 533–536.

Friedberg, B.C., 1985, *DNA repair*, Freeman, New York.

Gazzaniga, M.S., 1985, *The Social Brain*, Basic Books, New York.

Gedda, L.C. and Brenci, G., 1975, *Chronogénétique: l'hérédité du temps biologique*, Hermann, Paris.

Gensler, H.L., 1981, The effect of hamster age in UV-induced unscheduled DNA synthesis in freshly isolated lung and kidney cells, *Exper. Gerontol.* 16, 59–68.

Gepner, B. and Mestre, D., 2002, Rapid visual motion integration deficit in autism, *Trends Cogn. Sci.* 6, 455.

Germain, R.N., 2001, The art of the probable: system control in the adaptive immune system, *Science* 293, 240–245.

Gershon, H.L., 1979, Current status of age altered enzymes: alternative mechanisms, *Mech. Aging Develop.* 9, 189–196.

Gilinsky, A.S., 1984, *Mind and Brain*, Praeger Publishers, CBS, New York.

Goffman, E., 1973, *La mise en scène de la vie quotidienne*, Editions de Minuit, Paris.

Goguen, J.A., 1970, Mathematical representation of hierarchically organized systems. In: *Global Systems.* (E.O. Attinger, Ed.), Karger, Basel, 65–85.

Goguen, J.A., 2003, *Musical qualia, context, time and emotion*, Online, URL: cogprints.ecs.soton.ac.uk.

Goldbeter, A., 1990, *Rythmes et chaos dans les systèmes biochimiques et cellulaires*, Masson, Paris.

Goldstein, J., 1999, Emergence as a construct: history and issues, *Emergence* 1 (1), 49–72.

Goldstein, S., 1978, Human genetic disorders which feature accelerated aging. In: *The Genetics of Aging* (E.L. Schneider, Ed.), Plenum, New York, 171–224.

Goldstein, S. and Moerman, E.J., 1976, Defective proteins in normal and abnormal human fibroblasts during aging in vitro. In: *Cellular Ageing: Concepts and Mechanisms, Part 1* (R.G. Cutler, Ed.), Karger, New York, 24–43.

Gosgnach, S. *et al.*, 2006, V1 spinal neurons regulate the speed of vertebrate locomotor outputs, *Nature* 440, 215–219.

Gray, J.W., 1989, The category of sketches as a model for algebraic semantics. In: *Categories in Computer Science and Logic* (J.W. Gray and A. Scedrov, Eds.), American. Mathematics. Society., Providence R.I.

Grothendieck, A., 1961, Catégories fibrées et descente, *Séminaire Géométrie Algébrique* 4, I.H.E.S., Bures-sur-Yvette.

Grothendieck, A. and Verdier, J.I., 1963–1964, *Théorie des topos.* In: *SGA 4*, Lecture Notes in Math. 269, Springer, Berlin (reprint, 1972).

Guitart, R., 2000, *Evidence et Etrangeté*, P.U.F., Paris.

Haithcock, E., Dayani, Y., Neufeld, E., Zahand, A.J., Feinstein, N., Mattout, A., Gruenbaum, Y. and Liu, J., 2005, From the cover: age-related changes of nuclear architecture in Caenorhabditis elegans, *Proc. Natl. Acad. Sci. USA* 102, 16690–16695.

Hall, R., 1989, *L'excellence industrielle*, InterEditions, Paris.

Hameroff, S. and Penrose, R., 1995, Orchestrated reduction of quantum coherence in brain microtubules. In: *Scale in Conscious Experience: Is the Brain too Important to be Left to Specialist to Study?* (J. King and K.H. Pribram, Eds.), Lawrence Erlbaum, Mahevah, N.J., 341–274.

Harley, C.B. and Goldstein, S., 1980, Retesting the commitment theory of cellular aging, *Science* 207, 191–193.

Harman, D., 1986, Free radical theory of aging: role of free radical reactions in the origination and evolution of life, aging and disease processes. In: *Modern Trends in Aging Research*, Coll. INSERM-EURAGE, John Libbey Eurotex LTD, Vol. 147, Paris, 77–83.

Hartwell, L.H., Hopfield, J.J., Leibler, S. and Murray, A.W., 1999, From molecular to modular cell biology, *Nature* 492, C47–52.

Hauw, J.-J., Delaere, P. and Duyckaerts, C., 1987, La sénescence cérébrale: données morphologiques. In: *Le vieillissement cérébral normal et pathologique*, Maloine, Paris, 139–156.

Hayflick, L., 1977, The cellular basis for biological aging. In: *Handbook of the Biology of Aging* (C.B. Finch and L. Hayflick, Eds.), van Nostrand-Reinhold, New York.

Hebb, D.O., 1949, *The Organization of Behaviour*, Wiley, New York.

Hegel, G.W., 1807, Phänomenologie des Geistes. In: *La phénoménologie de l'esprit* (Trad. Hyppolite), Aubier, Paris (reprint 1939–1941).

Herrlich, H., 1979, Initiazl and final completions, *Lecture Notes in Mathematics.* 719, Springer, Berlin, 137–149.

Hespos, S.J. and Spelke, E.S., 2004, Conceptual preceptors of language, *Nature* 430, 453–456.

Hoebe, K., *et al.*, 2003, Identification of *Lps2* as a key transducer of MytD88-independent TIR signalling, *Nature* 424, 743–748.

Hofstadter, D., 1985, *Gödel, Escher, Bach.*, InterEditions, Paris.

Holliday, R., 1987a, The inheritance of epigenetic defects, *Science* 238, 163–170.

Holliday, R., 1987b, X-chromosome reactivation, *Nature* 327, 663.

Holliday, R., Huschtscha, L.L. and Kirkwood, T.B.L., 1981, Cellular aging: further evidence for the commitment theory, *Science* 213, 1505–1507.

Hopfield, J.J., 1982, Neural networks and physical systems, *Proc. Natl. Acad. Sci. USA* 79, 2554–2558.

Hopfield, J.J. and Tank, D.W., 1985, Neural computation of decisions in optimization problems, *Biol. Cybern.* 52, 141–152.

Horvath, T.B. and Davis, K.L., 1990, Central nervous system disorders in aging. In: *Handbook of the Biology of Aging* (E.L. Schneider and J.W. Rowe, Eds.), van Nostrand-Reinhold, New York, 307–329.

Houdé, O., 1992, *Catégorisation et développement cognitif*, PUF, Paris.

Howard, N., 1971, *Paradoxes of Rationality: Theory of Metagames and Political Behavior*, MIT Press, Cambridge, Mass.

Hubel, D.H. and Wiesel, T.N., 1962, Receptive fields ..., *J. Physio.* 160 (1), 106–154.

Husserl, E., 1904, *Vorlesungenzur Phänomenologie des innere Zeitbewuszseins*. In: *Leçons pour une phénomenologie de la conscience intime du temps*, P.U.F., Paris (reprint 1964).

Jackson, F., 1986, What Mary did not know? *J. Phil.* 83(5), 291–295.

Jacob, F., 1970, *La logique du vivant*, Gallimard, Paris.

Jacob, F., Brenner, S. and Cuzin, F., 1963, On the regulation of DNA replication in bacteria, *Cold Spring Harbor. Symp. Quant. Biol.* 28, 239–245.

James, W., 1890, *Principles of Psychology*, H. Holt and Co., New York.

Jeannerod, M., 1983, *Le cerveau-machine*, Fayard, Paris.

Josephson, B., 1998, Extendibility of activities and the design of the nervous system. In: *Proceedings Third International Conference on Emergence ECHO III* (G. Farre, Ed.), Helsinki.

Kainen, P.C., 1990, Functorial cybernetics of attention. In: *Neurocomputers and Attention II* (A. Holden and A. Kryukov, Eds.), Manchester University Press, Manchester, Chap. 57.

Kamiuchi, S., Saijo, M., Citterio, E., de Jager, M., Hoeijmakers, J. and Tanaka, K., 2002, Translocation of Cockayne syndrome group A protein to the nuclear matrix: possible relevance to transcription-coupled DNA repair, *Proc. Natl. Acad. Sci. USA* 99, 201–206.

Kampis, G., 1991, *Self-Modifying Systems in Biology and Cognitive Science*, Pergamon Press, Oxford.

Kan, D.M., 1958, Adjoint functors, *Trans. Am. Math. Soc.* 89, 294–329.

Kandel, E.R. and Schwartz, J.H., 1982, Molecular biology of learning: modulation of transmitter release, *Science* 218, 433–442.

Kant, E., 1790, *Critique of Judgement* (translated with Introduction and Notes by J.H. Bernard. 1914), Macmillan, London.

Karmiloff-Smith, A., 1992, *Beyond Modularity: A Developmental Perspective on Cognitive Science*, Mass. Inst. of Technology, Cambridge, Mass.

Kay, M.M.B., 1991, Drosophila to bacteriophage to erythrocyte: the erythrocyte as a model for molecular and membrane aging of terminally differentiated cells, *Gerontology* 37 (1–3), 5–32.

Kennes, B.J., Devière, J., Brohée, D., de Maertelaer, V. and Neve, P., 1986, The effect of age, sex and health status on the distribution and functions of human blood mononuclear subpopulations. In: *Modern trends in aging research*, Coll. INSERM-EURAGE, John Libbey Eurotex LTD, Vol. 147, Paris, 231–234.

Kim, J., 1998, *Mind in a Physical World: An Essay on the Mind-Body Problem and Mental Causation*, M.I.T. Press, Cambridge MA.

Kirkpatrick, S., Gelatt, C.D. and Vecchi, M.P., 1983, Optimization by simulated annealing, *Science* 220, 671–680.

Kirkwood, T.B.L., 1996, Human senescence, *Bioessays* 18, 1009–1016.

Kirkwood, T.B.L., Austad, S.N., 2000, Why do we age? *Nature* 408, 233–237.

Kitano, H., 2002, Computational systems biology, *Nature* 420, 206–210.

Klir, G.J., 1969, *An Approach to General System Theory*, van Nostrand, New York.

Klir, G.J., 1985, *Architecture of General Systems Problem Solving*, Plenum, New York.

Koechlin, E., Basso, G., Pietrini, P., Penzer, S. and Grafman, J., 1999, The role of the anterior prefrontal cortex in human cognition, *Nature* 399, 148–151.

Koestler, A., 1965, *Le cri d'Archimède*, Calmann-Lévy, Paris.

Köhler, W., 1947, *Gelstat Psychology*, Liveright, New York.

Kuhn, T.S., 1970, *La structure des révolutions scientifiques*, Flammarion, Paris.

Laborit, H., 1983, *La Colombe Assassinée*, Grasset, Paris.

Laborit, H., 1987, Comportement et vieillissement, in *Le vieillissement cérébral normal et pathologique*, Maloine, Paris, 44–60.

Lakatos, I., 1970, Falsification and the methodology of scientific programs. In: *Criticism and the Growth of Knowledge* (I. Lakatos and A. Musgrave, Eds.), Cambridge University Press, Cambridge.

Lamar, M. and Resnik, S.M., 2004, Aging and prefrontal functions: dissociating orbitofrontal and dorsolateral abilities, *Neurobiol. Aging* 25 (4), 553–558.

Lamour, Y., 1991, Démences de type Alzheimer: de l'anomalie génétique à la lésion structurale. *La Revue de Gériatrie*, Supplément no. 3, 15–28.

Landry, E., 1998, Category theory: the language of mathematics, Paper presented at the *Sixteenth Biennial Meeting of the Philosophy of Science Association*, October, Kansas City, Missouri.

Laszlo, E., 1989, *La cohérence du reel*, Gauthier-Villars, Bordas, Paris.

Lautman, A., 1938, *Essai sur les notions de structure et d'existence en mathématiques*, Hermann, Paris.

Lavie, G., Reznick, A.Z. and Gershon, D., 1982, Decreased proteins and puromycinyl peptide degradation in livers of senescent mice, *Biochem. J.* 202, 47–51.

Lawvere, F.W., 1963, Functorial semantics of algebraic theories, Dissertation, Columbia University, New York.

Lawvere, F.W., 1965, Metric Spaces, Generalized Logic, and Closed Categories, Rend. Sem. mat. fis. 43, 135–166.

Lawvere, F.W., 1966, An elementary of the theory of set. In: *Proceeding Conference on Categorical Algebra, La Jolla* (S.D. Harrison, S. Mac Lane and H. Rohrl, Eds.), Springer, Berlin.

Lawvere, F.W., 1972, Introduction: toposes, algebraic geometry and logic, *Lecture Notes in Mathematics* 274, Springer, Berlin, 1–12.

Lawvere, F.W., 1980, Toward the description in a smooth topos of the dynamically possible motions and deformations of a continuous body, *Cahiers Top.et Géom. Dif.* XXI (1), 377–392.

Lawvere, F.W., Schanuel, S.H., (Eds.) 1980, Categories in Continuum Physics, *Lecture Notes in Mathematics* 1174, Springer, Berlin.

LeBel, C.P. and Bondy, S.C., 1992, Oxidative damage and cerebral aging, *Prog. Neurobiol.* 38, 601–609.

Lecomte de Noüy, P., 1936, *Le temps et la vie*, Gallimard, Paris.

Le Goff, J., 1988, L'Histoire nouvelle. In: *La nouvelle histoire*, Editions Complexe, Paris, 62–63.

Le Moigne, J.-L., 1990, *La modélisation des systèmes complexes*, Dunod, Paris.

Levin, J. and Miller, J., 1996, Broadband neural encoding in the cricket cercal sensory system enhanced by stochastic resonance, *Nature* 380, 165–168.

Levi-Strauss, C., 1962, *La pensée sauvage*, Plon, Paris.

Linsker, R., 2005, Improved local learning rule for information maximization and related applications, *Neural Networks* 18 (3), 261–265.

Little, J.W. and Mount, D.W., 1982, The SOS regulatory system of *E. coli, Cell* 29, 11–22.

Lorenz, E.N., 1963, Deterministic nonperiodic flow, *J. Atmospheric Sci.* 20, 130–141.

Lorenz, K., 1973, *L'envers du miroir*, Champs Flammarion, Paris.

Louie, A.H., 1983, Categorical system theory, *Bull. Math. Biol.* 45, 1029–1072.

Lunca, M., 1993, *An Epistemological Programme for Interdisciplinarisation*, ISOR, Utrecht.

Macieira-Coelho, A., Liepkalns, C.I. and Puvion-Dutilleul, F., 1986, Genome reorganization during senescence of dividing cells, *Modern trends in aging research*, Coll. INSERM-EURAGE, John Libbey Eurotex LTD, Vol. 147, Paris, 41–54.

Mackie, J.I., 1974, *The Cement of the Universe*, Oxford University Press, Oxford.

Mac Lane, S., 1971, *Categories for the Working Mathematician*, Springer, Berlin.

Mac Lane, S., 1986, *Mathematics Form and Function*, Springer, Berlin.

Mac Lane, S., 1992, The protean character of mathematics. In: *The Space of Mathematics* (A. Echeverra, A. Ibarra and A. Mormann, Eds.), de Gruyter, New York, 3–12.

Mac Lane, S., 1997, The PNAS way back then, *Proc. Natl. Acad. Sci. USA* 94, 5983–5985.

Maenhaut-Michel, G., 1985, Mechanism of SOS induced targeted and untargeted mutagenesis in *E. coli, Biochimie* 67, 365–390.

Mahner, M. and Bunge, M.A., 2001, Function and functionalism: a synthetic perspective, *Phil. Sci.* 68, 75–94.

Mahner, M., Kary, M., 1997, What exactly are genomes, genotypes and phenotypes? And what about phenomes? *J. Theor. Biol.* 186, 55–63.

Malsburg (von der), C., 1995, Binding in models of perception and brain function, *Curr. Opin. Neurobiol.* 5, 520–526.

Malsburg, C. (von der) and Bienenstock, E., 1986, Statistical coding and short-term synaptic plasticity. In: *Disordered Systems and Biological Organization*, NATO ASI Series 20, Springer, Berlin, 247–272.

Mandelbrot, B., 1975, *Les objets fractals, forme, hasard et dimension*, Flammarion, Paris.

Marchais, P., 2003, *L'activité psychique*, L'Harmattan, Paris.

Marchais, P., 2004, L'angoisse et l'anxiété. Variations conceptuelles. Ouverture à la théorie des categories, *Annales Médico Psychologiques* 162, 196–202.

Marquis, J-P., 1995, Category theory and the foundations of mathematics: philosophical excavations, *Synthese* 103, 421–447.

Marquis, J-P., 1996, A critical note on Bunge's 'System Boundary' and a New Proposal, *Int. J. Gen. Sys.* 24 (3), 245–255.

Martin, G.M., Sprague, C.A. and Epstein, C.J., 1970, Replicative life span of cultivated human cells. Effect of donor's age, tissue and genotype, *Lab. Invest.* 23, 86–92.

Matsuno, K., 1989, *Protobiology: Physical Basis of Biology*, CRC Press, Boca Raton.

Maturana, H. and Varela, F., 1973, Autopoietic systems: a characterization of the living organization. In: *Autopoiesis and Cognition: the Realization of the Living*, Vol. 48 (H. Maturana and F. Varela, Eds.), Boston Studies in the Philosophy of Science, Reidel, Boston.

Mayr, E., 1976, *Evolution and the Diversity of Life, Essays*, Belknap Press of Harvard University, Cambridge.

Merleau-Ponty, M., 1945, *Phénoménologie de la perception*, Ed. Gallimard, Paris.

Meynadier, J., 1980, *Précis de physiologie cutanée*, La Porte Verte, Paris.

Miller, A.K.H., Alston, R.L. and Corsellis, J.A.N., 1980, Variation with age in the volumes of grey and white matter in the cerebral hemispheres of man, measurements with an image analyser, *Neuropath. Appl. Neurobiol.* 6, 119–132.

Miller, R.A., 1990, Aging and the immune response. In: *Handbook of the biology of aging*, 3rd ed., Academic Press, New York, 157–180.

Miltner, W., Braun, C., Arnold, M., Witte, H. and Taub, E., 1999, Neurobiology striving for coherence, *Nature* 397, 434–436.

Mingolla, E., 2003, The units of form vision. In: *Proceeding of the International Conference on theoretical Neurobiology* (A. Singh, Ed.), NBRC, New Delhi, 115–128.

Minsky, M., 1986, *The Society of Mind*, Simon and Schuster, New York.

Monod, J., 1970, *Le hasard et la nécessité*, Ed. du Seuil, Paris.

Moreau, P.L., 1985, Role of *E. coli* RecA protein in SOS induction and post-replication repair, *Biochimie* 67, 353–364.

Morin, E., 1977, *La Méthode*, Editions Seuil, Paris.

Moschetto, Y., De Backer, M., Pizieu, O., Bouissou, H., Pieraggi, M.T. and Rufian, M., 1974, Evolution de l'élastine d'aortes humaines au cours du vieillissement, *Paroi artérielle* 2, 161–177.

Moulin, T., 1986, Présentation sommaire des relateurs arithmétiques, *Cahiers Systema* 12, 25–113.

Muller, F.A., 2001, Sets, classes and categories, *Brit. J. Phil. Soc.* 52, 539–573.

Nagel, T., 1974, What is it like to be a bat? *The Philosophical Rev.* LXXXIII(4), 435–450.

Nemoto, S. and Finkel, T., 2004, Ageing and the mystery at Arles, *Nature* 429, 149–152.

Neuman (von), J., 1966, *The Theory of Self-Reproducing Automata*, University of Illinois Press, Chicago.

Niemeyer, P. and Peck, J., 1997, *Exploring Java*, O'Reilly, London.

Nietzsche, F., 1888, Götzen-Dämmerung. Translated in Crépuscule des idoles (Hemery, Trans.), Idées/Gallimard, Paris, 1974.

Novak, B. and Tyson, J.J., 1993, Numerical analysis of a comprehensive model of M-phase control in *Xenopus* oocyte extracts and intact embryos, *J. Cell Sci.* 106, 1153–1168.

Ohrloff, C. and Hockwin, O., 1986, Superoxide dismutase (SOD) in normal and cataractous human lenses. In: *Modern Trends in Aging Research*, Coll. INSERM-EURAGE, John Libbey Eurotex LTD, Vol. 147, 365–371.

Orgel, L.E., 1963, The maintenance of the accuracy of protein synthesis and its relevance to aging, *Proc. Natl. Acad. Sci. USA*, 49, 517.

Orgel, L.E., 1970, The maintenance of the accuracy of protein synthesis and its relevance to aging, correction, *Proc. Natl. Acad. Sci. USA* 67, 1476.

Pacifici, R.E. and Davies, K.J.A., 1991, Protein, lipid and DNA repair systems in oxidative stress: the free-radical theory of aging revisited, *Gerontology* 37 (1), 166–180.

Patarnello, S. and Carnevali, P., 1989, A neural network model to stimulate a conditioning experiment, *Intern. J. Neural Syst.* 1 (1), 47–53.

Paton, R.C., 1997, Glue, verb and text metaphors in Biology, *Acta Biotheoretica* 45, 1–15.

Paton, R.C., 2001, Structure and context in relation to integrative biology, *BioSystems*, Special edition in memory of Michael Conrad.

Paton, R.C., 2002, Process, structure, and context in relation to integrative biology, *BioSystems*, 64, 63–72.

Pavlov, I.P., 1927, *Conditioned Reflexes: An Investigation of the Physiological Activity of the Cerebral Cortex* (G.V. Anrep, Ed. and trans.), Dover, New York (reprint 1960).

Penrose, R., 1992, *L'esprit, l'ordinateur et les lois de la Physique*, InterEditions, Paris.

Piaget, J., 1940, Die geistige Entwicklung des Kindes, *Juventus Helvetica* 1. Translated in: *Six etudes de psychlogie*, Denoël-Gonthier, Paris, 1964.

Piaget, J., 1967, *Biologie et connaissance*, Gallimard, Paris.

Pierce, C.S., 1903, Abduction and induction. In: *Philosophical Writings of Pierce* (J. Buchler, Ed.), Dover Publications, New York, 150–156.

Poggio, T. and Bizzi, E., 2004, Generalization in vision and motor control, *Nature* 431, 768–774.

Poincaré, H., 1951, *Œuvres de Henri Poincaré*, Gauthier-Villars, Paris.

Popper, K., 1972, *Objective Knowledge, An Evolutionary Approach*, Clarendon Press, Oxford.

Pourcel, C., 1986, *Systèmes automatisés de production*, Cepadues-Ed., Toulouse.

Pribram, K.H., 1971, *Languages of the Brain*, Prentice Hall, Englewood Cliffs, NJ.

Pribram, K.H., 2000, Proposal for a quantum physical basis for selective learning. In: *Proceedings ECHO IV* (G. Farre, Ed.), Washington, 1–4.

Prigogine, I. and Glansdorff, P., 1971, *Structure, stabilité et fluctuations*, Masson, Paris.

Prigogine, I. and Stengers, I., 1982, *La nouvelle alliance*, Gallimard, Paris.

Pritchard, R.M., 1978, Control of DNA replication in bacteria. In: *DNA Synthesis: Present and Future* (I. Molyneux and M. Koyama, Eds.), Plenum Press, New York, 1–26.

Quine, W., 1960, *Word and Object*, M.I.T. Press, Cambridge, MA.

Radman, M., 1975, SOS repair hypothesis: phenomenology of an inducible DNA repair which is accompanied by mutagenesis. In: *Molecular Mechanisms for Repair of DNA* (Hanawalt and Setlow, Eds.), Plenum Press, New York, 355–367.

Raff, M.C., 1992, Social controls on cell survival and cell death, *Nature* 356, 397–399.

Rapoport, A., 1947, Mathematical theory of motivational interaction of two individuals, Parts I and II. *Bull. Math. Biophys.* 9, 17–28 and 41–61.

Rapoport, A., 1985, Application of game-theoretic concepts in Biology, *Bull. Math. Bio.* 47 (2), 167–192.

Rashevsky, N., 1967, Organismic sets. Outline of a general theory of biological and sociological organisms, *Bull. Math. Biophys.* 29, 139–152.

Rashevsky, N., 1968, Organismic sets II. Some general considerations, *Bull. Math. Biophys.* 30, 163–174.

Rashevsky, N., 1969, Hierarchical organization in Automata theoretic models of the central nervous system. In: *Information Processing in the Nervous System* (K.N. Leibovic, Ed.), Springer, New York, 21–35.

Rashevsky, N., 1972, Some relational cell models: the metabolic-repair system. In: *Foundations of Mathematical Biology*, Vol. 2 (R. Rosen, Ed.), Academic Press, New York, 217–253.

Rattan, S.I.S., Cavallius, J., Hartvigsen, G.K. and Clark, B.F.C., 1986, Amounts of active elongation factor 1α and its activity in livers of mice during aging. In: *Modern Trends in Aging Research*, Coll. INSERM-EURAGE, John Libbey Eurotex LTD, Vol. 147, Paris, 135–139.

Reeves, H., 1988, *Patience dans l'azur, l'évolution cosmique*, Edition du Seuil, Paris.

Renan, E., 1876, Dialogues et fragments philosophiques. In: *Œuvres complètes*, Vol. 1, Calmann-Levy, Paris (reprint 1947).

Ribeill, G., 1974, *Tensions et mutations sociales*, P.U.F., Paris.

Richardson, A. and Semsei, I., 1987, Effect of aging on translation and transcription, *Rev. Biol. Res. Aging* 3, 467–483.

Richardson, G.S., 1990, Circadian rythms and aging. In: *Handbook of the Biology of Aging*, 3rd ed., Academic Press, New York.

Rigby, B.J., Mitchell, T.W. and Robinson, M.S., 1977, Oxygen participation in the *in vivo* and *in vitro* aging of collagen fibers, *Biochem. Biophys. Res. Com.* 79, 400–405.

Riggs, B.L. and Melton, L.J., 1986, Involutional osteoporosis, *New Engl. J. Med.* 314, 1676–1696.

Robert, A.M., Boniface, R. and Robert, L., 1977, Biochemistry and pathology of basement membranes. In: *Frontiers of Matrix Biology*, Vol. 7 (A.M. Robert and L. Robert, Eds.), Karger, Basel.

Robert, L., 1983, *Mécanismes cellulaires et moléculaires du vieillissement*, Masson, Paris.

Robert, L., 1989, *Les horloges biologiques*, Flammarion, Paris.

Robert, L., 1990, Théories du vieillissement biologique. In: *Interdisciplinarité en Gérontologie* (A. Hebert, Ed.), Maloine, Paris.

Rodriguez, E., George, N., Lachaux, J.-P., Martinerie, J., Renault, B. and Varela, F., 1999, Perception's shadow: long-distance synchronization of humain brain activity, *Nature* 397, 430–433.

Rosch, E., 1973, Natural categories, *Cog. Psychol.* 4, 328–350.

Rosen, R., 1958a, The representation of biological systems from the standpoint of the theory of categories, *Bull. Math. Biophys.* 20, 245–260.

Rosen, R., 1958b, A relational theory of biological systems, *Bull. Math. Biophys.* 20, 317–341.

Rosen, R., 1959, A relational theory of biological systems II, *Bull. Math. Biophys.* 21, 109–128.

Rosen, R., 1969, Hierarchical organization in automata theoretic models of the central nervous system. In: *Information Processing in the Nervous System* (K.N. Leibovic, Ed.), Springer, New York, 21–35.

Rosen, R., 1978, Feedforwards and global system failure: a general mechanism for senescence, *J. Theor. Biol.* 74, 579–590.

Rosen, R., 1981, Pattern generation in networks. In: *Progress in Theoretical Biology*, Vol. 6 (R. Rosen, Ed.), Academic Press, New York, 161–209.

Rosen R., 1985a, Organisms as causal systems which are not mechanisms. In: *Theoretical Biology and Complexity*, Academic Press, New York, 165–203.

Rosen, R., 1985b, *Anticipatory Systems*, Pergamon, New York.

Rosen, R., 1986, *Theoretical Biology and Complexity*, Academic Press, New York.

Rumelhart, D.E., Hinton, G.E., Williams, R.J., 1986, Learning representations by back-propagating errors, *Nature* 323, 533.

Russell, B., 1949, *The Analysis of Mind*, George Allen and Unwin, London.

Russell, B., 1971, *La méthode scientifique en Philosophie*, Payot, Paris.

Sacks, O., 2004, Les instantanés de la conscience, *La Recherche* 374, 31–38.

Salthe, S.N., 1985, *Evolving Hierarchical Systems: Their Structure and Representation*, Columbia University Press, New York.

Samuel, P., 1948, On universal mappings and free topological groups, *Bull. A.M.S.* 54, 591–598.

Sapolsky, R.M., 1990, The adrenocortical axis. In: *Handbook of the Biology of Aging* (E.L. Schneider and J.W. Rowe, Eds.), van Nostrand, New York, 330–346.

Schempp, W.J., 1998, *Magnetic Resonance Imaging*, Wiley-Liss, New York.

Schneider, T.D., 1991, Theory of molecular channels, *J. Theor. Biol.* 148, 83–137.

Serres, M., 1969, *Hermes I: La Communication*, Ed. de Minuit, Paris.

Shanley, D.T. and Kirkwood, T.B., 2000, Caloric restriction and aging: the life-history analysis, *Evolution Int. J. Org. Evolution* 54(3), 740–750.

Shannon, C.E. and Weaver, W., 1949, *The Mathematical Theory of Communication*, University of Illinois Press, Urbana.

Shimokawa, I., Higami, Y., Hubbard, G.B., Mcmahan, C.A., Masoro, E.J. and Byung Pal Yu, 1993, Diet and the suitability of the male fischer 344 rat as a model for aging research, *J. Gerontol.* 48, B27–32.

Shmookler, R.J., Riabowol, K.T., Jones, R.A., Ravishanker, S. and Goldstein, S., 1986, Extrasomal circular DNAs from senescent human fibroblasts contain an over-representation of Alu1 and Kpn1 interspersed repeats, which may reflect genetic instability. In: *Modern Trends in Aging Research*, Coll. INSERM-EURAGE, John Libbey Eurotex LTD, Vol. 147, Paris, 31–39.

Signoret, J.-L., 1987, Le vieillissement du cerveau est-il modulaire? In: *Le vieillissement cérébral normal et pathologique*, Maloine, Paris, 61–72.

Simon, H.A., 1974, *La science des systèmes, Science de l'artificiel* (Transl. Le Moigne), Epi Ed., Paris.

Sinex, F.M., 1977, The molecular genetics of aging. In: *Handbook of the Biology of Aging* (C.B. Finch and L. Hayflick, Eds.), van Nostrand, New York, 639–665.

Singer, W., 2003, Oscillations and synchrony – Time as coding space in neuronal processing. In: *Proceeding of the International Conference on Theoretical Neurobiology* (A. Singh, Ed.), NBRC, New Delhi, 29–49.

Skinner, B.F., 1938, *The Behavior of Organisms: An Experimental Analysis*, D. Appleton-Century Co., New York.

Smale, S., 1967, Differentiable dynamical systems, *Bull. Am. Math. Soc.* 73, 747–817.

Smith-Sonneborn, J., 1990, Aging in Protozoa. In: *Handbook of the Biology of Aging* (E.L. Schneider and J.W. Rowe, Eds.), van Nostrand, NewYork, 24–44.

Stach, S., Benard, J. and Giurfa, M., 2004, Local-feature assembling in visual pattern recognition and generalization in honeybees, *Nature* 429, 758–761.

Stefanacci, L., 2003, Chaos to emotion: amygdala studies, from behaviour to genes. In: *Proceeding of the International Conference on Theoretical Neurobiology* (A. Singh, Ed.), NBRC, New Delhi, 19–28.

Stewart, I., 2004, Networking opportunity, *Nature* 427, 602–604.

Stryker, M.P., 1989, Is grand-mother an oscillation? *Nature* 338, 297.

Sutton, J.P., Beis, J.S. and Trainor, L.E.H., 1988a, Hierarchical model of memory and memory loss, *J. Phys. A: Math. Gen.* 21, 4443–4454.

Sutton, J.P., Beis, J.S. and Trainor, L.E.H., 1988b, A hierarchical model of neocortical synaptic organization, *Math. Comput. Modelling* 11, 346–350.

Sutton, R.S. and Barto, A.G., 1998, *Reinforcement Learning*, M.I.T., Cambridge, MA.

Sweatt, J.D. and Kandel, E.R., 1989, Persistent and transcriptionally-dependent increase in protein phosphorylation in long-term facilitation, *Nature* 339, 51–54.

Szilard, L., 1959, On the nature of the aging process, *Proc. Natl. Acad. Sci. USA* 45, 30–45.

Teilhard de Chardin, P., 1955, *Le phénomène humain*, Editions du Seuil, Paris.

Teodorova, S., 1985, Principle of invariance and criterion of evolution in biological structures, *Revue de Bio-Math.* 89, 17–30.

Thom, R., 1974, *Modèles Mathématiques de la Morphogenèse,* Union Générale d'Edition, Coll. 10/18, Paris.

Thom, R., 1988, *Esquisse d'une Sémiophysique*, InterEditions, Paris.

Thomas, N., 1999, Are theories of imagerie theories of imagination?, *Cog. Sci.* 23, 207–245.

Thouvenot, J., 1985, Régulations, asservissements en biogestion. Application aux systèmes en croissance, *Revue Bio-Math.* 89 (1), 1–16.

Tinbergen, N., 1951, *The Study of Instinct*, Clarendon Press, Oxford.

Tononi, G., Sporns, O. and Edelman, G.M., 1999, Measures of degeneracy and redundancy in biological networks, *Proc. Natl. Acad. Sci. USA* 96, 3257–3262.

Toulouse, G., Dehaene, S. and Changeux, J.-P., 1986, Spin glass model of learning by selection, *Proc. Natl. Acad. Sci. USA* 83, 1695–1698.

Trifunovic, A., *et al.* 2004, Premature ageing in mice expressing defective mitochondrial DNA polymerase, *Nature* 429, 417–422.

Uexküll, J.V., 1956, *Mondes animaux et monde humain*, Ed. Gonthier, Paris.

Ullmo, J., 1993, *La pensée scientifique moderne*, Champs Flammarion, Paris.

Usher, M. and Donnelly, N., 1998, Visual synchrony affects binding and segmentation in perception, *Nature* 394, 179–182.

Valen (van), L., 1983, A new evolutionary law, *Evol. Theor.* 1, 1–30.

Vallée. R., 1986, Subjectivité et systèmes. In: *Perspectives systémiques*, Actes Colloque de Cerisy (B. Paulré, Ed.), L'Interdisciplinaire, Limonest, 44–53.

Vallée, R., 1995, *Cognition et Système*, L'Interdisciplinaire, Limonest.

Vanbremeersch, J.-P., Ehresmann, A.C. and Chandler, J.L.C., 1996, Are interactions between different timescales a characteristic of complexity? *Actes du Symposium ECHO*, Ehresmann, Amiens.

Varela, F.J., 1989, *Autonomie et connaissance*, Editions du Seuil, Paris.

Verzar, F., 1957, Aging of connective tissue, *Gerontol* 1, 363–378.

Vogel, E.H., Castro, M.E. and Saavedra, M.A., 2004, Quantitative models of Pavlonian conditioning, *Brain Res. Bull.* 63 (3), 173–202.

Waddington, C.H., 1940, *The Strategy of the Genes.*, Allen and Unwin, London.

Wallace, R., 2004, Consciousness, cognition and the hierarchy of contexts: extending the global neuronal workspace, Online URL at http://www.cogprints.ecs.soton.ac.uk.

Walters, R.E.C., 1991, *Categories and Computer Science*, Cambridge University Press, Cambridge, UK.

Warner, M.W., 1982, Representations of (M, R)-systems by Categories of Automata, *Bull. Math. Biol.* 44, 661–68.

Watslawick, P., Helminck Beavin, J. and Jackson, D.D., 1967, *Pragmatics of Human Communication*, Norton, New York.

Webster, G.C., 1985, Protein synthesis in aging organisms. In: *Molecular Biology of Aging: Gene Stability and Gene Expression* (R.S. Sohal, L.S. Birnbaum and R.G. Cutler, Eds.), Raven Press, New York.

Wehr, M. and Laurent, G., 1996, Odour encoding by temporal sequences of firing in oscillating neural assemblies, *Nature* 384, 162–166.

Whitehead, A.N., 1925, Lowell Lectures. In: *Science and the Modern World,* 1947 (reprint), Free Press, New York.

Whiten, A., Goodall, J., McGrew, W.C., Nishida, T. and Reynolds, V., 1999, Cultures in chimpanzees, *Nature* 399, 682–685.

Wiener, N., 1948, *Cybernetics or Control and Communication in the Animal and the Machine*, Hermann, Paris.

Wiesenfeld, K. and Moss, F., 1995, Stochastic resonance and the benefits of noise, *Nature* 373, 33–36.

Williams, G.C., 1957, Pleiotropy, natural selection and the evolution of senescence, *Evolution* 11, 498–411.

Winfree, A.T., 2001, *The Geometry of Biological Time*, 2nd ed.,Springer, New York.

Wittgenstein, L., 1953, *Philosophical Investigations*, Blackwell, Oxford.

Wojtyk, R.I. and Goldstein, S., 1980, Fidelity of protein synthesis does not decline during aging of cultured human fibroblasts, *J. Cell Physiol.* 103, 299–303.

Wright, J.W. and Harding, J.W., 2004, The brain angiotensin system and extra-cellular matrix molecules in neural plasticity, learning and memory, *Prog. Neurobiol.* 72 (4), 263–293.

Zeeman, E.C., 1977, *Catastrophe Theory, Selected Papers*, Addison-Wesley, New York.

Zhang, Li., Tao, H., Holt, C., Harris, W. and Poo, M., 1998, A critical window for cooperation and competition among developing retinotectal synapses, *Nature* 395, 37–43.

List of Figures

Fig. 1.1   A graph.
Fig. 1.2   A path.
Fig. 1.3   A category.
Fig. 1.4   Difference between a graph and a category.
Fig. 1.5   Commutative diagrams.
Fig. 1.6   Functors.
Fig. 1.7   A free object.
Fig. 1.8   The perception field.
Fig. 1.9   Operating field.
Fig. 1.10 The graph G of routes.
Fig. 1.11 The category of paths.
Fig. 1.12 The category of travel times.
Fig. 1.13 Categories associated to integers.
Fig. 2.1   A pattern.
Fig. 2.2   A collective link.
Fig. 2.3   Role of the indices.
Fig. 2.4   Colimit of a pattern.
Fig. 2.5   Comparison of the sum and the colimit.
Fig. 2.6   Oxygenation of the haemoglobin tetramer.
Fig. 3.1   P-factors of a link to cP.
Fig. 3.2   Links correlated by a zigzag.
Fig. 3.3   A perspective.
Fig. 3.4   A cluster between patterns.
Fig. 3.5   Composition of clusters.
Fig. 3.6   A (Q, P)-simple link.
Fig. 3.7   Composition of simple links.
Fig. 3.8   Comparison between a pattern and a sub-pattern.
Fig. 3.9   A representative sub-pattern.
Fig. 3.10 Connected patterns.
Fig. 3.11 Complex link.
Fig. 3.12 A hierarchical category.
Fig. 3.13 A ramification of length 2.
Fig. 3.14 Different ramifications.
Fig. 3.15 A reducible 2-iterated colimit.
Fig. 4.1   Image of a pattern and of a collective link.
Fig. 4.2   Deformation of a colimit by a functor.
Fig. 4.3   Preservation of a colimit.
Fig. 4.4   Distributed link and limit of a pattern.
Fig. 4.5   A pro-cluster.
Fig. 4.6   Complexification with respect to an option.
Fig. 4.7   Absorption and elimination.

Fig. 4.8   Binding of patterns.
Fig. 4.9   Forcing of colimits.
Fig. 4.10   Construction of the simple links.
Fig. 4.11   Two successive complexifications.
Fig. 4.12   Emergent complex link.
Fig. 5.1   Transition between two configurations.
Fig. 5.2   An evolutive system.
Fig. 5.3   Components of an evolutive system.
Fig. 5.4   Different interconnections between components.
Fig. 5.5   Stability span.
Fig. 5.6   Complex identity.
Fig. 5.7   Variations of the different spans.
Fig. 5.8   Fibration associated to an evolutive system.
Fig. 6.1   A schematic view of a memory evolutive system.
Fig. 6.2   Landscape of a co-regulator.
Fig. 6.3   Selection of a procedure.
Fig. 6.4   Anticipated landscape.
Fig. 6.5   Evaluation.
Fig. 6.6   Complete step of a co-regulator.
Fig. 6.7   Structural temporal constraints.
Fig. 6.8   A business enterprise.
Fig. 7.1   Different causes of fractures.
Fig. 7.2   Different types of dyschrony.
Fig. 7.3   Interactions between two heterogeneous co-regulators.
Fig. 7.4   Fractures and their repairs.
Fig. 7.5   Aging curves.
Fig. 7.6   Pathology and recovery during aging.
Fig. 8.1   Formation of a record.
Fig. 8.2   Procedural memory.
Fig. 8.3   Effector record of a limit of procedures.
Fig. 8.4   Activator link.
Fig. 8.5   Formation of a new procedure.
Fig. 8.6   Storing of a procedure in the memory.
Fig. 8.7   Formation of an E-concept.
Fig. 8.8   E-universal link from a record to its E-concept.
Fig. 8.9   The functor E-concept.
Fig. 8.10   Colimit of E-concepts.
Fig. 9.1   Neurons and synapses.
Fig. 9.2   A synaptic path.
Fig. 9.3   Graph of neurons.
Fig. 9.4   A cat-neuron.
Fig. 9.5   Cluster and simple link between cat-neurons.
Fig. 9.6   A cat-neuron of level 2.
Fig. 9.7   Partial record.
Fig. 9.8   Internal E-record.

Fig. 9.9   Activation of a procedure.
Fig. 9.10   Record of S.
Fig. 9.11   Activator link.
Fig. 9.12   Formation of a procedure.
Fig. 9.13   Conditioning.
Fig. 10.1   E-trace of a record.
Fig. 10.2   Different traces of a record.
Fig. 10.3   E-concept.
Fig. 10.4   Instances of an E-concept.
Fig. 10.5   Concept for several attributes.
Fig. 10.6   Archetypal core and its fans.
Fig. 10.7   **AC** and **Exp**.
Fig. 10.8   Extension of the archetypal core.
Fig. A.1   A multi-colimit.

Index

abduction, 343
absorption, 119
activated domain, 340
activator link, 259, 313
actual landscape, 190
actual present, 194
adjoint functor, 32
admissible procedure, 184, 262
aging, 235
ambiguous image, 356
anticipated landscape, 194
anticipatory system, 228
Archetypal Core, 332
arrows, 23
aspect, 189
assembly of neurons, 290
associativity, 27

base functor, 35
based hierarchy, 134
basic procedure, 256
binding of a pattern, 57, 119
binding link, 57
binding of a cluster, 84
Binding Problem, 59, 296
blind-sight, 347
broad complexification, 129
business enterprise, 208

cascade of re-synchronizations, 222, 237
category, 25, 27
category associated to a labelled graph, 40
category constructed by generators and relations, 39
Category of categories, 46

Category of evolutive systems, 173
Category of graphs, 46
Category of neurons, 292
Category of particles and atoms, 90
Category of sets, 45
cat(egory)-neuron, 291, 295
cat-neuron of level k, 298
causality, 141, 227, 229
characteristic of complexity, 141
classical conditioning, 316
classification, 120
classifier of a pattern, 114
classifying complexification, 119
classifying option, 118
cluster, 81, 296
coherent assembly, 59
colimit, 57
colimits of graphs, 67
colimits of sets, 66
collective binding link, 57
collective link, 53
command of a procedure, 254
communication, 178
comparison functor, 196
comparison link, 63, 86
complex identity, 165
complex link, 94
complex switch, 90
complexification, 118
Complexification Theorem, 128
complexity order, 105
component of an evolutive system, 157
component of a pattern, 52
composite cluster, 82
composite of links, 27
concept, 278, 328
conditioning, 316
configuration category, 33, 151

connected patterns, 90–91
consciousness, 339, 345
conscious processes, 338
conscious units, 338
consolidated memory, 303
continuity span, 168
coordination neuron, 289
co-product, 62
co-regulator, 177, 184, 262, 306
correlated links, 79
correlated by a zigzag, 76
Cosmic Evolutive System, 162
crystals, 107

decoherence, 163
decomposition, 61
degeneracy of the neuronal code, 297
Degeneracy Principle, 92
de-synchronization, 217
diagram, 30
dialectics between co-regulators, 222
difference functor, 189
discrete category, 43
distinguished link of a pattern, 52
distributed link, 114
drives, 315
dyschrony, 198, 216

E-concept, 273, 324
ecosystem, 207
edges, 23
efficient cause, 141, 227
effector record, 255, 312
effectors, 184
E-internal record, 307
E-invariance class, 271, 323
elementary topos, 26
elimination, 119
Emergence Problem, 132, 138
Emergence Process, 139
Emergentist Monism, 302
Emergentist Reductionism, 138
empirical memory, 305
entropy, 108
episode of presence, 187

equivalence relation, 43
E-trace, 271, 323
E-universal link, 273
evaluating co-regulators, 318
evaluation process, 194, 267
evolutive functor, 155
evolutive system, 154, 173
evolutive system of cat-neurons, 301
evolutive system of neurons, 292
experiential memory, 334
(extended) Hebb Rule, 300
extended landscape, 341

fan, 333
fibration, 171
field of an object, 35
field of a pattern, 79
final causation, 229
final object, 36
forcing, 123
formal cause, 141
fracture, 198, 214
free category, 46
free object, 32
full sub-category, 30
functionally equivalent, 34, 125
functor, 31

gene transcription, 204
global landscape, 340
graph, 23
greatest lower bound, 115
Grothendieck topology, 333
group, 45
groupoid, 28
groupoid of pairs, 43

Hard Problem, 345
Hebb Rule, 290
heterogeneous co-regulators, 219
hierarchical category, 95
hierarchical evolutive system, 161
homologous patterns, 91
homomorphism, 24
hyperstructure, 357

identity, 28
identity functor, 31
image, 110
immune system, 205
indices of a pattern, 51
IndK, 83
information, 178
initial date, 155, 157
initial object, 36
insertion, 31
insertion cluster, 82
instances of a concept, 273, 278, 324, 329
intentional co-regulator, 338
intentional net, 340
intermingling of causes, 142
internal E-record, 249, 309
interplay among the procedures, 201
intersection, 115
invariance class, 329
inverse, 28
isomorphism, 28, 32
iterated colimit, 99
Iterated Complexification Theorem, 133

journal, 210

knowledge, 280

labelled in a monoid, 40
landscape of a co-regulator, 189
language, 348
large categories, 45
learning, 290
least upper bound, 67
limit of a pattern, 114
link of a category, 28
link of an evolutive system, 157
living system, 148
local colimit, 355
local section, 172
long-term memory, 182

MacNeille completion, 130
mechanical system, 147
mechanism, 226
mediated link, 74
memory, 180, 182, 184, 304
Memory Evolutive Neural System (MENS), 14, 303, 307
memory evolutive system, 184
Mind–Brain Problem, 302
mixed complexification, 119
Mixed Complexification Theorem, 130
mixed option, 118
models, 4, 149, 224
monoid, 44
morphisms, 28
multi-colimit, 354
multifold object, 90
Multiplicity Principle, 92, 297

net of co-regulators, 184, 306
n-complex link, 135
neuron, 287
n-multifold, 108

objects of a category, 28
operant conditioning, 318
operating field of an object, 36
operating field of a pattern, 56
Operating Reductionism, 64
operative procedure, 201, 266
operon, 204
opposite category, 29
option, 118
option associated to a procedure, 192
organisms, 226

partial functor, 34
(partial) order, 43
partial record, 249, 308
path, 24
path of length n, 24
pattern, 51
pattern of commands, 254
peaks of consciousness, 344, 357
perception field, 35

perceptual categorization, 322
perspective for a pattern, 77
perspective for a co-regulator, 190
P-factor of a link, 75
phase space, 148
p(re-)o(rdered) set, 43
pre-order, 43
preservation of a colimit, 113
pro-cluster, 116
pro-multifold objects, 117
pro-simple link, 117
procedural memory, 252, 257, 305
procedure, 254, 312
procedure-concept, 330
product, 114
projection distributed link, 114
propagation delay, 170, 187
prospection process, 344

qualia, 345
quasi-crystals, 107
quotient category, 38

ramification of length k, 100
recall, 245, 249, 311
recall of a concept, 330
receptors, 186
record, 246, 249, 310
Reduction Theorem, 104
Reductionism Problem, 103
reflexes, 315
relation, 43
renewal span, 167–168
replication with repair of DNA, 231
representative sub-pattern, 88
re-synchronization, 217, 222, 237
retrograde propagation, 198
retrospection process, 342

self-consciousness, 348
semantic memory, 278, 305, 328
shift, 274, 330

short-term memory, 303
simple link, 84, 296
sketch, 51
SOS system, 232
source of an arrow, 23
stability span, 164
standard changes, 117
stimulus-reward memory, 319
strategy, 118
strength of a synapse, 287
strongly connected patterns, 90
structural and temporal constraints of a
    co-regulator, 198
sub-category, 29
sub-pattern, 86
sum, 62
synapse, 287
synchronous assembly of neurons, 290

target, 23
teleonomy, 229
Theory of Aging, 234, 237
Theory of Sketches, 46
time lag, 188
time scale, 150
time scale of a co-regulator, 184
transition functor, 152
transport network, 68
treatment units, 306

underlying graph, 27
universal completion, 130
universal problems, 32
universal property, 57

value-dominated memory, 319
vertices, 23

weight, 39

zigzag, correlating links, 76